Power Systems

More information about this series at http://www.springer.com/series/4622

Krzysztof Sozański

Digital Signal Processing in Power Electronics Control Circuits

Second Edition

Springer

Krzysztof Sozański
Institute of Electrical Engineering
University of Zielona Góra
Zielona Góra
Poland

ISSN 1612-1287 ISSN 1860-4676 (electronic)
Power Systems
ISBN 978-1-4471-7331-1 ISBN 978-1-4471-7332-8 (eBook)
DOI 10.1007/978-1-4471-7332-8

Library of Congress Control Number: 2017938127

Printed on acid-free paper

This Springer imprint is published by Springer Nature
The registered company is Springer-Verlag London Ltd.
The registered company address is: 236 Gray's Inn Road, London WC1X 8HB, United Kingdom

This book is dedicated to my dear parents Maria and Kazimierz, and my darling children, Anna, Mateusz and Andrzej.

Preface

Preface to the Second Edition

The first edition of the book was one of the elements of my habilitation (a qualification above a Ph.D., which is a necessary step for obtaining the title of a professor in Poland and other European countries), and as a result it was subsequently very carefully reviewed by four reviewers. The habilitation was a success. Using the insightful comments from my reviewers, I have improved the current version and eliminated some typographic errors that were initially overlooked.

This book is based on the first edition of the book and besides some changes to the previous material, there are some new elements I include in the new edition. In particular, I add more MATLAB listings, which would further complement and enrich the text by providing worked examples of the solutions proposed in the book.

More figures are added and all the existing ones are revised and in color. My new solution for a shunt active power filter is also included in the new edition. All those changes are introduced in the Chapters from the first edition.

This book also includes a new Chapter considering selected problems of simulation of power electronic systems together with digital control circuits. These simulations are conducted using MATLAB and PSIM. In the case of the PSIM programs, the considered methods use C code for describing the digital control algorithm.

I have written this book in his endeavor to abide by the following maxim *nulla dies sine linea* ↔ *nie ma dnia bez kreski* ↔ *not a day without a line drawn* (Apelles, Greek painter, flourished 4th century bc). However, this is not always easily achieved.

Zielona Góra, Poland
February 2017

Krzysztof Sozański

Preface to the First Edition

Power electronics circuits are becoming increasingly important in the modern world due to the rapid progress in developments of microelectronics in areas such as microprocessors, digital signal processors, memory circuits, complementary metal-oxide-semiconductors, analog-to-digital converters, digital-to-analog converters and power semiconductors—especially metal-oxide-semiconductor field-effect transistors and insulated gate bipolar transistors.

Specifically, the development of power transistors has shifted the range of applications from a few amperes and hundreds of volts to several thousands of amperes and a few kilovolts, with a switching frequency measured in millions of hertz. Power electronics circuits are now used everywhere: in power systems, industry, telecommunications, transportation, commerce, etc. They even exist in such modern popular devices as digital cameras, mobile phones, and portable media players. Power electronics are also used in micropower circuits, especially in energy harvesting circuits.

In the early years of power electronics, in the sixties and seventies, analog control circuits were most commonly used, meaning that only the simplest control algorithms could be applied. Some years later, in the eighties and early nineties, hybrid control circuits were used, which consisted of both analog and digital components. In subsequent years there followed a slow transition to fully digitalized control systems, which are currently widely used and enabled the application of more complex digital signal processing algorithms.

In this book the author considers signal processing, starting from analog signal acquisition, through its conversion to digital form, methods of its filtration and separation, and ending with pulse control of output power transistors. The author has focused on two applications for the considered methods of digital signal processing: an active power filter and a digital class-D power amplifier.

Both applications require precise digital control circuits with very high dynamic range of control signals. Therefore, in the author's opinion these applications will provide very good illustrations for the considered methods. In this book the author's original solutions for both applications are presented. In the author's opinion the adopted solutions can also be extended to other power electronics devices.

The discussion of the first application, APF, starts with the analysis of first harmonic detectors based on: IIR filter, wave digital filters, sliding DFT and sliding Goertzel, moving DFT. Following that, author's implementation of classical control circuits based on $p-q$ algorithm is presented. Next, the dynamics of APF is considered. Dynamic distortion of APF makes it impossible to fully compensate line harmonics. In some cases, the line current *THD* ratio for systems with APF compensation can reach a value of a dozen or so percent. Therefore the author has dealt with this problem by proposing APF models suitable for analysis and simulation of this phenomenon. For predictable line current changes, it is possible to develop a predictable control algorithm to eliminate APF dynamics compensation errors. In the following sections the author's modification using a predictive circuit to eliminate dynamic compensation errors is described. In this book control circuits with filter banks which allow the selection of compensated harmonics are described. The considered filter banks are based on sliding DFT, sliding Goertzel, moving DFT and $p - q$ algorithm.

For unpredictable line current changes the author has developed a multirate APF. The presented multirate APF has a fast response for sudden changes in the load current. So, using multirate APF, it is possible to decrease the *THD* ratio of line current even for unpredictable loads.

The second application is a digital class-D amplifier. Both APFs and the amplifiers are especially demanding in terms of the dynamics of processed signals. However, in the case of a class-D amplifier, the dynamics reach 120 dB, which results in high requirements for the type of algorithm used and its digital realization. The author has proposed a modulator with a noise shaping circuit for a class-D amplifier. Interpolators are also considered that allow for the increasing of the sampling frequency whilst maintaining a substantial separation of signal from noise. The author also presents an original analog power supply voltage fluctuation compensation circuit for the class-D amplifier. The class-D amplifier with digital click modulation is given special consideration too. Finally, two-way and three-way loudspeaker systems, designed by the author, are presented, where the signal from input to output is digitally processed.

The great majority of the presented methods and circuits is the original work of the author. Listings from MATLAB or in C language are attached to some of the considered algorithms to make the application of the algorithms easier. The presented methods and circuits can be successfully applied to the whole range of power electronics circuits.

The issues concerning digital signal processing are relatively widely described in the literature. However, in the author's opinion, there are very few publications combining digital signal processing and power electronics, due to the fact that these two areas of knowledge have been developed independently over the years. The author hopes that this book will, to some extent, bridge the gap between digital signal processing and power electronics. This book may be useful for scientists and engineers who implement control circuits, as well as for students of electrical engineering courses. It may also be of some value to those who create new topologies and new power electronics circuits, giving them some insights into possible control algorithms.

Zielona Góra, Poland
December 2012

Krzysztof Sozański

Acknowledgements

I would like to thank the reviewers of my habilitation (in alphabetical order) Andrzej Dzielinski from Warsaw University of Technology, Boguslaw Grzesik from Silesian University of Technology, Gliwice, Adam Kempski from the University of Zielona Góra, Jan Mucko from University of Science and Technology, Bydgoszcz.

Many thanks also to my English consultants, Peter Preston from the University of Zielona Góra and my daughter Anna Sozanska who is studying at the University of Cambridge. I hope that I successfully applied all of your comments and corrections.

I would also like to thank everyone who supported me during the writing of this book.

Contents

1 Introduction 1
1.1 Power Electronics Systems.......... 1
1.2 Digital Control Circuits for Power Electronics Systems.......... 3
1.2.1 Analog Versus Digital Control Circuit 5
1.2.2 Causal and Non-causal Digital Circuits.......... 5
1.2.3 LTI Discrete-Time Circuits.......... 6
1.2.4 Digital Filters 8
1.2.5 Hard Real-Time Control Systems 10
1.2.6 Sampling Rate 11
1.2.7 Simultaneous Sampling.......... 12
1.2.8 Number of Bits.......... 12
1.3 Multirate Control Circuits 13
1.4 Active Power Filters 14
1.5 Digital Class-D Power Amplifiers 16
1.6 Symbols of Variables.......... 18
1.7 What Is in This Book 18
References 20

2 Analog Signals Conditioning and Discretization.......... 23
2.1 Introduction 23
2.2 Analog Input 23
2.2.1 Galvanic Isolation.......... 23
2.2.2 Common Mode Voltage 25
2.2.3 Isolation Amplifiers 27
2.3 Current Measurements.......... 32
2.3.1 A Resistive Shunt.......... 32
2.3.2 Current Transformers 33
2.3.3 Transformer with Hall Sensor.......... 35
2.3.4 Current Transformer with Magnetic Modulation.......... 38
2.3.5 Current Transducer with Air Coil.......... 38
2.3.6 Comparison of Current Sensing Techniques.......... 41

2.4 Selected Parameters of Digital Control Circuit . . . 41
2.5 Total Harmonic Distortion . . . 43
2.6 Sampling of Analog Signal . . . 45
2.6.1 Synchronization of Sampling Process . . . 47
2.6.2 Maximum Signal Frequency Versus Signal Acquisition Time . . . 49
2.6.3 Errors in Multichannel System . . . 50
2.6.4 Amplitude and Phase Errors of Sequential Sampling A/D Conversion . . . 52
2.6.5 Sampling Clock Jitter . . . 54
2.7 Signal Quantization . . . 57
2.7.1 Dynamic Range of Signal . . . 59
2.7.2 Signal Headroom . . . 59
2.7.3 Noise Shaping Technique . . . 60
2.7.4 Dither . . . 63
2.7.5 Propagation of Quantization Noise . . . 65
2.7.6 Effective Number of Bits . . . 66
2.8 A/D Converters Suitable for Power Electronics Control Circuits . . . 68
2.8.1 A/D Converter with Successive Approximation . . . 69
2.8.2 A/D Converter with Delta Sigma Modulator . . . 70
2.8.3 Selected Simultaneous Sampling A/D Converters . . . 70
2.8.4 ADS8364 . . . 71
2.8.5 AD7608 . . . 72
2.8.6 ADS1278 . . . 73
2.8.7 ADS8568 . . . 74
2.8.8 A/D Conventer of TMS320F28335 . . . 75
2.8.9 A/D Converters of TMS320F2837xD . . . 77
2.9 Conclusions . . . 77
References . . . 78

3 Selected Methods of Signal Filtration and Separation and Their Implementation . . . 83
3.1 Introduction . . . 83
3.2 Digital Filters . . . 84
3.2.1 Digital Filter Specifications . . . 84
3.2.2 Finite Impulse Response Digital Filters . . . 85
3.2.3 Infinite Impulse Response Digital Filters . . . 87
3.2.4 Design of Digital IIR Filters . . . 90
3.3 Lattice Wave Digital Filters . . . 94
3.3.1 Comparison of Classical IIR Filter and Lattice Wave Digital Filter . . . 96
3.3.2 Realization of LWDF . . . 97

3.4 Modified Lattice Wave Digital Filters 100
3.4.1 First-Order Sections 101
3.4.2 Second-Order Sections 104
3.5 Linear-Phase IIR Filters 107
3.5.1 Example of a Linear-Phase IIR Filter 111
3.5.2 Comparison of FIR and LF IIR 114
3.6 Multirate Circuits 115
3.6.1 Signal Interpolation 115
3.6.2 Signal Decimation 119
3.6.3 Multirate Circuits with Wave Digital Filters 121
3.6.4 Interpolators with Linear-Phase IIR Filters 123
3.7 Digital Filter Banks 126
3.7.1 Strictly Complementary Filter Bank 128
3.7.2 DFT Filter Bank 129
3.7.3 Sliding DFT Algorithm 131
3.7.4 Sliding Goertzel Algorithm 136
3.7.5 Moving DFT Algorithm 136
3.7.6 Wave Digital Lattice Filter Bank 139
3.8 Implementation of Digital Signal Processing Algorithms 146
3.8.1 Basic Features of the DSP 148
3.9 Selected Microcontrollers Suitable for Power Electronics Control Circuits 153
3.9.1 TMS320F28335 155
3.9.2 TMS320F2837xD 157
3.9.3 Digital Signal Processor—TMS320C6xxx Family 157
3.9.4 Digital Signal Processors—SHARC Family 159
3.10 Conclusions 161
References 161

4 Selected Simulation Methods and Programs for Power Electronics Circuits 167
4.1 Introduction 167
4.2 Simulation Using MATLAB® 169
4.2.1 DC and AC Analysis of Analog Circuits 169
4.2.2 DC and AC Nodal and Loop Analysis of Circuits 174
4.2.3 Transient Analysis of Analog Circuits 177
4.2.4 Simulation of Power Electronics System Together with Digital Control Circuit 180
4.2.5 Simulation of the Power Electronics System Using Simulink® 183
4.3 Simulation Using PSIM 184
4.3.1 Simulation Using the Typical PSIM Blocks 184
4.3.2 Simulation Using C Code 185
4.3.3 Simulation with AC Analysis 190
4.3.4 Simulation to Hardware Implementation 192

4.4 Conclusions . . . 196
References . . . 196

5 Selected Active Power Filter Control Algorithms . . . 199
5.1 Introduction . . . 199
5.2 Control Circuit of Shunt APFs . . . 200
5.2.1 Synchronization . . . 201
5.3 Simulation of APF . . . 203
5.3.1 Simulation of APF Using MATLAB . . . 204
5.3.2 Simulation of APF Using PSIM . . . 206
5.4 APF Control with First Harmonic Detector . . . 206
5.4.1 Control Circuit with Low-Pass 4-Order Butterworth Filter . . . 208
5.4.2 Control Circuit with Low-Pass 5-Order Butterworth LWDF . . . 209
5.4.3 Control Circuit with Sliding DFT . . . 210
5.4.4 Control Circuit with Sliding Goertzel . . . 215
5.4.5 Control Circuit with Moving DFT . . . 215
5.5 The Control Circuit for the Shunt APF Based on $p-q$ Algorithm . . . 218
5.6 Shunt APF Classical Control Circuit . . . 222
5.6.1 High-Pass IIR Filter . . . 226
5.6.2 Improved High-Pass Filter . . . 227
5.6.3 DC Bank Voltage Controller . . . 229
5.6.4 The Remaining Part of the $p-q$ Algorithm . . . 229
5.6.5 Output Current Controller . . . 230
5.6.6 Modernized Digital Controller for the APF . . . 231
5.7 Dynamics of Shunt APF . . . 233
5.7.1 Methods of Reducing APF Dynamic Distortion . . . 233
5.7.2 Control Circuit . . . 236
5.7.3 APF Output Current Ripple Calculation . . . 240
5.7.4 Simulation of APF Control Circuit . . . 243
5.8 Predictive Control Algorithm for APF . . . 244
5.8.1 Advance Time T_A . . . 245
5.8.2 Experimental Results for Steady-State . . . 248
5.8.3 Step Response of APF . . . 248
5.9 Selected Harmonics Separation Methods Suitable for APF . . . 254
5.9.1 Control Circuit with MDFT . . . 255
5.9.2 Control Circuit with $p-q$ Algorithm . . . 255
5.10 Multirate Shunt APF . . . 258
5.10.1 Analog Input Circuit . . . 260
5.10.2 The Output Inductors . . . 264
5.10.3 APF Simulation Results . . . 264

5.11 Multirate Shunt APF with Prediction 270
5.12 Conclusions 271
References 272

6 Digital Signal Processing Circuits for Digital Class-D Power Amplifiers 277
6.1 Introduction 277
6.2 Digital Class-D Power Amplifier Circuits 278
6.3 Modulators for Digital Class-D Power Amplifiers 281
6.3.1 Oversampled Pulse Width Modulator 284
6.4 Basic Topologies of Control Circuits for Digital Class-D Power Amplifiers 285
6.4.1 Open Loop Amplifiers 285
6.4.2 Amplifiers with Digital Feedback for Supply Voltage 287
6.4.3 Amplifiers with Analog Feedback for Output Pulses . . . 287
6.4.4 Amplifiers with Digital Feedback 290
6.5 Supply Units for Class-D Power Amplifiers 291
6.6 Click Modulation 293
6.7 Interpolators for High Quality Audio Signals 295
6.7.1 Single Stage Interpolators 295
6.7.2 Multistage Interpolators 296
6.8 Class-D Audio Power Amplifiers 300
6.8.1 Digital Crossovers 302
6.9 Loudspeaker Measurements 304
6.10 Class-D Power Amplifier with Digital Click Modulator 309
6.10.1 Digital Crossovers 310
6.10.2 Realization of Digital Click Modulator 314
6.10.3 Experimental Results 318
6.11 Digital Audio Class-D Power Amplifier with TAS5508 DSP . . . 321
6.11.1 TAS5508-5121K8EVM 322
6.11.2 Three-Way Digital Crossover 326
6.11.3 Experimental Results 326
6.12 Conclusions 329
References 330

7 Conclusion 333
7.1 Summary of Results 333
7.2 Future Work 335
References 336

Index 337

Abbreviations and Symbols

Abbreviations

AC	Alternating current
A/D	Analog-to-digital converter
ALU	Arithmetic-logic unit
APF	Active power filter
av	Average value of signal
BPF	Band-pass filter
BSF	Band-stop filter
CM	Click modulator also called zero position coding
D/A	Digital-to-analog converter
D/t	Digital-to-time converter
DAI	Digital audio interface
dB	Decibel, $20\log(U_2/U_1)$, $10\log(P_2/P_1)$
DC	Direct current
DFT	Discrete Fourier transform algorithm
DPWM	Digital pulse width modulation
DSM	Delta sigma modulator
DSP	Digital signal processor
e.g.	For example (exempli gratia—Latin)
EMI	Electromagnetic interference
etc.	And other things, or and so forth (et cetera—Latin)
FFT	Discrete fast Fourier transform algorithm
FIR	Finite impulse response digital filter
FLOPS	Floating-point operations per second
FPGA	Field programmable gate array
HPF	High-pass filter
i.e.	This is (id est—Latin)
IC	Integrated circuit
IGBT	The insulated gate bipolar transistor
IIR	Infinite impulse response digital filter

IIS	Inter-IC sound, integrated interchip sound, or IIS, is an electrical serial bus interface standard used for connecting digital audio devices together
$\mathrm{Im}(x), \Im(x)$	Imaginary part of x
IPS	Instructions per second, MIPS
IPT	Instantaneous power theory
LBWDF	Lattice bireciprocal wave digital filter
LC	LC circuit, circuit composed of capacitor and inductor
LogChrip	Logarithmic chirps signal
LPF	Low-pass filter
LR	Linkwitz–Riley filter
LSB	Last significant bit
LTI	Linear time-invariant circuit (system)
LWDF	Lattice wave digital filter
MAC	Multiplication and accumulation, special arithmetic operation of DSP
MDFT	Moving discrete Fourier transform algorithm
MIPS	Million instructions per second
MLS	Maximal length sequence signal
MOSFET	Metal–oxide–semiconductor field-effect transistor
MSB	Most significant bit
MWDF	Modified wave digital filters
PCB	Printed circuit board
PCM	Pulse code modulation
PDF	Probability density function
PLL	Phase locked loop or phase lock loop circuit
PWM	Pulse width modulation
QMF	Quadrature mirror filter
$\mathrm{Re}(x), \Re(x)$	Real part of x
RLC	Circuit, circuit composed of resistor, capacitor and inductor
rms	Root mean square
S/PDIF	Sony/Philips digital interconnect format (more commonly known as Sony Philips digital interface)
SA	Successive approximation
SC	Strictly complementary digital filter bank
SDFT	Sliding discrete Fourier transform algorithm
SGDFT	Sliding Goertzel discrete Fourier transform algorithm
SH	Sample–and–hold circuit, sampling circuit
SOS	Second-order section
SPS	Sample per seconds
TR	Time reversal
U2	Two's complement binary code
WDF	Wave digital filters
ZePoC	Zero position coding also called click modulation
μC	Microcontroller
μP	General-purpose microprocessor

Symbols

C	Capacitance
δ_p	Ripple in the passband
δ_z	Ripple in the stopband
f_c	Filter crossover (cutoff) frequency
f_s	Sampling rate, sampling frequency
f_k	Power transistor switching frequency
f_M	Line voltage frequency
f_p	Passband frequency
f_z	Stopband frequency
$H(s)$	Analog transfer function
$H(z)$	Digital transfer function
$i_C, i_C(t)$	Instantaneous value of APF compensation current
$i_C(n)$	Discrete signal represents APF compensation current
$i_M, i_M(t)$	Instantaneous value of line current, power line current
$i_M(n)$	Discrete signal represents power line current
$i_L, i_L(t)$	Instantaneous value of load current
$i_L(n)$	Discrete signal represents load current
L	Inductance
N	Length of sample block
N_M	Number of samples per line voltage period
R	Signal oversampling ratio
R	Resistance
M	Signal decimation ratio
$SINAD$	Signal to noise and distortion ratio
SNR	Signal to noise ratio
THD	Total harmonic distortion
T_M	Line voltage period
T_s	Sampling period
Q_g	MOSFET total gate charge
Q_{rr}	Diode reverse recovery charge
$u_{C1}, u_{C1}(t)$	Instantaneous value of capacitor voltage
$X(s)$	Analog signal transmittance
$X(z)$	Digital signal transmittance
Z	Impedance
Z_C	APF compensation circuit output impedance
Z_M	Power line impedance
Z_L	Load impedance
z^{-1}	Unit delay operator

Chapter 1
Introduction

1.1 Power Electronics Systems

Over the past 30 years, the area of power electronics has expanded greatly. This has been caused by the extensive development of microelectronics in areas such as microprocessors, digital signal processors, memory circuits, complementary metal-oxide-semiconductors (CMOS), analog-to-digital (A/D) converters, digital-to-analog (D/A) converters and power semiconductors—especially metal-oxide-semiconductor field-effect transistors (MOSFET) and insulated gate bipolar transistors (IGBT). Specifically, the development of power transistors has shifted the range of applications from a few amperes and hundreds of volts to several thousand amperes and a few kilovolts with a switching frequency measured in millions of hertz. Another area of application of power electronics circuits are micropower circuits and particularly energy harvesting circuits. Hence, power electronics circuits are now used everywhere, in power systems, industry, telecommunications, transportation, commerce, etc. They even exist in such modern popular devices as smartphones, tablets, laptops, digital cameras, mobile phones, and portable media players. The background of power electronics is described by many authors for example Mohan [31], Erickson [17], Bose [11], Trzynadlowski [55] etc.

Power electronics is quite a difficult field of science and technology, requiring extensive knowledge of related fields. These include areas such as power systems, electrical machines, signal processing, analog and digital control, electronics, electromagnetic compatibility, solid-state electronics, embedded software design, circuit theory, circuit simulation, electromagnetic theory, thermal design, etc. These aspects are shown in Fig. 1.1. In order to produce properly functioning power electronics devices, all these aspects should be included. Often, the omission of one of these factors causes a system malfunction. This is due to the fact that a power electronics system converts energy and any errors are revealed in energy dissipation, which leads to the disruption or destruction of components.

A simplified diagram of a power electronics system is shown in Fig. 1.2. The system makes the conversion of electric power from input to load. The power input is typically from a power system, electrochemical battery, solar cells or fuel cells, etc. The input power can be direct current (DC) or alternating current (AC): single-phase, two-phase, three-phase or more. The output power depends on the load.

K. Sozański, *Digital Signal Processing in Power Electronics Control Circuits*,
Power Systems, DOI 10.1007/978-1-4471-7332-8_1

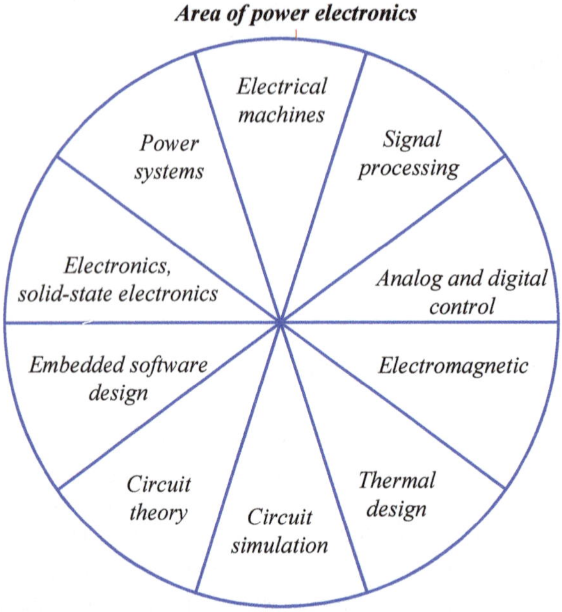

Fig. 1.1 Multidisciplinary nature of power electronics circuit

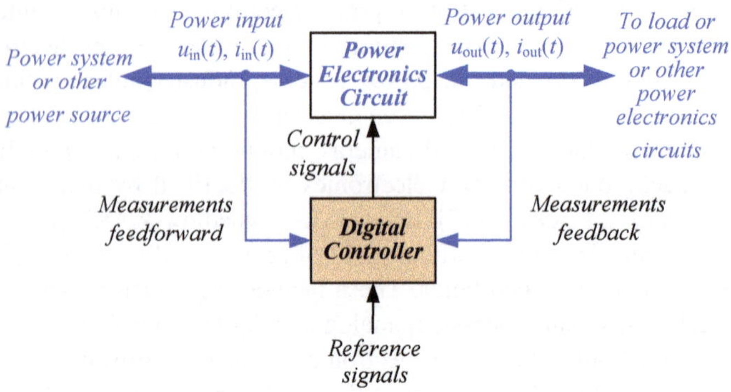

Fig. 1.2 A power electronics system

The power electronics circuit is monitored by a controller which compares the output power (feedback) and input power (feedforward) with the reference value needed to achieve the desired results. The controller can be designed to use either an analog or digital technique. Today the digital approach is most common, while the analog approach is reserved only for simple power electronics systems. The power flows through an electronics system from source to output. The output can be connected to a load or to another power system or to other power electronics circuits, etc. However, the power electronics circuit may be reversed and then the energy can flow from output to source.

1.2 Digital Control Circuits for Power Electronics Systems

The most common power electronics system element is the inverter. Single-phase or three-phase (or even more in multi-phase or in multilevel devices) inverters are frequently used parts of power electronics systems, in such devices as AC and DC motor drives, uninterruptible power supplies, harmonic compensators, DC power supplies, controlled rectifiers, AC and DC power transmission systems, smart grid etc.

A simplified block diagram of an exemplary three-phase inverter with digital controller is shown in Fig. 1.3. The inverter consists of six power IGBT transistors Q_1, Q_2, Q_3, Q_4, Q_5, Q_6, controlled by drivers with galvanic isolation. This galvanic isolation should have low input-to-output capacitance and should be highly resistant to a high output voltage slew rate, in the range from 10 to 30 kV/μ s. One of the inverter legs consists of two transistors Q_1 and Q_4, which are connected to the load Z_{L1} through the LC low-pass filter (L_{f1}, C_{f1}) for suppressing pulse width modulation (PWM) components. The inverter is controlled by a digital signal processor (DSP) which performs the control algorithm. The control DSP function can be realized using one of the following devices: general purpose microprocessors, microcontrollers, advanced microprocessors and microcontrollers, digital signal processors and programmable digital devices, etc. As with transistor control signals, analogue signals representing currents and voltages are galvanically isolated too. The analog signals are converted into the digital form by an A/D converter. The algorithm controls the output signal value using controlling output transistors by pulses generated by the same kind of pulse modulator. In the early days of digital control circuits, the control algorithms were very simple and represented only digital versions of analog control circuits with a single sampling frequency.

Over time, control algorithms have turned into highly advanced and more complicated solutions. Currently, control circuits designed to increase signal dynamic range use a few sampling frequencies. Digital circuits using different sampling frequencies are called multirate digital circuits. There are not many comprehensive publications dealing with digital control circuits for power electronics systems, though it is possible to find a lot of limited discussion in many conference papers. The background of digital control circuits is described by Astrom and Wittenmark [6], Williamson [60], Kazimerkowski et al. [25, 26], Buso and Mattavelli [12], some aspects of power and signal processing are discussed by Bollen et al. [10]. Also, some problems of control circuits for active power filters (APF) are described by the author [42, 46–50]. The crucial first aspects to be considered during the digital control circuit design include:

- control system functions,
- control algorithm,
- sampling rate,

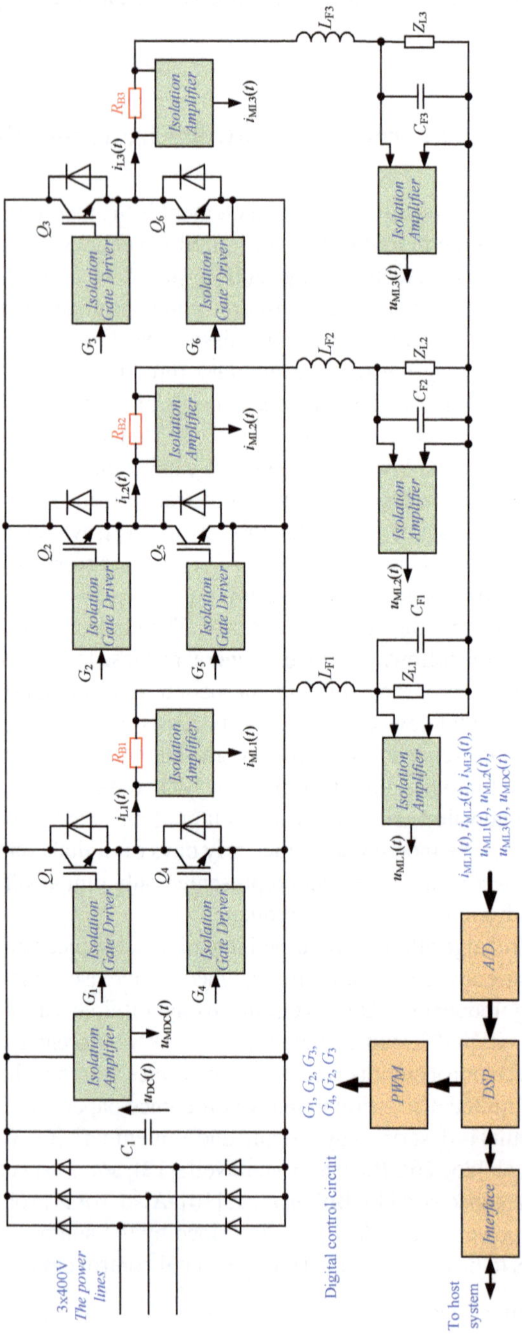

Fig. 1.3 Exemplary three-phase inverter with digital controller

- number of bits,
- type of digital circuit for realization,
- fixed-point or floating-point implementation of control algorithm.

Discussion of these problems is found in this book.
Examples of digital control circuit applications include active power filters and digital class-D power amplifiers. Both applications require precise digital control circuits with very high dynamic range of control signals. Therefore, in the author's opinion, these applications will provide very good illustrations for the considered methods. This book presents the author's original solutions for both applications. In the author's opinion, the adopted solutions can also be extended to other electronics devices.

1.2.1 Analog Versus Digital Control Circuit

Historically the control circuit of a power electronics device was analog. Therefore in the literature, even today, a lot of digital control circuits are described using analog transfer functions $H(s)$. This may be acceptable if the sampling frequency f_s and power transistor switching frequency f_k are much bigger than the frequency of the higher component of the band of interest. Figure 1.4a shows a magnitude frequency response of an analog circuit and Fig. 1.4b shows a magnitude frequency response of its digital representation. The relation between analog frequency and digital frequency for the most commonly used bilinear transform is nonlinear, and especially for high frequencies near $f_s/2$ the frequency characteristic is compressed. The frequency response of an analog circuit is spanned between zero and infinity and it should be compressed from zero to $f_s/2$ for the digital domain. Therefore characteristics of analog and digital circuits are different especially near $f_s/2$. This problem is considered in Chap. 3. In the author's opinion, this is the most appropriate way to consider a digital circuit in contradistinction to an analog circuit. This helps to avoid errors and instability in high-frequency components. Another problem concerns algorithm computation, for which simulation research should use the same arithmetic resolution as in a real control circuit. This will help to avoid unexpected associations with the instability caused by the limited resolution of the arithmetic. These problems are also considered in Chap. 3. The problems of digital signal processing are described in many publications, and as basic books ("bricks") the following books can be recommended [7, 15, 29, 33, 34, 36–38, 59, 61, 62]. The author presents some selected solutions of digital signal processing useful for power electronics circuits.

1.2.2 Causal and Non-causal Digital Circuits

A circuit (system) for which the output signals at any instant depends only on the past or/and present values of the input signals is called a causal system (circuit) e.g.,

Fig. 1.4 Frequency response of equivalent analog and digital circuits: **a** analog, **b** digital for $f_s = 10$ Hz

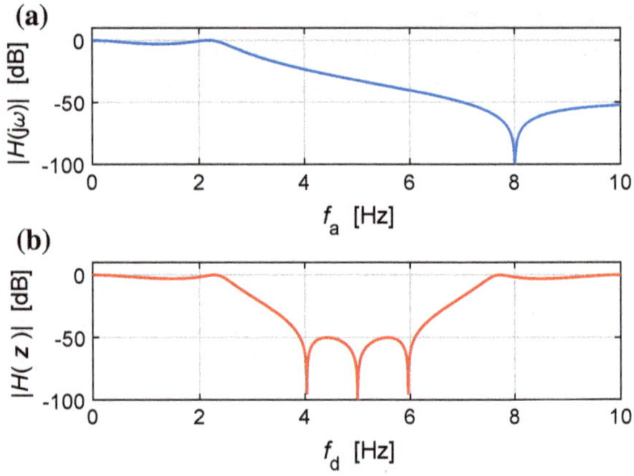

$y(n) = x(n) - 0.3x(n-1) + 0.1x(n-3)$. All real-time physical systems are causal because time only moves forward.

A circuit (system) for which the output at any instant depends also on future values of the input signals, is called a non-causal system (circuit) e.g., $y(n) = x(n) - 0.5x(n+1) + 0.1x(n+3)$, $y(n) = x(-n)$, $y(n) = y(n^2)$. A non-causal circuit (system) is also called a non-realizable circuit (system). A circuit (system) depending only on future input signal values is an anti-causal circuit (system), e.g. $y(n) = x(n+1)$.

This book mainly considers causal circuits (systems), because they are easier to work with and understand, and because most practical systems are causal in nature. However, non-causal circuits in some applications are very attractive e.g., the realization of a linear-phase IIR filter uses a non-causal zero-phase IIR filter. Similarly, the author uses a non-causal circuit in the control circuit for an APF.

1.2.3 LTI Discrete-Time Circuits

This book mainly considers linear time-invariant (LTI) discrete-time circuits. Let $x(n)$ be the discrete input signal, $y(n)$ be the discrete output signal and $h(n)$ be a discrete impulse response from the discrete unit impulse $\delta(n)$, which is the impulse response. Discrete impulse $\delta(n)$ is also called Kronecker delta

$$\delta(n) = \begin{cases} 1 & n = 0 \\ 0 & n \neq 0 \end{cases} . \tag{1.1}$$

It plays the same role as the Dirac delta in continuous-time circuits. A block diagram of the LTI discrete-time circuit is shown in Fig. 1.5. Time domain LTI discrete-time circuits can be described by the following equations, for a non-causal circuit,

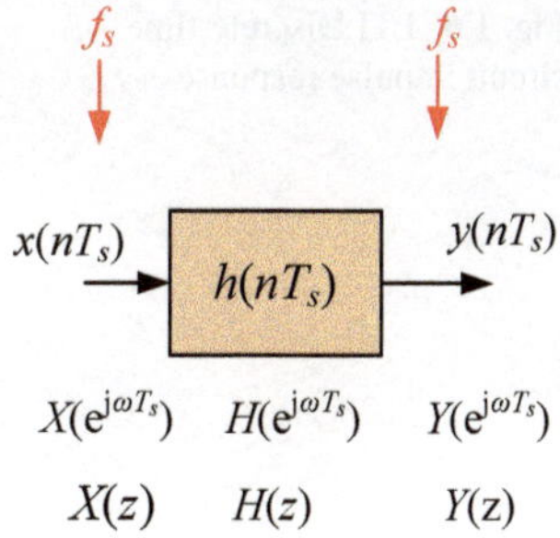

Fig. 1.5 LTI Discrete-time circuit

$$y(n) = \sum_{k=-\infty}^{\infty} h(k)x(n-k) \quad , \tag{1.2}$$

and for a casual circuit,

$$y(n) = \sum_{k=0}^{\infty} h(k)x(n-k) \quad . \tag{1.3}$$

For a frequency domain LTI discrete-time circuit

$$Y(e^{j\omega T_s}) = H(e^{j\omega T_s})X(e^{j\omega T_s}) \quad , \tag{1.4}$$

where: $X(e^{j\omega T_s})$, $Y(e^{j\omega T_s})$—Fourier transformation of discrete signals, $H(e^{j\omega T_s})$—LTI discrete-time circuit frequency transfer function, and

$$j\omega T_s = j2\pi/f_s \quad . \tag{1.5}$$

A LTI discrete circuit may also be described by a Z-transform

$$Y(z) = H(z)X(z) \quad , \tag{1.6}$$

where: $X(z)$, $Y(z)$—Z-transform of discrete signals, $H(z)$—transfer function. Therefore for LTI discrete-time circuits it is possible to write relations as

$$x(n) \leftrightarrow x(nT_s) \leftrightarrow X(e^{j\omega T_s}) \leftrightarrow X(z) \tag{1.7}$$

For a description of LTI discrete-time circuits it is important to know their impulse response. Substituting for input signal discrete impulse $x(n) = \delta(n)$, it is possible to calculate the circuit impulse response

$$y(n) = \sum_{k=-\infty}^{\infty} h(k)\delta(n-k) = h(n) \quad , \tag{1.8}$$

Fig. 1.6 LTI Discrete-time circuit impulse response

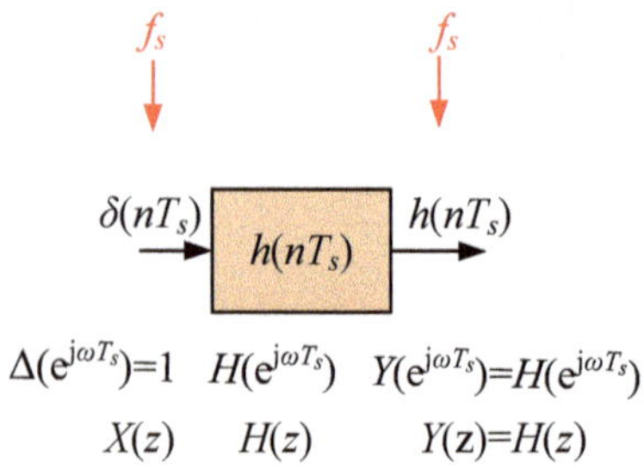

and for frequency domain ($\Delta(e^{j\omega T_s}) = 1$)

$$Y(e^{j\omega T_s}) = H(e^{j\omega T_s}) \overbrace{X(e^{j\omega T_s})}^{1} = H(e^{j\omega T_s}) \quad . \tag{1.9}$$

A block diagram of a LTI discrete circuit for impulse response is depicted in Fig. 1.6 A highly detailed description of LTI discrete circuits can be found in many books, e.g.: Oppenheim et al. [33], Rabiner and Gold [38], Proakis and Manolakis [37], Mitra [30], Zielinski [61], Chen [13], Wanhammar [59], Venezuela and Constantindes [57], Orfanidis [34], Tantaratana [54].

As can be seen in the Eqs. (1.2, 1.3, 1.8), in digital signal processing, the multiply-accumulate (MAC) operation is a basic operation that computes the product of two numbers and adds that product to an accumulator.

1.2.4 Digital Filters

In engineering practice, there is often a common problem of the presence of noise in digital measuring signals. If the frequency of the noise is higher than the frequency of the signal, in the first approach, the most common remedy is the use of averaging. Typically $N + 1$ signal samples are summed and the result is divided by the number of samples $N + 1$. A diagram of such a circuit for calculating the average of four current samples, also called the moving average, is shown in Fig. 1.7. The input signal is stored in the input buffer and the output signal is stored in the output buffer, but only four ($N + 1$) current input samples should be stored. A diagram of a N-order moving average circuit is depicted in Fig. 1.8. The operation of such a circuit can be described by the equation

$$y(n) = \frac{1}{N+1} \sum_{k=0}^{N} x(n-k) \quad , \tag{1.10}$$

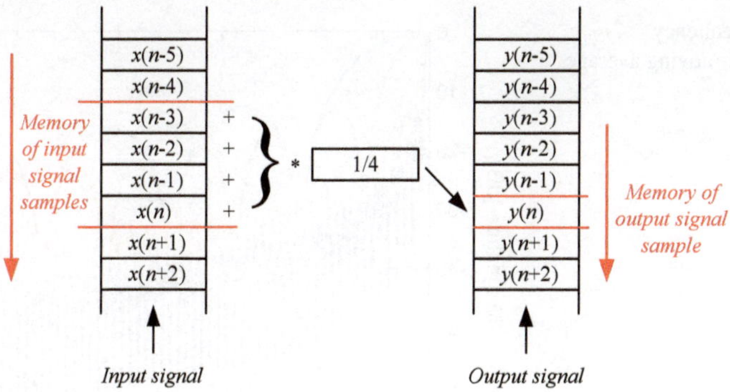

Fig. 1.7 3-Order moving average circuit

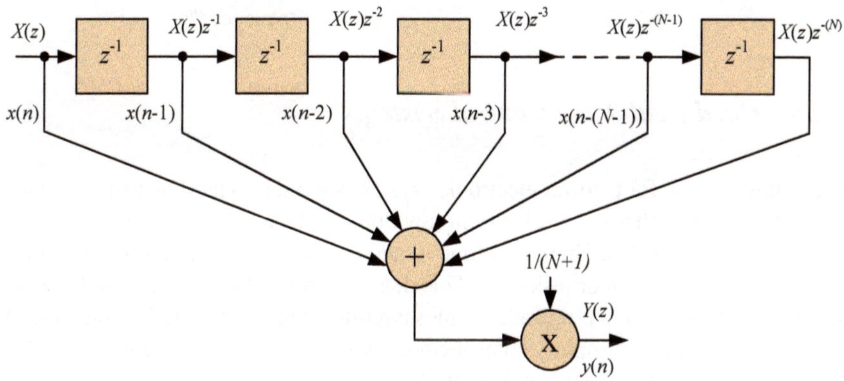

Fig. 1.8 Block diagram of N-order moving average circuit

where: $N + 1$—number of signal samples.
The N-order moving average transfer function can be described by the equation

$$H(z) = \frac{Y(z)}{X(z)} = \frac{1}{N+1} \sum_{k=0}^{N} z^{-k} \quad . \tag{1.11}$$

Figure 1.9 shows the frequency characteristics for the moving average of the orders: 8, 16, 64, 128. As can be seen from the graph, this method does not allow the obtaining of very high attenuation. Achieving greater attenuation is possible thanks to the use of more complex circuits than the moving average. These problems are further described in Chap. 3, with special attention being paid to the wave digital filter (WDF), which is very suitable for the realisation of low resolution arithmetics [18, 19, 21].

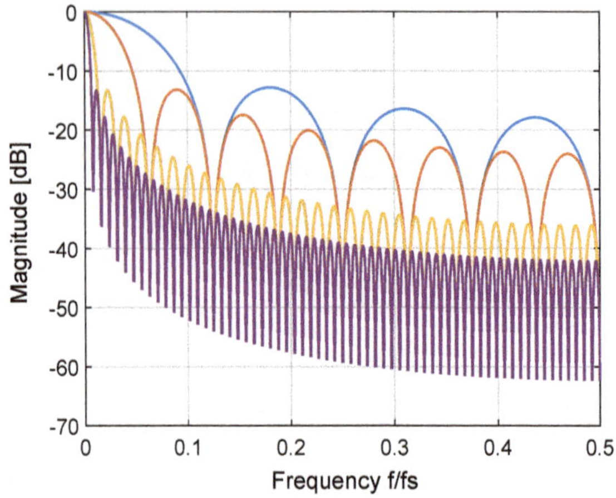

Fig. 1.9 Frequency responses of moving average circuits

1.2.5 Hard Real-Time Control Systems

The control circuit for a power electronics system should be a hard real-time system. What this means is that a control system function (hardware, software, or a combination of both) is considered hard real-time if, and only if, it has a hard deadline for the completion of an action or task [35]. This deadline must always be met, otherwise, the task has failed and in power electronics circuits there is a high risk of damage. A block diagram of an exemplary control circuit with one analog input and one analog output is shown in Fig. 1.10. Analog input signal $x(t)$ is converted into digital form with sampling rate f_s, and is then processed by a digital signal processor (DSP). Finally, output signal $y(n)$ is converted by a PWM modulator to pulses controlling output inverter switches S_1 and S_2. In this circuit all digital circuits have the same sampling frequency f_s. A typical timing diagram for such a control circuit for power electronics is depicted in Fig. 1.11. During the sampling period T_s all operations (calculation, conversion, communication etc.) must be completed. A control system may have more than one hard real-time task as well as other nonreal-time tasks. This is possible, as long as the system can properly schedule these tasks in such a way that the hard real-time tasks always meet their deadlines. The control system should ensure that the system still operates under worst-case response time to events. The control circuit should also be stable, even under transient overload when the system is overloaded by events and it is impossible to meet all deadlines, yet the deadlines of selected critical tasks must still be guaranteed.

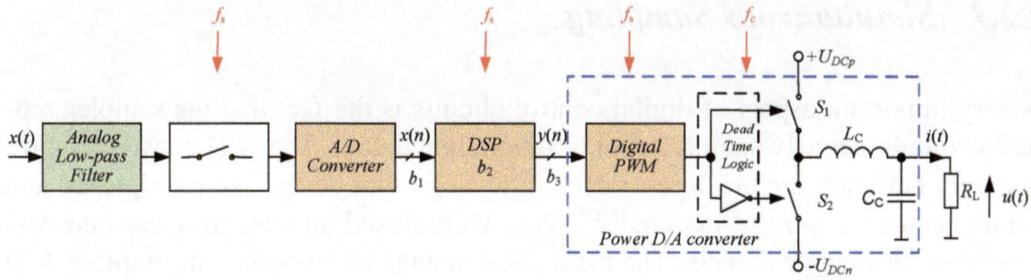

Fig. 1.10 Block diagram of exemplary control circuit for power electronics system

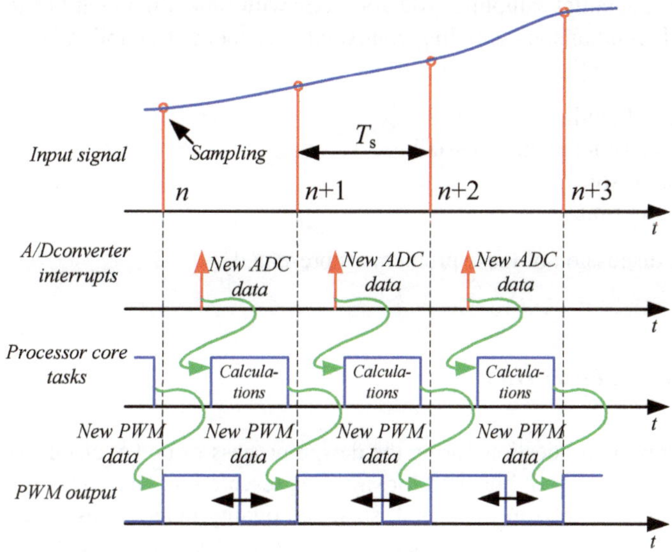

Fig. 1.11 Timing diagram for power electronics control circuit

1.2.6 Sampling Rate

A continuous analogue signal is sampled at discrete intervals, $T_s = 1/f_s$, which must be carefully chosen to ensure an accurate representation of the original analogue signal. It is clear that the more samples taken (i.e., faster sampling rates), the more accurate the digital representation, while if fewer samples are taken (i.e., lower sampling rates), a point is reached where critical information about the signal is actually lost. In the classical system for analogue signal band $0 \cdots f_b$, half of the sampling frequency $f_s/2$ is only a little bit higher than f_b. A rule of sampled data systems is that the input signal's spectrum gets folded around a frequency one-half that of the sampling clock. An ideal anti-aliasing filter would pass all signals in the band of interest and block all signals outside of that band. The quality of the anti-aliasing filter is the major factor in the signal-to-noise ratio (*SNR*). Problems of choosing the correct sampling rate are discussed in Chaps. 2 and 3.

1.2.7 Simultaneous Sampling

A very important aspect of digital control circuits is the fact that the samples represent simultaneously sampled data or time-aligned data. The A/D converters have two very common architectures, which in analog input design are multiplexed and simultaneous sampling solutions [27, 28]. Multiplexed architectures use one A/D converter for many channels. The main disadvantage of a sequential sampling A/D converter is the time error between channel samples. The best solution is to use a simultaneous sampling A/D converter, however, if it is not possible to use such a converter, the sequential sampling A/D converter with time alignment has to be used. Benefits of simultaneous sampling compared to sequential sampling include:

- less jitter,
- higher bandwidth,
- less channel-to-channel crosstalk,
- less phase error,
- less settling time.

In Chap. 2 discussion of these problems is presented.

1.2.8 Number of Bits

Another important question during the design process of power electronics control circuits is the number of bits used for signal representation. During the quantization process noise is added to the signal, so signal quality deteriorates. The quality of signal can be described by the signal to noise ratio (*SNR*) [24]. Signal-to-noise ratio can be determined by

$$SNR_{dB} = 10 \log_{10}\left(\frac{P_x}{P_n}\right) \quad , \tag{1.12}$$

where: P_x—power of signal, P_n—power of noise.
The following aspects of signal acquisition and processing are considered in this book:

- current and voltage measurements,
- signal galvanic isolation,
- choice of sampling rate,
- choice of number of bits,
- discussion of sequential and simultaneous sampling,
- synchronization with line voltage,
- signal filtration and separation.

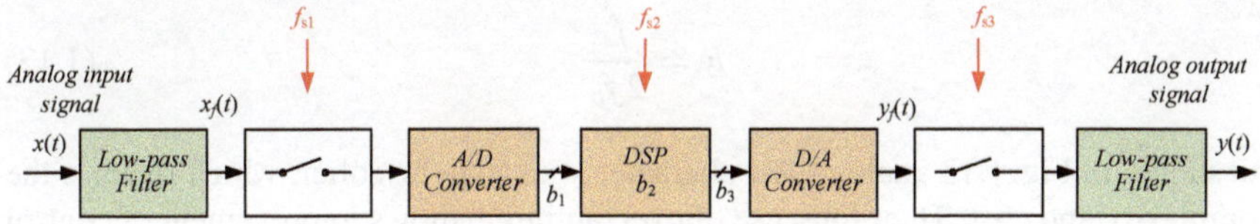

Fig. 1.12 The digital multirate control circuit with analog input and analog output

Fig. 1.13 The oversampling

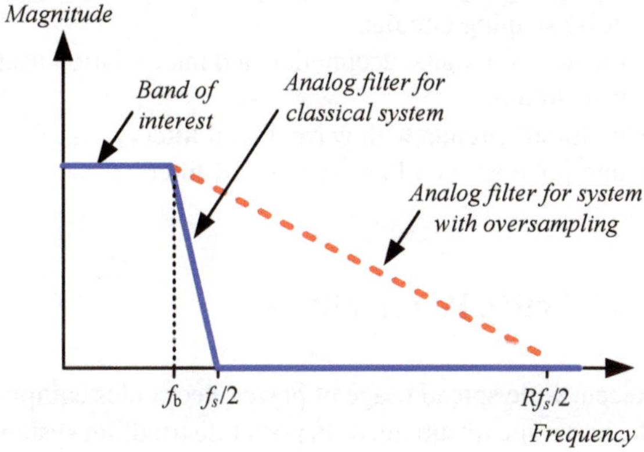

1.3 Multirate Control Circuits

Today most control circuits are built using multirate circuits. The most common multirate circuits are signal interpolators and decimators, used for changing the signal sampling rate. Multirate circuits are described by Crochiere and Rabiner [14], Vaidyanathan [56], Flige [20], Proakis and Manolakis [37] and by many others. An exemplary digital multirate control circuit with analog input and analog output is depicted in Fig. 1.12. The analog input signal $x(t)$, after passing through a low-pass anti-aliasing filter, is sampled with frequency f_{s1} and afterward is converted to digital form with b_1 bits resolution by an A/D converter. The digital signal processor performs the algorithms with sampling rate f_{s2} and resolution b_2 bits. The algorithm output signal is converted to analog form by a D/A converter with the sampling rate f_{s3} and with the resolution b_3 bits. In power electronics systems the output D/A converter is usually a pulse width modulator producing power transistor controlling pulses. In the past, digital systems worked at sampling speeds close to the maximum signal frequency. This resulted in very high requirements for analog input and output filters. At the present time the sampling rate can be easily and inexpensively increased, so the requirements for the analog input and output filter can be much lower. This phenomenon is shown in Fig. 1.13. A signal with band $0-f_b$ is sampled at a frequency much greater than $2f_b$. This technique is called oversampling. Oversampling ratio R is determined from the equation

$$R = \frac{f_s}{2f_b} \quad . \tag{1.13}$$

As shown in Fig. 1.13 analog filter characteristics may be softer, which reduces the complexity and cost. Therefore, use of oversampling allows improvement of control system performance. This book considers the following multirate circuit problems:

- increasing signal dynamic range by oversampling in A/D and D/A conversion,
- noise shaping circuits,
- methods for signal decimation and interpolation useful for power electronics control circuits,
- multirate circuits with wave digital filters,
- interpolators with linear-phase IIR filters.

1.4 Active Power Filters

Recent wide spread usage of power electronics equipment has caused an increase in the harmonic disturbances in power distribution systems. The control by transistors, AC power thyristors, thyristors and other semiconductor switches is widely applied to feed electric power to electrical loads, such as furnaces, computer power supplies, adjustable speed drives, etc. The nonlinear loads draw harmonic and reactive power components of current from the AC mains. In three-phase systems, they can also cause unbalance and draw excessive neutral currents. Reactive power burden, unbalance and injected harmonics cause a poor power factor, a low power system efficiency and excessive neutral currents.

Conventionally, passive LC filters and capacitors have been used to eliminate line current harmonics and to increase the power factor. However, in some practical applications, in which the amplitude and the harmonic content of the distortion power can vary randomly, this conventional solution becomes ineffective.

To suppress these harmonics, an active power harmonic compensator should be used, which is also called an active power filter (APF). The APF can be connected in series or in parallel with the supply network. The series APF is appropriate for the harmonic compensation of a large capacity diode rectifier with a DC link capacitor. The shunt APF (called also: parallel APF or current-fed APF) permits compensation of the harmonics and asymmetries of the mains currents caused by nonlinear loads. The idea of a shunt APF was introduced by Gyugyi and Strycula in 1976 [23]. Problems with APFs are discussed in many publications, among others [1, 3–5, 9, 53]. The shunt APF compensates the harmonics and asymmetries of the line current, caused by nonlinear loads. Two versions of harmonic compensation circuits with shunt APF are depicted in Fig. 1.14, where Z_M represents the mains power line impedance, Z_L represents nonlinear load and $e_M(t)$, the mains power line (source) voltage. Figure 1.14a shows an APF without feedback (with unity gain) and Fig. 1.14b presents an APF with feedback. Due to its better stability, in this book APFs without

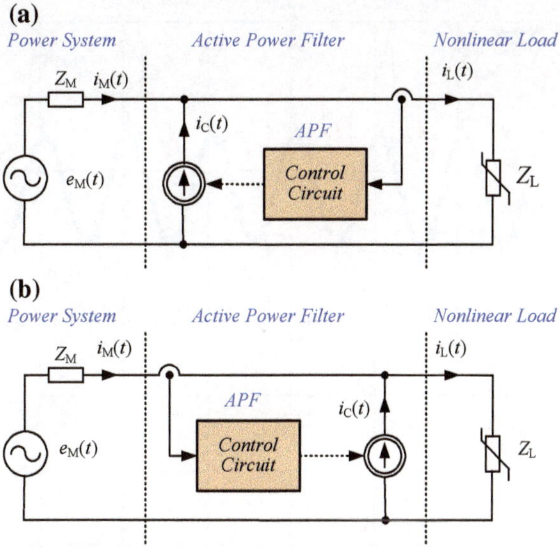

Fig. 1.14 Harmonic compensation circuit with current-fed active power filters: **a** without feedback (with unity gain), **b** with feedback

feedback are considered (Fig. 1.14a). The shunt APF injects AC power current $i_C(t)$ to cancel the main AC harmonic content. The line current $i_M(t)$ is the result of summing the load current $i_L(t)$ and the compensating current $i_C(t)$

$$i_M(t) = i_L(t) - i_C(t). \tag{1.14}$$

Three-phase shunt APFs are one of the best devices for compensating:

- harmonics,
- reactive power,
- asymmetries,
- voltage sag and swell—for shunt APF with additional DC energy bank.

Classical control algorithms of APFs are widely described in the literature, among which may be mentioned [1, 3–5, 9]. Experimental waveforms of compensation circuit with shunt APF load current i_L, line current i_M, and their spectra are shown in Fig. 1.15. When the value of load current changes rapidly, as in current i_L in Fig. 1.15, the APF transient response is too slow and the line current i_s suffers from dynamic distortion. This distortion causes an increase of harmonic content in the line current, which is dependent on a time constant. In the APF currents shown in Fig. 1.15 the total harmonic distortion (THD) ratio is increased by more than twelve percent. Solutions to this problem are presented in Chap. 5. There are also considered the following APF control algorithm modifications:

- new control circuits with prediction circuit,
- control algorithm based on sliding discrete Fourier transform (SDFT) and sliding Goertzel (SGDFT), moving DFT,

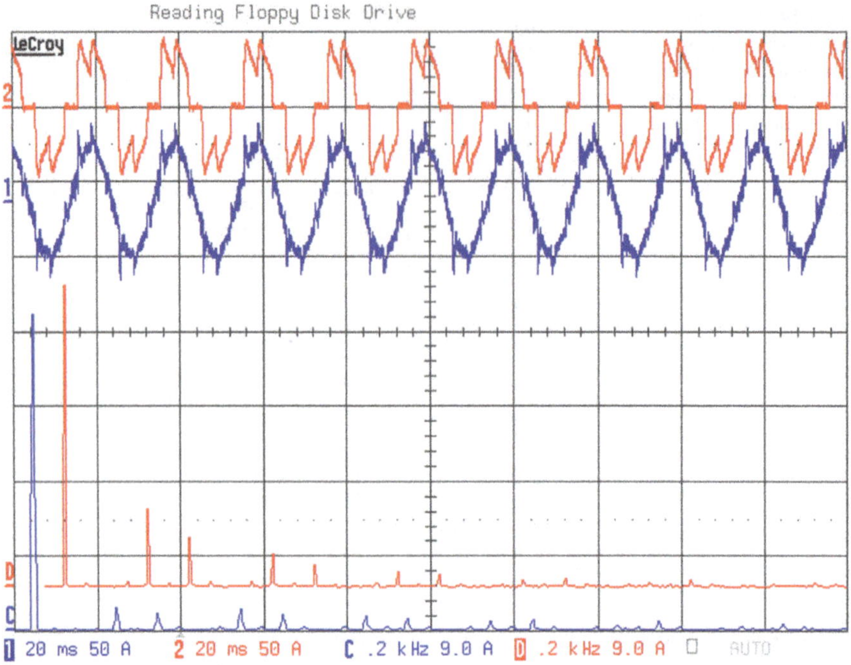

Fig. 1.15 Experimental waveforms in the shunt APF: load current $i_L(t)$—*red*, line current $i_M(t)$—*blue*, and theirs spectra

- filter bank solution for selective harmonics compensation,
- synchronization circuits,
- improving the dynamic range for load current much less than the APF nominal current,
- improving the dynamics of the APFs.

In this book the APF with novel modified output multirate shunt APF suitable for unpredictable loads [45] is described too.

1.5 Digital Class-D Power Amplifiers

In the past, the typical issues associated with audio power amplifiers [16, 40, 41] were not reported in the literature of power electronics, and despite the many similarities, both worlds lived parallel but in separated existences. However, the introduction of power class-D amplifiers, initially analog and then digital, encroached on both these areas. A power digital amplifier class-D is not much different from the power inverter or APF. The output power of audio power amplifiers is in the range of up to tens of watts in the homes, and in the range of up to kilowatts in the entertainment power

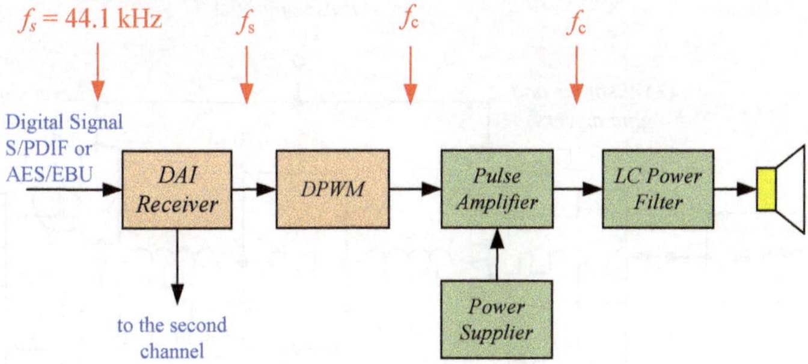

Fig. 1.16 Block diagram of digital class-D audio power amplifier

amplifier. The only significant difference is the dynamic range of processed signals, which for class-D amplifiers is more than 100 dB. In this respect, the author thinks it is appropriate to consider these problems.

Today there is an easy access to digital audio signal sources. Most audio signal sources are digital and have analog output solely because of conventional systems [8, 22, 32, 51, 52, 58]. However, it does not seem unreasonable to supply a digital signal directly to the loudspeaker. A block diagram of a digital power audio amplifier is shown in Fig. 1.16 [43, 50]. The digital audio input signal S/PDIF or AES/EBU (in the CD player standard, i.e., $b = 16$ bit with sampling rate $f_s = 44.1$ kHz) is divided into two channels left and right by a digital audio interface receiver (DAI). The next stage is a digital pulse width modulator (DPWM). The input signal is converted into signal pulses controlling the output power pulse amplifier. The loudspeaker is connected to a pulse amplifier through LC low-pass filter used for suppressing modulation harmonics. A typical audio band has a range of 20 Hz to 20 kHz. In a class-D power amplifier, depicted in Fig. 1.17, the output pulse power amplifier works as a one-bit D/A converter [50]. In the classical analog PWM circuit the conversion resolution is theoretically unlimited and the spectrum of modulation components depends only on the modulation method. In practice the conversion resolution is limited by component nonidealities [22, 39], while in DPWM resolution it is limited by the number of bits. Consideration of the following class-D amplifier problems is given in Chap. 6:

- PWM for open loop digital class-D audio power amplifier, noise shaping technique,
- harmonics distortion in open loop digital class-D amplifier,
- amplifier load—loudspeaker model for low, mid and high frequencies,
- digital class-D audio power amplifier with feedback from supply voltage, output pulse, output signal (after output filter),
- digital click modulator,
- influence of output DC source impedance on open loop class-D amplifier,
- power supply voltage ripple.

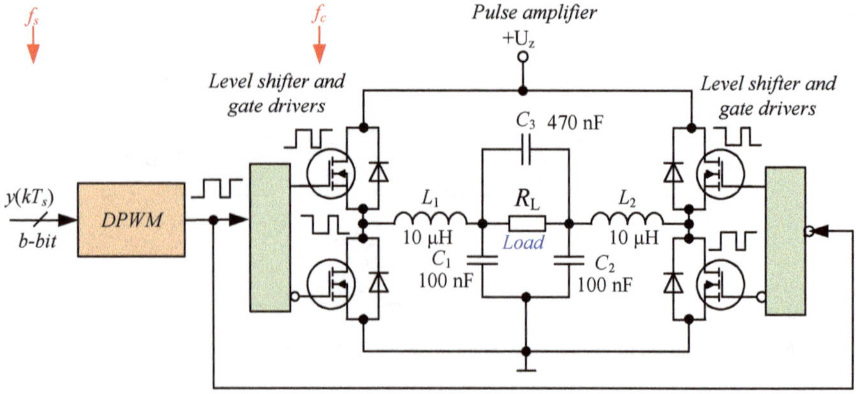

Fig. 1.17 Simplified diagram of a pulse amplifier width DPWM modulator

1.6 Symbols of Variables

In this book the symbols used for instantaneous values of variables that are a function of time are lowercase letters. The symbols u, i, and p respectively denote voltage, current and power. The following explicit notations are also used: $u(t) \equiv u, i(t) \equiv i$, $p(t) \equiv p$. For discrete instantaneous values of variables of time, notations used are: $u(nT_s) \equiv u(n)$, $i(nT_s) \equiv i(n)$, $p(nT_s) \equiv p(n)$, where: n—sample number, T_s—sampling period. Uppercase symbols U, I, P refer to their values computed from instantaneous values. Typically for DC current, uppercase denotes average value (avg) $U \equiv U_{avg}$, $I \equiv I_{avg}$, $P \equiv P_{avg}$ and for AC current, it denotes root mean square value (rms): $U \equiv U_{rms}$, $I \equiv I_{rms}$, $P \equiv P_{rms}$. Upper case symbols are also used for: Z-transform of signal $U(z)$, Laplace transform of signal $U(s)$ and spectrum of signal $U(\mathrm{j}\omega)$.

1.7 What Is in This Book

Chapter 2 covers the general problems of analogue signal acquisition from power electronics, such as sampling rate, number of bits, galvanic isolation, signal-to-noise ratio *SNR*, anti-aliasing filter, AC and DC current and voltage sensors, bandwidth of signal, signal range, jitter, and dither. Included is discussion of the influence of the above factors on signal dynamic range. This chapter discusses noise shaping circuits too. At the end of the chapter some selected A/D converters suitable for power electronics control circuits are also presented. Special attention is paid to simultaneous sampling A/D converters.

Chapter 3 presents methods of digital signal filtration and separation useful for power electronics control circuits. Moreover, taken into consideration are digital filters and filter banks, and algorithms that are especially useful for active power filters and class-D power amplifiers: sliding discrete Fourier transform (DFT), sliding Goertzel DFT and moving DFT, and linear-phase IIR filters. Wave digital filters are also considered there, which are especially useful for implementation using low-resolution arithmetics. Chapter 3 also includes problems of using digital filter banks in power electronics control circuits. Some consideration is given to DFT based filter banks and filter banks using lattice wave digital filters.

In Chap. 3 methods and circuits for signal changing sampling rate are also described, and interpolators and decimators useful for power electronics control circuits are considered. Special attention is paid to phase shifting by interpolator and decimator filters. Problems concerning the realization of the digital control algorithm are also described in Chap. 3. In addition, the following realizations of digital control circuits are considered: digital signal processors, microprocessors, microcontrollers, programmable digital circuits. In this chapter, these solutions are compared.

Selected aspects for simulation of power electronics circuit are considered in Chap. 4. Special attention is paid to simulation of power electronics circuits and power electronics circuits together with digital control circuits. The presented methods are accompanied by simulations using MATLAB® and PSIM. In the case of the PSIM program, the considered methods use C source for describing the digital control algorithm. In the case of the PSIM program, considered methods use C source for describing digital control algorithms.

Chapter 5 describes control circuits for shunt active power filters. Special attention is paid to improvement of the APF's dynamic range, as well as to the reduction of the APF's dynamic distortion. There is a presentation of the active power filter with modern control circuits. For predictable loads, a predictive control algorithm is used, and this allows for reduction of dynamic distortion. The usefulness of this method has been confirmed by experimental test results. However for unpredictable loads, a three-phase multirate APF with modified output inverter was designed. The simulation and the experimental test results of the considered APFs are presented and discussed.

A class-D digital power audio amplifier is presented in Chap. 6. The chapter starts with discussion of the open loop digital class-D power audio amplifier. Special attention is paid to the amplifier supply voltage influence on output signal quality. Next, amplifiers with feedback are considered: with supply voltage feedback, with output pulse feedback, with output signal feedback. In this chapter methods and circuits for changing signal sampling rate are described. Signal interpolators useful for class-D amplifier power circuits are considered. Special attention is paid to the phase shift introduced by the signal interpolator circuits. There are also examined signal interpolators using a two-path digital filter and linear-phase IIR filters. Special attention is paid to the linear-phase IIR filters using a lattice wave digital filter with extended dynamic range. In this chapter the pulse width modulator (PWM) with noise shaping technique and the modulator using click modulation are examined. Some simulation and experimental results are presented and discussed too.

Finally, Chap. 7 concludes the study of digital signal processing methods applied to power electronics control circuits.

References

1. Akagi H (1996) New trends in active filters for power conditioning. IEEE Trans Ind Appl 32(6):1312–1322
2. Akagi H, Kanazawa Y, Nabae A (1983) Generalized theory of instantaneous reactive power and its applications. Trans IEE Jpn 103(7):483–490
3. Akagi H, Kanazawa Y, Nabae A (1984) Instantaneous reactive power compensators comprising switching devices without energy storage components. IEEE Trans Ind Appl 1A–20(3):625–630
4. Akagi H, Watanabe EH, Aredes M (2007) Instantaneous power theory and applications to power conditioning. Wiley, Hoboken
5. Aredes M (1996) Active power line conditioners. PhD thesis, Technische Universitat Berlin
6. Astrom KJ, Wittenmark B (1997) Computer-controlled system. Theory and design. Prentice Hall Inc., New Jersey
7. Attia JO (1999) Electronics and circuit analysis using Matlab. CRC Press, Boca Raton
8. Baxandall PF (1959) Transistor sine-wave LC oscillators. In: ICTASD—IRE, pp 748–759
9. Benysek G, Pasko M (eds) (2012) Power theories for improved power quality. Springer, London
10. Bollen MHJ, Gu IYH, Santoso S, McGranaghan MF, Crossley PA, Ribeiro MV, Ribeiro PF (2009) Bridging the gap between signal and power. IEEE Sig Process Mag 26(4):12–31
11. Bose BK (2006) Power electronics and motor drives: advances and trends. Academic Press, Cambridge
12. Buso S, Mattavelli P (2015) Digital control in power electronics, 2nd edn. Morgan & Claypool, San Rafael
13. Chen WK (ed) (1995) The circuits and filters handbook. IEEE Press, Washington, D.C
14. Crochiere RE, Rabiner LR (1983) Multirate digital signal processing. Prentice Hall Inc., Upper Saddle River
15. Dabrowski A (ed) (1997) Digital signal processing using digital signal processors. Wydawnictwo Politechniki Poznanskiej, Poznan (In Polish)
16. Duncan B (1996) High performance audio power amplifier for music performance and reproduction. Newnes, Oxford
17. Erickson RW, Maksimovic D (2004) Fundamentals of power electronics. Kluwer Academic Publishers, Dordrecht
18. Fettweis A (1971) Digital filter structures related to classical filter networks. AEU 2:79–89 Band 25. Heft
19. Fettweis A (1986) Wave digital filters: theory and practice. Proc IEEE 74(2):270–327
20. Flige N (1994) Multirate digital signal processing. Wiley, Hoboken
21. Gazsi L (1985) Explicit formulas for lattice wave digital filters. IEEE Trans Circuits Syst 32(1):68–88
22. Goldberg JM, Sandler MB (1994) New high accuracy pulse width modulation based digital-to-analogue convertor/power amplifier. IEE Proc Circ Devices Syst 141(4):315–324
23. Gyugyi L, Strycula EC (1976) Active AC power filters. In: Proceedings of IEEE industry applications annual meeting, pp 529–535
24. IEEE (2011) IEEE Standard for terminology and test methods for analog-to-digital converters, IEEE Std. 1241-2010, IEEE, Technical report
25. Kazimierkowski M, Malesani L (1998) Current control techniques for three-phase voltage-source converters: a survey. IEEE Trans Ind Electron 45(5)

26. Kazimierkowski MP, Kishnan R, Blaabjerg F (2002) Control in power electronics. Academic Press, Cambridge
27. Kester W (2004) Analog-digital conversion. Analog Devices Inc., Norwood
28. Kester W (2005) The data conversion handbook. Newnes, Oxford
29. Lyons R (2004) Understanding digital signal processing, 2nd edn. Prentice Hall, Upper Saddle River
30. Mitra S (2006) Digital signal processing: a computer-based approach. McGraw-Hill, New York
31. Mohan N, Undeland TM, Robbins WP (1995) Power electronics, converters, applications and design. Wiley, Hoboken
32. Nielsen K (1998) Audio power amplifier techniques with energy efficient power conversion. PhD thesis, Departament of Applied Electronics, Technical University of Denmark
33. Oppenheim AV, Schafer RW (1999) Discrete-time signal processing. Prentice Hall, New Jersey
34. Orfanidis SJ (2010) Introduction to signal processing. Prentice Hall Inc., New Jersey
35. Oshana R (2005) DSP software development techniques for embedded and real-time systems. Newnes, Oxford
36. Owen M (2007) Practical signal processing. Cambridge University Press, Cambridge
37. Proakis JG, Manolakis DM (1996) Digital signal processing, principles, algorithms, and application. Prentice Hall Inc., Upper Saddle River
38. Rabiner LR, Gold B (1975) Theory and application of digital signal processing. Prentice Hall Inc., Upper Saddle River
39. Santi S, Ballardini M, Rovatti R, Setti G (2005) The effects of digital implementation on ZePoC codec. In: Conference proceedings, ECCTD—IEEE III, pp 173–176
40. Self D (2002) Audio power amplifier design handbook. Newnes, Oxford
41. Slone GR (1999) High-power audio amplifier construction manual. McGraw-Hill, New York
42. Sozański K (2012) Realization of a digital control algorithm. In: Benysek G, Pasko M (eds) Power theories for improved power quality. Springer, London, pp 117–168
43. Sozanski K (1999) Design and research of digital filters banks using digital signal processors, PhD thesis, Technical University of Poznan (in Polish)
44. Sozanski K (2004) Harmonic compensation using the sliding DFT algorithm. In: Conference proceedings, 35th annual IEEE power electronics specialists conference, PESC 2004, Aachen, Germany
45. Sozanski K (2007) The shunt active power filter with better dynamic performance. In: Conference proceedings, power tech 2007 conference, Lausanne, Switzerland
46. Sozanski K (2008) Improved shunt active power filters. Przeglad Elektrotechniczny (Electrical Review) 45(11):290–294
47. Sozanski K (2010) Digital realization of a click modulator for an audio power amplifier. Przeglad Elektrotechniczny (Electrical Review) 2010(2):353–357
48. Sozanski K (2015) Selected problems of digital signal processing in power electronic circuits. In: Conference proceedings SENE 2015, Lodz, Poland
49. Sozanski K (2016) Signal-to-noise ratio in power electronic digital control circuits. In: Conference proceedings: signal processing, algorithms, architectures, arrangements and applications—SPA 2016. Poznan University of Technology, pp 162–171
50. Sozanski K, Strzelecki R, Fedyczak Z (2001) Digital control circuit for class-D audio power amplifier. In: Conference proceedings, 2001 IEEE 32nd annual power electronics specialists conference—PESC 2001, pp 1245–1250
51. Streitenberger M, Bresch H, Mathis W (2000) Theory and implementation of a new type of digital power amplifiers for audio applications. In: Conference proceeding, ICAS 2000, vol I. IEEE, pp 511–514
52. Streitenberger M, Felgenhauer F, Bresch H, Mathis W (2002) Class-D audio amplifiers with separated baseband for low-power mobile applications. In: Conference proceeding, ICCSC'02. IEEE, pp 186–189
53. Strzelecki R, Fedyczak Z, Sozański K, Rusiński J (2000) Active power filter EFA1. Technical report, Instytut Elektrotechniki Przemyslowej, Politechnika Zielonogorska (in Polish)

54. Tantaratana S (1995) Design of IIR filters. In: Chen WK (ed) (1995) The circuits and filters handbook. IEEE Press
55. Trzynadlowski A (2010) Introduction to modern power electronics. Wiley, Hoboken
56. Vaidyanathan PP (1992) Multirate systems and filter banks. Prentice Hall Inc., Upper Saddle River
57. Venezuela RA, Constantindes AG (1982) Digital signal processing schemes for efficient interpolation and decimation. IEE Proc G 130(6):225–235
58. Verona J (2001) Power digital-to-analog conversion using sigma-delta and pulse width modulations. ECE1371 Analog Electronics II 2001(II):1–14
59. Wanhammar L (1999) DSP integrated circuit. Academic Press, Cambridge
60. Williamson D (1991) Digital control and implementation. Prentice Hall Inc., New Jersey
61. Zielinski TP (2005) Digital signal processing: from theory to application. Wydawnictwo Komunikacji i Lacznosci, Warsaw (in Polish)
62. Zielinski TP, Korohoda P, Rumian R (eds) (2014) Digital signal processing in telecommunication, Basics. Multimedia, Transmission, Wydawnictwo Naukowe PWN, Warsaw (in Polish)

Chapter 2
Analog Signals Conditioning and Discretization

2.1 Introduction

In control systems it is necessary to observe the work of the controlled object. This is usually achieved in power electronics systems by measuring currents and voltages [7, 112]. This chapter is devoted to the problems of processing the analog current and voltage signals to convert them into digital form in power electronics systems [89, 90]. The problems discussed in this chapter, the author believes, are essential for the proper operation of control systems.

2.2 Analog Input

Typical circuits of analog inputs for measuring voltages and currents are shown in Fig. 2.1. In the voltage measurement circuit (Fig. 2.1a), voltage from the voltage divider (R_1, R_2) goes through the amplifier input and the antialiasing filter to the sample–and–hold circuit (SH) and analog–to–digital converter (A/D). Finally digital signal $x_1(nT_s)$ corresponding to the measured voltage is sent to the processor control system. A similar process takes place in the current measurement circuit (Fig. 2.1b), where instead of a voltage divider, current transducer is applied. Current measurement issues are discussed in Sect. 2.3.

2.2.1 Galvanic Isolation

In low-power electronics systems it is possible to use measurement systems electrically coupled with the control system. This is often the cheapest and best option, especially if accuracy is taken into account. However, usually in high-power systems galvanic isolation in measurements is required. It should be noted that the

K. Sozański, *Digital Signal Processing in Power Electronics Control Circuits*,
Power Systems, DOI 10.1007/978-1-4471-7332-8_2

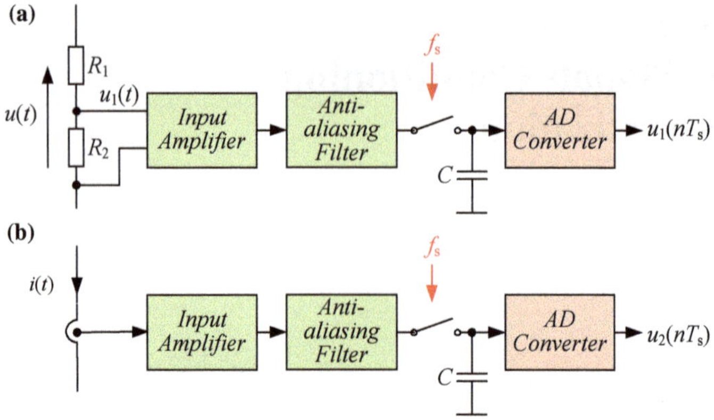

Fig. 2.1 Analog input of A/D converter: **a** voltage measurement, **b** current measurement

application of galvanic isolation always deteriorates the quality of the signal, but it is often necessary. A system with galvanic isolation is shown in Fig. 2.2, consists of two circuits, A and B, with signals between circuits are connected through an isolation barrier. The signals can be transmitted through an isolation barrier using optical wave, sound wave, radio wave, capacitive coupling, inductive coupling, or mechanical coupling (piezoelectrics). Circuits A and B are supplied by separate power supply units. Reasons for the use of galvanic isolation include:

- **Protection from high voltages**. Isolation provides a dielectric barrier that acts as an insulator against high voltages in power electronics circuits where higher power levels are required. It is also important for safety reasons.
- **Break ground loop**. Galvanic isolation breaking the ground currents (return path) of an electrical circuit to only one side of the barrier, enabling a noise-free environment for sensitive measurements on the other side. Galvanic isolation is increasing noise immunity.
- **Level translation**. Enabling noise-free data transfer between circuits that operate at different voltage rails is a common challenge for electronics designers.

Fig. 2.2 System with galvanic isolation

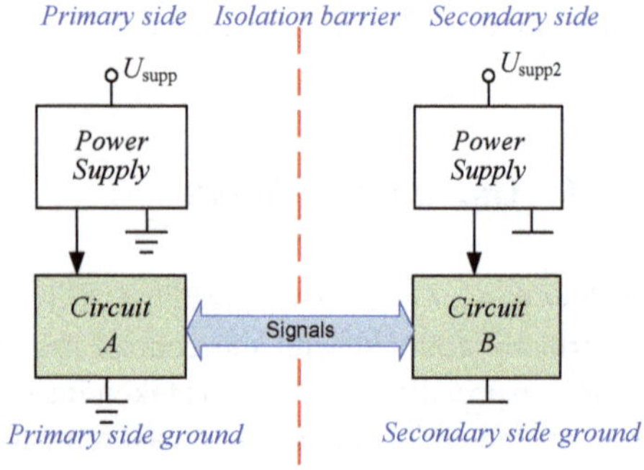

For example, the control gates of transistors in the inverter is a key issue in power electronics circuits. Although there are many non-isolated level shifters available to circumvent this problem, using an isolator provides several solid advantages. Isolators are the most noise-free and robust solution, and they prevent parasitic paths that may inadvertently switch devices on or off.

- **Protection from high common voltage**. Galvanic isolation protects parts of circuit working with different voltage level. This eliminates the need for a level shifter, however, requires the use of additional glvanically isolated power supply.

2.2.2 *Common Mode Voltage*

Measurements in power electronics systems are very difficult due to a very high slew rate of common mode voltage. The common mode voltage swings of several hundred volts in tens of nanoseconds are common in modern switching inverters. The isolation amplifier should be designed to ignore very high common mode transient slew rates (of at least from 10 to 25 kV/μs or more).

Figure 2.3 depicts an isolation amplifier connected to one leg of a power inverter. During the operation of the inverter, the potential of point A voltage changes from −500 to 500 V in a time equal to 100 ns. The slew rate of voltage change can be determined by the equation

$$SR = \frac{\mathrm{d}u(t)}{\mathrm{d}t} \approx \frac{\Delta U}{\Delta t} \ . \tag{2.1}$$

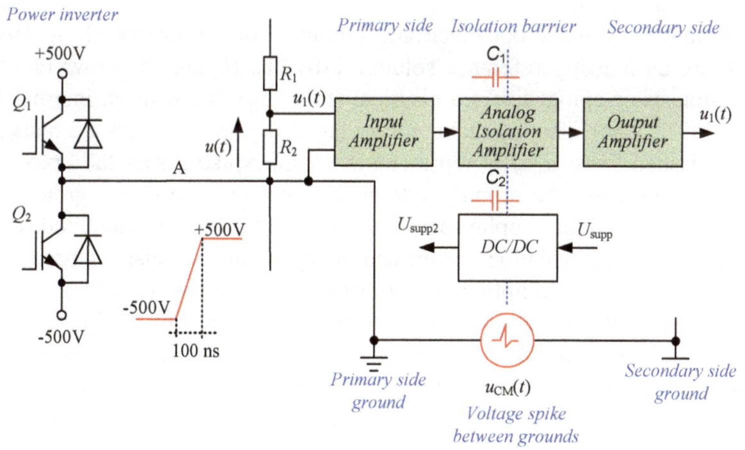

Fig. 2.3 Circuit for illustration of common mode voltage

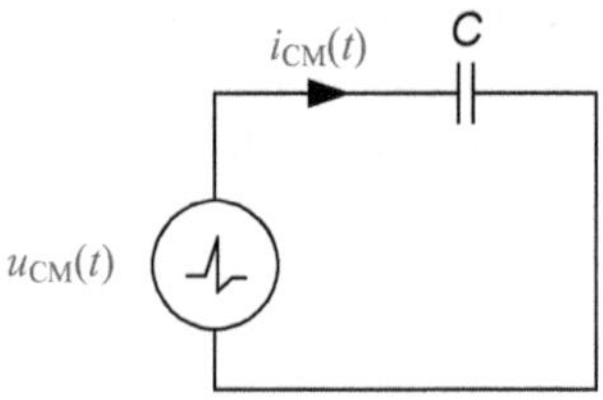

Fig. 2.4 Simplified circuit for common mode current

Capacitor C_1 represents the resultant common mode (parasitic) capacitance on the signal path, capacitor C_2 represents the resultant common mode (parasitic) capacitance in the DC/DC converter. The circuit of an isolation amplifier for a common mode (parasitic) current calculation can be simplified to the circuit shown in Fig. 2.4. Thus, the value of parasitic current can be calculated by equation

$$i_{CM}(t) = C\frac{\mathrm{d}u_{CM}(t)}{\mathrm{d}t} \approx C\frac{\Delta U_{CM}}{\Delta t} \quad , \tag{2.2}$$

where: C—resultant common mode (parasitic) capacitance.
As an example, for $SR = 10\,\mathrm{kV/\mu s}$ and $C = 10\,\mathrm{pF}$ common mode current the value is 10 mA. It is a too high value especially for sensitive electronics input circuits, such as optoelectronic devices.

Common mode rejection (CMR) is a measure of the ability of a device to tolerate common mode noise. Usually common mode rejection is specified as common mode transient rejection (CMTR). CMTR describes the maximum tolerable rate-of-rise (or fall) of a common mode voltage, usually given in kilovolts per microsecond (kV/μs). The specification for CMTR also includes the amplitude of the common mode voltage $u_{CM}(t)$ that can be tolerated.

For example, Avago's optoelectronic circuits (former produced by Hewlett Packard) are an industry reference solution [10, 11]. Figure 2.5 shows two types of single-transistor optocouplers a classical (Fig. 2.5a) and with an internal Faraday shield (Fig. 2.5c). The internal components of the optocouplers are marked by the green dashed lines. $u_{CM}(t)$ represents a voltage spike across the optocoupler isolation path between the primary side ground and secondary side ground. $E(t)$ represents a signal voltage applied across the input side. The common mode current $i_{CM}(t)$ flows through isolation barrier and changes output transistor base current $i_B(t)$. Figure 2.5b shows simplified waveforms for common mode voltage response for the classical optocoupler. This type of failure is avoided by adding an internal Faraday shield as shown in Fig. 2.5c. Referring to Fig. 2.5c, the parasitic distributed capacitances, C_1 and C_2 are shown across the LED anode to secondary side ground and LED cathode to secondary side ground. The common mode current flown through capacitor C_1 changes the LED current during common mode transients. For instance, if the LED is on, then during a positive transient (i.e., $\mathrm{d}u_{CM}(t)/\mathrm{d}t > 0$), the LED current will be decreased. For fast enough transients, this may turn the LED off. Figure 2.5d shown simplified waveforms for common mode voltage response for

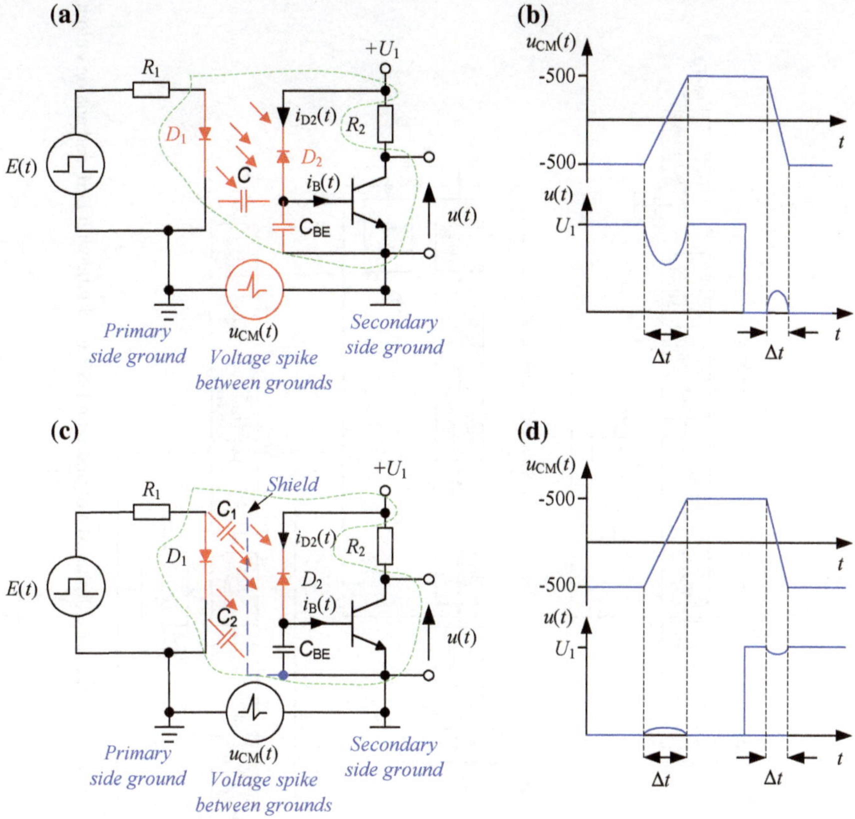

Fig. 2.5 Optocouplers: **a** a classical optocoupler, **b** simplified waveforms for common mode voltage response for classical circuit, **c** a optocoupler with shield, **d** simplified waveforms for common mode voltage response for circuit with shield

circuit with the internal Faraday shield. This type of failure is avoided by ensuring that C_1 and C_2 are small. Also, the printed circuit board and connections should be designed to minimize the parasitic capacity.

2.2.3 *Isolation Amplifiers*

Topologies of isolation amplifiers are shown in Fig. 2.6. A topology with analog isolation amplifier is depicted in Fig. 2.6a. In this solution solution is mostly used either an optical or electromagnetic separation. The cheapest one is optical separation with linearization. The advantage of such a system is the small signal delay. A typical example of solution with optical separation is the integrated circuit (IC),

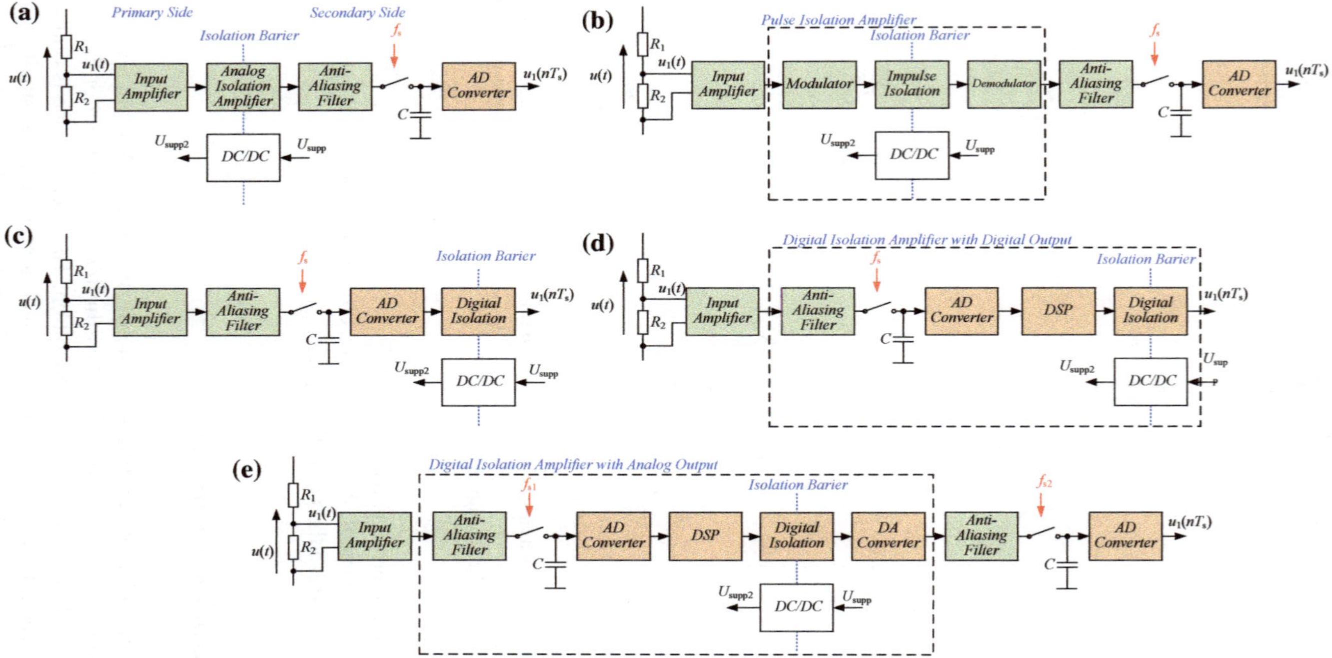

Fig. 2.6 Signal isolation technique typologies: **a** an analog solution, **b** a pulse solution, **c** a digital, **d** a digital with DSP, **e** a digital with DSP and analog output

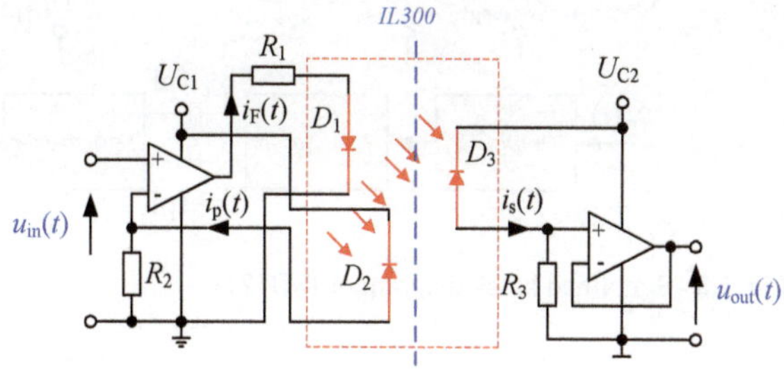

Fig. 2.7 Isolation amplifier with feedback IL300

IL300 from Vishay [107] (originally introduced by Siemens). The IC Fig. 2.7 consists of an LED D_1 irradiating an isolated feedback photodiode and an output photodiode D_3 in a bifurcated arrangement. The feedback photodiode D_2 captures a part of the LEDs flux and generates a control signal $i_p(t)$ that can be used to serve the LED drive current $i_F(t)$. This technique couples AC and DC signals and compensates for the LED's non-linear, time, and temperature characteristics. The IC achieves the following parameters: 0.01% servo linearity, wide bandwidth, >200 kHz, high gain stability, 0.005 %/°C typically.

The time and temperature stability of the coupler transfer function is insured by using matched PIN photodiodes D_2 and D_3 that accurately track the output flux of the LED. In alternative solutions the transformer is used, which allows the transfer of AC signals and DC signals by adding the Hall sensor. Primary circuits are supplied by an isolated DC/DC converter.

The most common used topology of the insulation circuit is shown in Fig. 2.6b. In this system an isolation pulse amplifier is applied. It consists of an input modulator, pulse signal isolator and demodulator. Such systems give high accuracy at a low price. In typical applications the modulation frequency is in the range of 10 kHz–1 MHz. The use of modulation prolongs the typical impulse response time and causes the occurrence of modulation components in the output signal. In applications with transformer isolation, the transformer can also be used to transfer energy to supply the primary side circuits.

Examples of industrial integrated circuits include: isolation amplifier with capacitive isolation ISO124, ISO121 [93, 96], Si8920 [83, 84], classical isolation amplifier with transformer isolation AD215 [4], isolation amplifier with optical isolation barrier HCPL-7800 [8].

An interesting solution has been applied in the ISO121, through the use of precision isolation amplifiers with duty cycle modulation-demodulation technique. The signal is transmitted digitally across a 2 × 1 pF differential capacitive barrier. Simplified block diagram of ISO121 is shown in Fig. 2.8. The IC achieves the following parameters: 0.01% max nonlinearity, bandwidth, 6 kHz, high gain stability and 150 μV/°C maximum offset voltage drift. The IC, shown in Fig. 2.9, is built into a ceramic barrier, where modulator and demodulator are placed at the ends and two 1 pF matched barrier capacitors are placed in the middle. This IC has excellent reliability.

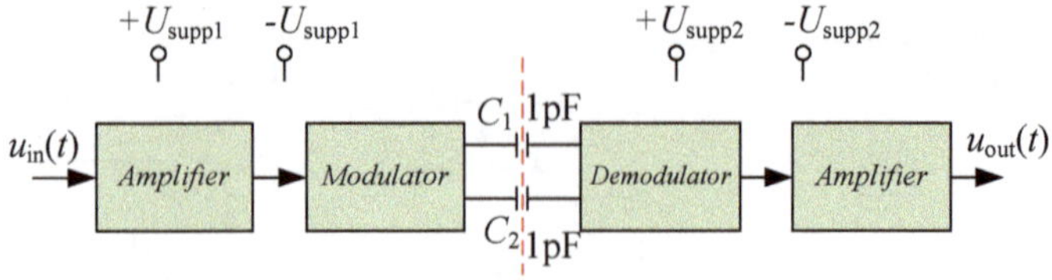

Fig. 2.8 Simplified block diagram of ISO121

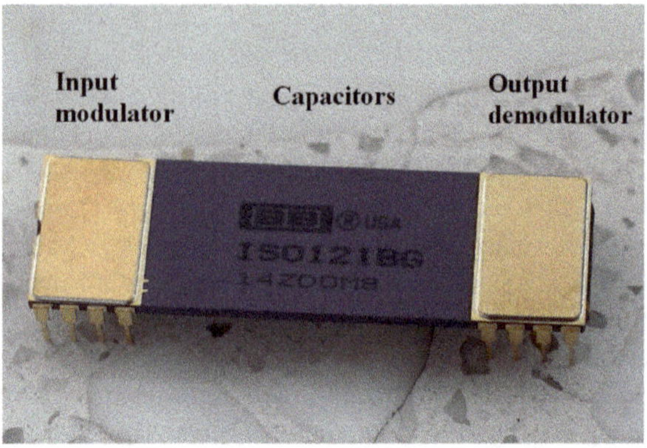

Fig. 2.9 Precision isolation amplifier with capacitor barrier isolation

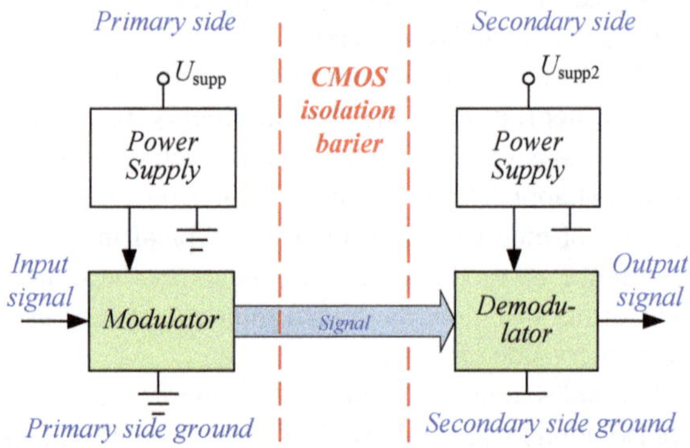

Fig. 2.10 Si8920 isolation amplifier with capacitor barrier isolation

The Si8920 is a low-cost galvanically isolated analog amplifier [83, 84] from Silicon Laboratories. The block diagram of the Si8920 IC is shown in Fig. 2.10. The low-voltage differential input ±100 mV, with nonlinearity: 0.1% full scale, is good for measuring voltage across a current shunt resistor or for any application where a sensor must be isolated from the control system. The amplifier has very high common-mode transient immunity: 75 kV/μs, it is important for power electrons application. Another advantage of this circuit is that it has low signal delay: 0.75 μs, especially important for feedback circuit. This technology enables higher performance, reduced variation with temperature and age, tighter part-to-part matching, and longer lifetimes

Table 2.1 Comparison of isolation amplifiers

Signal isolation technique	Accuracy	Delay	Power consumption	Relative cost	Typical signal ranges
Analog, optoelectronic isolation	0.5–5%, High nonlinearity and temperature drift	Low	Low	Low	DC to 200 kHz
Analog with transformer isolation	0.5–5%	Low	Low	Low	AC, 50–200 kHz
Pulse, optoelectronic isolation	0.01–0.5%	Moderate	Low	Low	DC to 100 kHz, output with signal containing residuals of the modulation component
Pulse with transformer isolation	0.01–0.5%	Moderate	Moderate	Low	DC to 300 kHz, output with signal containing residuals of the modulation component
Pulse with capacitive isolation	0.01–0.5%	Moderate	Moderate	Low	DC to 300 kHz, output with signal containing residuals of the modulation component
With A/D converter on primary side	Depending on the A/D converter, possible error correction	Low	Moderate-High	Moderate	DC to 100 kHz, difficult realization of simultaneous sampling
With A/D converter and DSP on primary side	Depending on the A/D converter	Moderate	High	High	DC to 100 kHz, difficult realization of simultaneous sampling
Digital transducer on primary side with analog output	Depending on the A/D and D/A converters, possible error correction	High	High	High	DC to 50 kHz, double signal conversion, difficult realization of simultaneous sampling

compared to other isolation technologies. Despite the use of typical plastic enclosure (DIL8) it supports up to 5 kV_{RMS} withstand voltage per UL1577 standard.

In the next solution the A/D converter was moved to the primary side. Hence signal to the secondary side is transmitted in digital form (Fig. 2.6c). Such a solution allows the elimination of errors introduced by the insulation system. But in the case of multi-channel systems it is difficult to synchronize the sampling moments. Optically isolated sigma delta modulator HCPL-7860 [9] with digital output is the simplest example of such a solution.

In the next circuit (Fig. 2.6d) is an added DSP to allow local measurement error correction and additional algorithms. In another topology (Fig. 2.6e) a digital isolation transducer is used. In this design the input signal is converted to digital form and in this form it is sent to the secondary side, then it is converted to analog from. In this configuration, as in the previous one, it is possible to correct an error. The disadvantages of this system are the high price and the double signal conversion causing signal delay. Comparison and summary of the galvanic isolation techniques are presented in Table 2.1.

2.3 Current Measurements

Current sensors are often used to provide essential information to power electronics control systems. Current transducers convert measured current to a proportional AC or DC voltage or milliamp signal. These devices should have extremely low insertion impedance. There are also current transducers with digital output and in the author's opinion they will play an increasing role in current measurement systems in future. There are several techniques that are typically used for measuring currents: sense resistor (resistive shunt), current transformer, current transformer with Hall effect sensor, current transformer with magnetic modulation and air coil.

2.3.1 A Resistive Shunt

A sense resistor is inserted in series with the load. Through Ohm's law, $U = IR$, we know the voltage drop across the resistor is proportional to the current. This system is very simple and provides very accurate measurements, given that the resistance value has a tight tolerance. To maintain low dissipation power voltage across sense resistors, the resistance value should be very low, therefore they also require a high quality amplifier, such as instrumentation amplifiers, to generate an exact signal. This

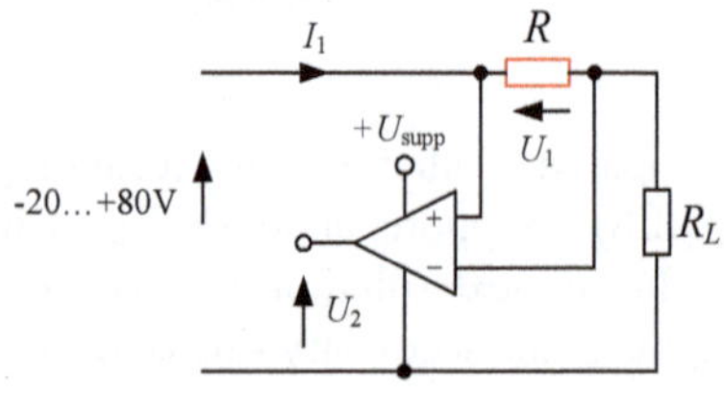

Fig. 2.11 Current measurement with high common mode amplifier

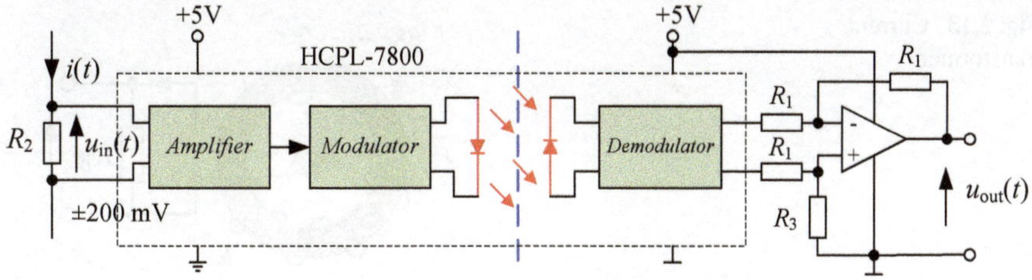

Fig. 2.12 Simplified diagram of HCPL-7800 current measurement circuit

kind of measurement does not provide galvanic isolation, so in some applications an isolation amplifier needs to be used. For larger currents sense resistors with high performance thermal packages have been used. In the last few years there has been extensive development of portable battery-powered electronic devices, and as a result a high demand for current measurement systems has been created. It is a typical way of measuring battery currents, thus, there has arisen a great demand for simple and inexpensive measurement systems. Figure 2.11 shows a measurement system with an amplifier with a unique high common mode ratio to allow for work above the positive supply voltage (U_{supp}) and under the negative supply voltage of amplifier. Common mode voltage of such amplifiers achieves −20–80 V. A typical voltage value U_2 is several hundred millivolts. Many manufacturers produce such circuits, for example: integrated circuit INA270 from Texas Instruments [97], AD8210 from Analog Devices [5] etc. Due to the voltage range, this kind of amplifier is also widely used in automotive applications. For current sensing in power electronics circuits the HCPL-7800 family isolation amplifier was designed [8]. The HCPL-7800 utilizes delta sigma modulator converter technology, chopper stabilized amplifiers, and a fully differential circuit topology. Figure 2.12 depicts a simplified diagram of a current measurement circuit. In a typical implementation, currents flow through an external resistor and the resulting analog voltage drop is sensed by the HCPL-7800. The HCPL-7800 input voltage range is equal to ± 200 mV. A differential output voltage is created on the other side of the HCPL-7800 optical isolation barrier. This differential output voltage is proportional to the input current and can be converted to a single-ended signal by using an operational amplifier as shown in Fig. 2.12. The HCPL-7800 was designed to ignore very high common-mode transient slew rates (of at least 10 kV/μs). A similar solution can be achieved using newer IC Si8920 [83, 84] from Silicon Laboratories, which was described in the previous section.

2.3.2 Current Transformers

Current transformers are relatively simple and passive (self-powered) devices, and do not require driving circuitry to operate. They are two-wire components with voltage or current output. The primary current (AC) will generate a magnetic field. The field

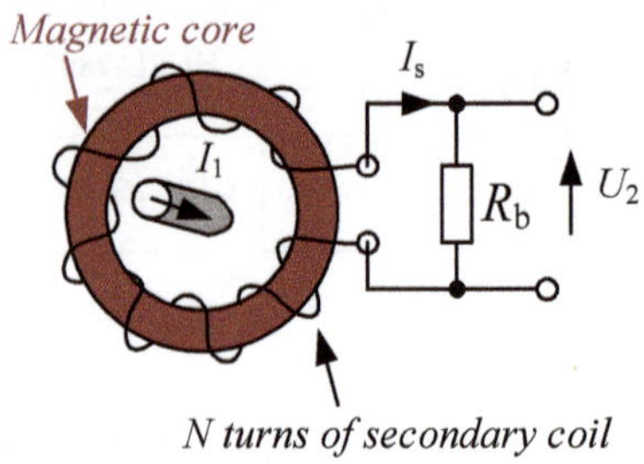

Fig. 2.13 Current transformer

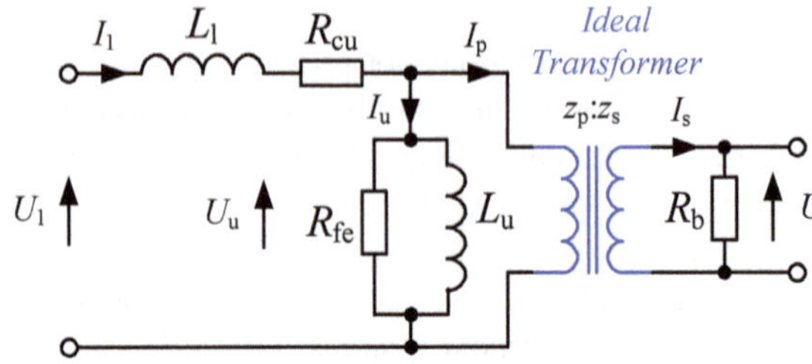

Fig. 2.14 Simplified equivalent circuit of current transformer

is concentrated by magnetic core. The secondary coil is coupled with the primary coil by magnetic field. This is the principle that governs all transformers. Current transformers are designed to measure AC current and typically work between 20 and 400 Hz, although some units will work in the kilohertz range. Inductive current transducers are available in both solid-core and split-core configurations. For an ideal transformer, the secondary current magnitude is proportional to ratio of the primary number of turns z_p (typically from one to several) to secondary number of turns z_s (typically thousands), thus, the secondary current I_s for an ideal transformer can be calculated by the equation

$$I_s = \frac{z_p}{z_s} I_p \quad . \tag{2.3}$$

The current transformer is shown in Fig. 2.13. The secondary current is then sensed through a sense resistor R_b to convert the output into a voltage U_2.

For a real transformer this equation is more complicated. Figure 2.14 shows a simplified version of a low frequency current transformer model of electric circuit. This model is called the high side equivalent circuit model because all parameters have been moved to the primary side of the ideal transformer. In this circuit: L_l—resultant windings leakage inductance, R_{cu}—resultant windings resistance, R_{fe}—resistance which represents power losses in transformer core (mainly due to hysteresis), L_u—main transformer inductance, magnetizing inductance.

Hence, for this transformer model secondary current should be calculated by the equation

$$I_s = \frac{z_p}{z_s}(I_1 - I_u) \quad . \tag{2.4}$$

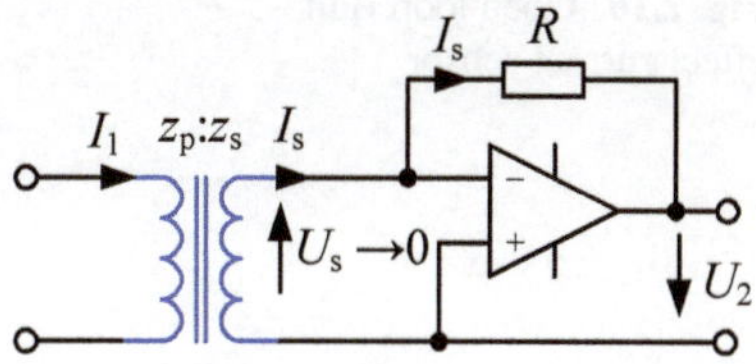

Fig. 2.15 Current converter with minimized current transformer output voltage

Magnetizing current i_u will give an error in the current transformation of the secondary side. In order to reduce the error, voltage at the output should be very small. Another way to decrease magnetizing current is by increasing the core size. Figure 2.15 shows a circuit which minimizes current transformer output voltage and hence reduces current error.

The presented current transformer model has sufficient accuracy only for low frequency use, while for high frequency use it has to be more complicated for example in [17]. Current transformers are one of the simplest and relatively cheap solutions for current measurements but they have one major disadvantage in that they can not transform DC. Therefore, in applications that require measurements of the DC current, other techniques should be used.

2.3.3 *Transformer with Hall Sensor*

A simplified diagram of an open loop Hall effect current sensor is depicted in Fig. 2.16. The sensor measures DC, AC and complex current waveforms while providing galvanic isolation. The Hall effect current sensor consists of three basic components: the magnetic core, the Hall effect sensor, and signal conditioning circuitry. The Hall sensor is located in the magnetic core gap. The magnetic flux created by the primary current I_1 is concentrated in a magnetic circuit and measured in the air gap using a Hall sensor.

$$U_H = \frac{k}{d} I_C B + U_{off} \quad . \tag{2.5}$$

where: k—the Hall constant of the conducting material, d—the thickness of the sensor, I_C—constant current, B—magnetic flux density, and U_{off} is the offset voltage of the Hall generator in the absence of an external field. Such an arrangement is referred to as a Hall generator and the ratio k/dI_C is generally described as the Hall generator sensitivity. The output signal from the Hall device is then conditioned to provide an exact representation of the primary current at the output. The relation between primary current and the magnetic flux density, B, is non linear. Therefore, a magnetic core in the linear region is used. Within the linear region of the hysteresis loop of the material used for the magnetic circuit, the magnetic flux density, B, is proportional to the primary current, I_1, and the Hall voltage, U_H, is proportional to the magnetic flux density. Therefore the output of the Hall generator is proportional

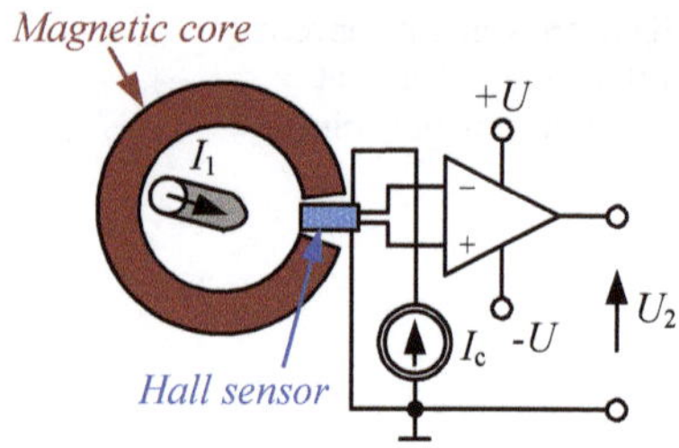

Fig. 2.16 Open loop Hall effect current sensor

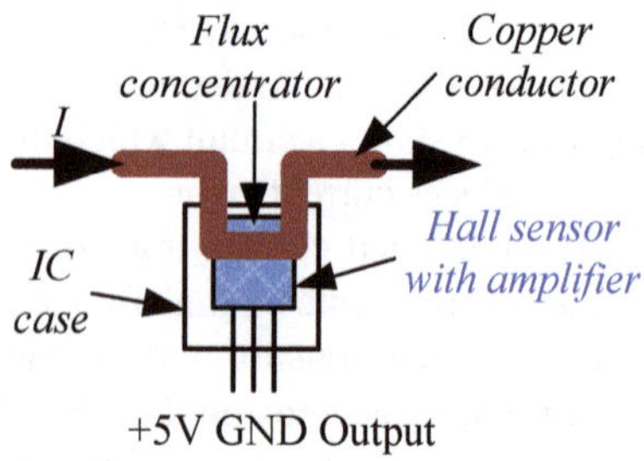

Fig. 2.17 IC current sensor from Allegro MicroSystems

to the primary current, plus the Hall offset voltage, U_{off}. The advantages of open loop transducers include: low cost, small size, low weight, low power consumption and very low insertion power losses. The accuracy is limited by the combination of:

- DC offset at zero current (hall generator, electronics, or remanent magnetization (remanence) of core ferromagnetic material)
- gain error (current source, hall generator, core gap)
- linearity (core material, hall generator, electronics)
- big influence of temperature changes
- output noise
- bandwidth limitation (attenuation, phase shift, current frequency)

This type of sensor is widely used, and among the producers are: LEM Components, ABB, Honeywell, Allegro Microsystems, ChenYoung, etc. Especially noteworthy is Allegro's sensor, which is constructed in the form of a monolithic integrated circuit with flux concentrator [2, 3, 27]. A typical schematic diagram of such sensor is shown in Fig. 2.17.

An example of the fully integrated circuit is Hall effect-based linear current sensor ACS752SCA-100 from Allegro MicroSystems [2], with 3 kV_{RMS} voltage isolation and a low-resistance current conductor (130 μΩ) and single +5 V supply on the secondary side. The current sensor is shown in Fig. 2.18, on the right side are two primary current terminals. The sensor has a primary sensed current range from −100 to +100 A, this current is converted to the output voltage signal with a sensitivity of 20 mV/A.

To reduce magnetic core and Hall sensor errors a closed loop topology was introduced. Simplified diagram of a closed loop Hall effect current sensor is depicted

Fig. 2.18 IC current sensor ACS752SCA-100

in Fig. 2.19. In a closed loop topology, the Hall sensor drives the output amplifier current to a secondary coil, which will generate a magnetic flux to cancel the primary current magnetic flux. So the resultant flux should be equal to zero. The secondary current, which is proportional to the primary current by the secondary coil ratio, can then be measured as voltage across a sense resistor. By keeping the resultant flux in the core at zero, the errors associated with offset drift, sensitivity drift and saturation of the magnetic core will also be significantly decreased. Closed-loop Hall effect current sensors also provide the shortest response times. However, in such devices, the nominal secondary coil current from several milliamps to hundreds of milliamps, thus power consumption is much higher in closed loop Hall sensor devices than in open loop topologies. In the closed loop configuration the maximum current magnitude is limited by a finite amount of compensation current in the device. The closed loop topology is widely used, for example Lem, ABB, and this type of transducer is widespread in industrial applications, with many manufacturers now supplying it. Good examples are the above mentioned typical industrial current transducers: LA 55-P [56] from LEM Components, ESM1000 from ABB [1], CSNB121 from Honeywell [42], CYHCS-SH from ChenYoung [25], etc. Figure 2.20 shows closed

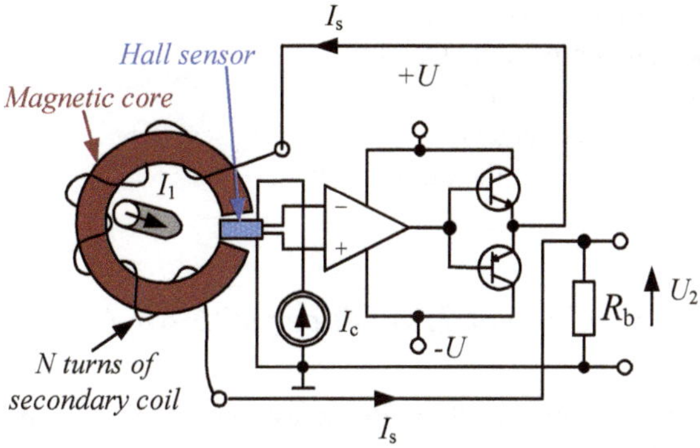

Fig. 2.19 Closed loop Hall effect current sensor

Fig. 2.20 Closed loop Hall current sensors in APF

loop Hall effect current sensors LA 205 [57] mounted on the current rail in APF EFA1 [92].

2.3.4 Current Transformer with Magnetic Modulation

Certain power electronics applications such as: medical equipment, meter or accessories for measuring equipment, require a precision current transducer. In order to eliminate the shortcomings associated with Hall effect, sensors are also developed with magnetic modulation topologies allowing the measurement of a DC component. These topologies have one, two or three magnetic cores, for example, the high precision current transducer ITB 300-S from LEM Components [80]. Features of current transformers with magnetic modulation include:

- high global accuracy,
- high linearity <1 ppm,
- high temperature stability,
- low cross-over distortion,
- wide frequency range,
- low noise on the output signal.

2.3.5 Current Transducer with Air Coil

A Rogowski coil allows the measurement of alternating current (AC). The principle of Rogowski coil operation is well described by Ray and Davis in [76–78]. The

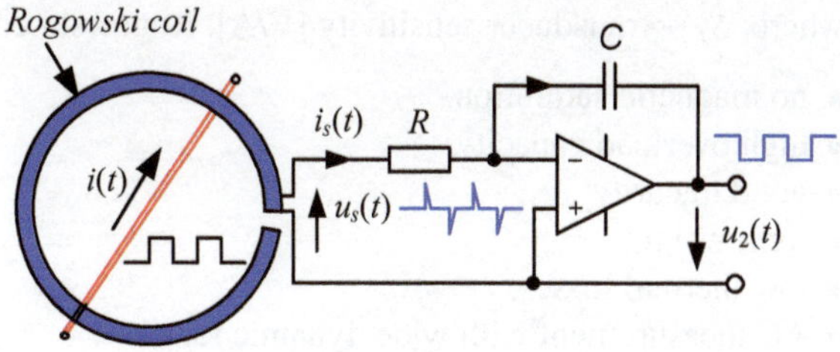

Fig. 2.21 Rogowski coil measurement circuit

Rogowski coil measurement circuit is depicted in Fig. 2.21. It consists of a helical coil of wire wrapped around a straight conductor whose current $i(t)$ is to be measured, since the voltage that is induced in the coil is proportional to the rate of change (derivative) of current in the straight conductor. The coil output signal is connected to an integrator circuit, so that the achieved output signal is proportional to the current. The voltage induced in the coil is given by the equation

$$u_s(t) = \mu_0 N M \frac{di(t)}{dt} = S \frac{di(t)}{dt} \quad , \tag{2.6}$$

where: μ_0—permeability of free space, N—turns/meter, M—cross-sectional area [m^2], S—coil sensitivity [Vs/A]. In next the stage the coil voltage is integrated by the integrator, so that transducer output voltage is

$$u_2(t) = -\frac{1}{RC} \int u_s(t) dt = -S_T i(t) \quad , \tag{2.7}$$

Fig. 2.22 Rogowski coil in the laboratory

where: S_T—transducer sensitivity [V/A]. Rogowski coil features are:

- no magnetic saturation,
- high overload capacity,
- good linearity,
- light weight,
- low thermal losses,
- AC measurement with wide dynamic range.

Figure 2.22 shows Rogowski coil and integrator in the laboratory. The free end is normally inserted into a socket adjacent to the cable connection but can be unplugged to enable the coil to be looped around the conductor or device carrying the current to be measured. Application of an induction coil without magnetic core allows for the elimination of errors associated with non-linear magnetic material. But the lack of a magnetic field concentrator causes the measuring system to be very sensitive to

Table 2.2 Comparison of current sensing techniques

Current-sensing technique	Galvanic isolation	Accuracy	Power dissipation	Relative cost	Typical current ranges
Transformer	Yes	0.1–5%	Low-Moderate	Low	Up to 15 kA, AC
Sensing resistor	None	0.01–5% mainly depends on resistor tolerance	Moderate-High	Low	Up to 500 A, DC-100 kHz
Sensing resistor with high common mode amplifier	None	0.01–5%	Moderate-High	Low	Up to 200 A, DC-100 kHz, common voltage from −20 to 100 V
Open loop Hall effect sensor	Yes	5–10%	Low	Low	Up to 15 kA, DC-50 kHz
Closed loop Hall effect sensor	Yes	1-5%	Moderate-High	Moderate-High	Up to 15 kA, DC-200 kHz
Transformer with magnetic modulation	Yes	0.001–0.5%	High	High	Up to 700 A, DC-500 kHz
Rogowski coil—one air coil	Yes	1–2%	Low	Low	Up to 10 kA, 10 Hz–100 kHz
Two air coils	Yes	0.5–1%	Low	Low	Up to 10 kA, 10 Hz–100 kHz

external interference fields. The use of two inductive coils allows for the partial elimination of this phenomenon [55]. The induced voltages from coils are then integrated in order to obtain both amplitude and phase information for the measured current. In this solution, in comparison to the Rogowski coil, accuracy is independent of the position of the cable in the aperture and of external fields.

2.3.6 Comparison of Current Sensing Techniques

Table 2.2 presents the basic features of current transducers considered in this chapter. The discussed transducers have analog outputs, but it should be noted that currently the development of transducers with digital output are under development. They can be equipped with their own processors, allowing for compensation of some errors. Parallel to an industrial solution of current sensing techniques Hartman et al. [40] designed an alternative which consists of a wideband current transformer and a demagnetizing circuit. The sensor concept is capable of measuring AC currents with dc offset, having periodic zero crossings, as given in power-factor-corrected (PFC) circuits.

2.4 Selected Parameters of Digital Control Circuit

An example of a power electronic circuit with an open-loop multirate digital control circuit is depicted in Fig. 2.23. The circuit may be a part of the control system for DC/DC, DC/AC, AC/DC and AC/AC converters or active power filters etc. [46, 47, 89, 90]. An open system was chosen because of its simpler analysis. In the closed-loop circuit, the impact of feedback should be taken into account.

Analog input signal $x(t)$ (Fig. 2.23) is converted to digital form $x(n)$ by an A/D converter with sampling ratio f_{s1} and b_1-bit resolution [89, 90]. In the next stage, a digital control algorithm using DSP is executed. The algorithm is calculated with b_2-bit resolution and sampling ratio f_{s2}. Finally, output control signal $y(n)$ is transferred to a digital PWM with b_3-bit resolution and sampling ratio $f_s 3$. The PWM controls output power switches S_1 and S_2. The switches work with switching frequency f_C. The digital PWM with two power switches S_1 and S_2 and an analog output filter L_C, C_C works as a digital-to-analog (D/A) power converter. It converts energy from a direct current source (DC) to the output. Typically the main issue is the quality of the output voltage and current. Therefore, the following signal parameters should be considered:

- f_{s1}, f_{s2}, f_{s3}—signal sampling ratios,
- b_1, b_2, b_3—signal resolutions (in bits),
- f_C—transistor switching frequency,
- *THD*—total harmonic distortion ratio,

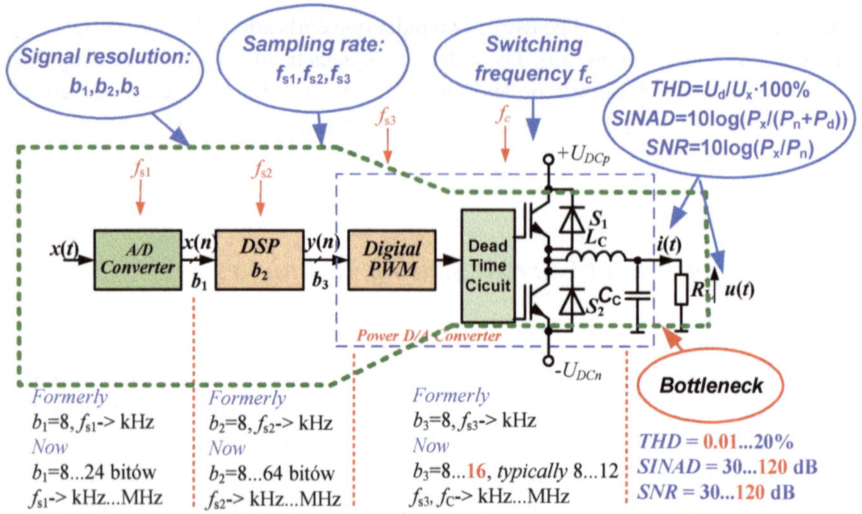

Fig. 2.23 An example of a power electronic circuit with an open-loop multirate digital control circuit

- *SNR*—output signal-to-noise ratio
- *SINAD*—output signal to noise and distortion ratio.

Finally, the required *SINAD* value of output current or voltage of power electronics circuit are dependent on the application. For example, for battery charger *SINAD* value equal to 30 dB is sufficient. However, for the high-quality audio power amplifier, the *SINAD* value should be bigger than 100 dB. As we can see the spread of parameters of digital control systems for power electronics circuits is very high. Nowadays there is a huge offer of A/D converters and DSPs, so it is easy to choose circuits with adequate parameters: b_1, b_2, f_{s1} and f_{s2}. The only considerable limitation may be the price of those circuits. The last stage, with digital PWM, is still a "bottleneck" of the whole system, especially for high resolution, for example, high-quality power audio amplifier with parameters: $f_{s3} = 44.1$ kHz and $b_3 = 16$ bit, the clock frequency of PWM counters can be calculated using the equation

$$f_h = f_{s3} 2^{b3} \approx 2.8\,\text{GHz} \quad . \tag{2.8}$$

The calculated value of clock frequency is too high for ordinary digital circuits, therefore, resolution of digital PWM should be reduced. However, it will result in deterioration of the signal-to-noise ratio. A solution to this problem is shown in this chapter.

2.5 Total Harmonic Distortion

Total harmonic distortion (*THD*) ratio is a form of nonlinear distortion in circuits in which harmonics (signals whose frequency is an integer multiple of the input signal) are generated. A wide class of nonlinear circuits can be described by the equation

$$y(t) = a_1 x(t) + a_2 x(t)^2 + a_3 x(t)^3 + \cdots + a_k x(t)^k \quad . \tag{2.9}$$

In linear circuits only a_1 coefficient is nonzero. For example a nonlinear circuit described by equation $y(t) = x(t) - 0.2x(t)^3 + 0.15x(t)^5 + 0.11x(t)^7 - 0.05x(t)^9$ the response for DC input signal in the range from -1 to 1 is shown in Fig. 2.24. MATLAB® program for the nonlinear circuit simulation is shown in Listing 2.1.

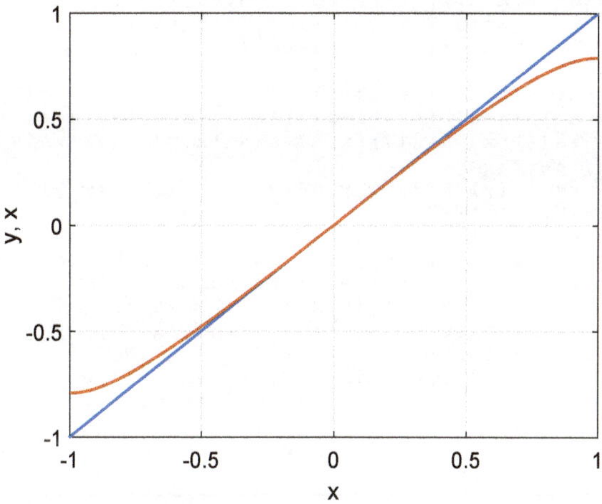

Fig. 2.24 The static characteristics of nonlinear circuit, response for input signal range from -1 to 1

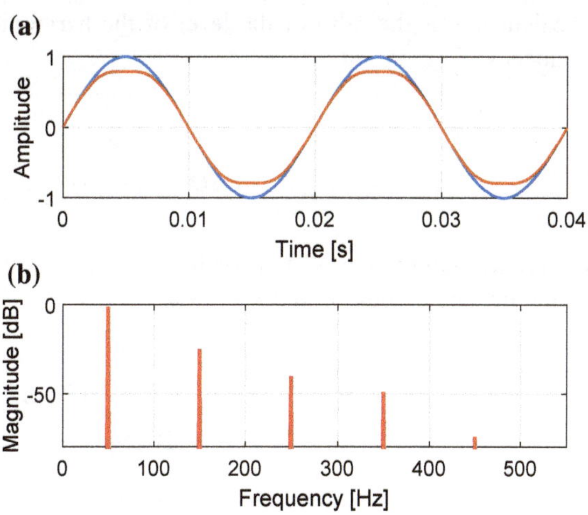

Fig. 2.25 The nonlinear circuit response for sinusoidal signal, $f = 100$ Hz: **a** the input and the output signal waveforms, **b** spectrum of the output signal

The response for unity amplitude sinusoidal signal, $f = 100\,\text{Hz}$, is depicted in Fig. 2.25. The waveforms of input and output signal are presented in Fig. 2.25a. The output signal distortion causes generation of signal harmonics. The spectrum of the output signal is shown in Fig. 2.25b.

Listing 2.1 Nonlinear circuit

```
clear all; close all;
roz_fon=18; grub_lin=2;
N=2^12; % number of samples
fs=12800; % sampling frequency
f1=50; f_1k=round(f1/(fs/N))*fs/N; % line frequency
A1=1;
t=(0:N-1)/fs; % time vector
a=[1 0 -0.2 0 0.15 0 -0.11 0 -0.05]; % coefficients
%% ----- Response for DC -----
x=((0:N-1)/N)*2-1; % input signal -1...1
y=a(1)*x+a(2)*x.^2+a(3)*x.^3+a(4)*x.^4+a(5)*x.^5 + ...
  +a(6)*x.^6+a(7)*x.^7+a(8)*x.^8+a(9)*x.^9;
plot(x,x,x,y,'LineWidth',grub_lin);
        set(gca,'FontSize',[roz_fon],'FontWeight','d'),
        xlabel('x'); ylabel('y,␣x'); grid on;
%% ----- Response for sinusoidal signal 100 Hz -----
x=A1*sin(2*pi*f_1k*t); % input sinusoidal signal
y=a(1)*x+a(2)*x.^2+a(3)*x.^3+a(4)*x.^4+a(5)*x.^5+...
  +a(6)*x.^6+a(7)*x.^7+a(8)*x.^8+a(9)*x.^9;
% spectrum
w_y=fft(y)/N*2; w_y_dB=20*log10(abs(w_y)+eps);
f=(0:N-1)*fs/N;
subplot(211);
        plot(t,x,t,y,'LineWidth',grub_lin); grid on;
  set(gca,'FontSize',[roz_fon],'FontWeight','n'),
  set(gca,'Xlim',[0 2/f_1k]); xlabel('Time␣[s]');
  ylabel('Amplitude'); title('(a)');
subplot(212);
        plot(f,w_y_dB,'r','LineWidth',grub_lin+1); grid on;
        set(gca,'Ylim',[-80 0]);set(gca,'Xlim',[0 11*f_1k]);
        set(gca,'FontSize',[roz_fon],'FontWeight','n'),
        xlabel('Frequency␣[Hz]'); ylabel('Magnitude␣[dB]');
        title('(b)');
```

THD ratio is measured in percents or in decibels (dB), and harmonic distortion is calculated as the ratio of the level of the harmonic to the level of the original frequency

$$THD = \frac{\sqrt{\sum_{k=2}^{N} U_k^2}}{U_1}. \tag{2.10}$$

where: U_1—amplitude of first (or fundamental) harmonic, U_k—amplitude of k-th harmonic, and expressed in dB

$$THD_{dB} = 20\log\frac{\sqrt{\sum_{k=2}^{N} U_k^2}}{U_1}. \tag{2.11}$$

In power electronics systems other distortion factors [13, 41] are also used, one of which is weighted harmonic distortion ratio in which the importance of harmonics decreases as the frequency increases

$$WTHD = \frac{\sqrt{\sum_{k=2}^{K}\left(\frac{U_k}{k}\right)^2}}{U_1} . \tag{2.12}$$

2.6 Sampling of Analog Signal

Sampling is an essential part of signal processing. It enables A/D transformation of the analog signal to occur. However useful this process is, some precautions need to be taken to ensure that the output signal is not changed significantly. Therefore, while sampling the analog signal at discrete intervals, $T_s = 1/f_s$, the signal sampling frequency f_s (also called signal sampling speed) must be carefully chosen to ensure an accurate representation of the original analog signal. It is evident that the more samples are taken (faster signal sampling rates), the more accurate the digital representation is. Hence if fewer samples are taken (lower sampling rates), a point is reached where critical information about the signal is actually lost. The discussion in this chapter focuses on periodic and uniform sampling. An example of a sample and hold circuit (SH) (also called sampling circuit or track and hold (TH) circuit) is shown in Fig. 2.26 and an illustration of an analog sinusoidal signal sampling process is presented in Fig. 2.27.

The history of the mathematical background of sampling theory goes back to the 1920s, when it was founded by Nyquist [65, 66] of Bell Telephone Laboratories. This original work was shortly supplemented by Hartley [39] and Whittaker. These papers formed the basis for the pulse code modulation (PCM) work and were followed in 1948 by Shannon who wrote a paper on communication theory [82]. The sampling

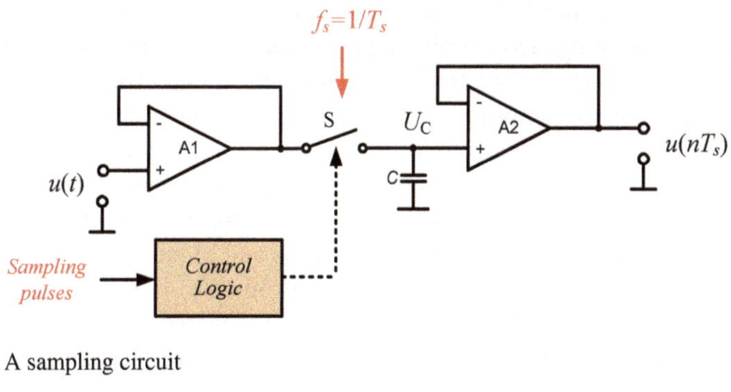

Fig. 2.26 A sampling circuit

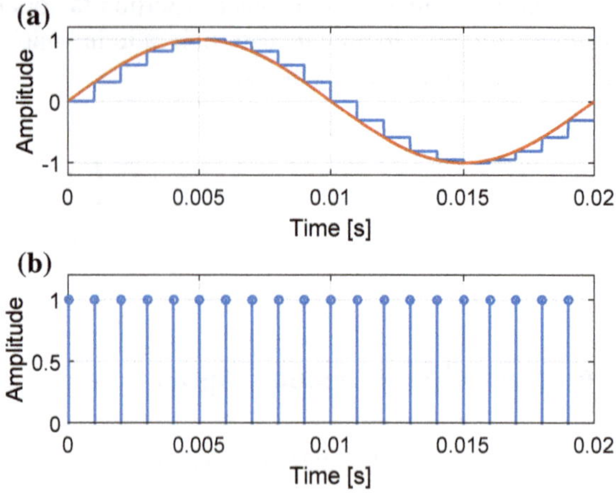

Fig. 2.27 Illustration of analog sinusoidal signal sampling process: **a** input signal and sampled signal, **b** sampling pulses

theory was discovered independently by Kotielnikov from the Soviet Union in 1933 [51]. Simply stated, the sampling theory criteria require that the sampling frequency must be at least twice the highest frequency contained in the analog signal, otherwise the information about the analog signal will be lost. If the sampling frequency is less than twice the maximum analog signal frequency, a phenomenon known as aliasing will occur. It is a source of spurious signals occurring within the anti-aliasing filter as a result of the high signal bandwidth available in today's ADCs. The main purpose of using an anti-aliasing filter is to limit the input signal's bandwidth to eliminate high-frequency components. Therefore it is required that in data sampling systems the input signal's spectrum frequency must not exceed one-half of the sample clock frequency. An ideal anti-aliasing filter would pass all signals within the band of interest and block all signals from outside of that band. It is the quality of the anti-aliasing filter which is the major factor in the signal-to-noise ratio (SNR). The SNR typically expressed in decibel, can be determined by the Eq. 1.12.

When the half sampling frequency $f_s/2$ is only slightly higher than f_b for signal band $0 - f_b$ (as happens in the classic system), the anti-aliasing has to have very sharp amplitude characteristics and a high damping factor in the stop-band. This results in the anti-aliasing filter being very complicated and expensive. Currently, the development of integrated circuit (IC) manufacturing technology has led to the fact that fast digital circuits are freely available and cheap. Therefore, in modern systems the value of the sampling rate f_s can be much higher than f_b (this is called oversampling), so the requirements for the anti-aliasing filter are much lower. The spectra of a sampling signal process with aliasing occurrence are shown in Fig. 2.28.

The background of digital signal processing useful for A/D and D/A conversion is described by Oppenheim and Schafer [67], Proakis and Manolakis [74], Lyons [58], Rabiner and Gold [75], Zolzer [110, 111] and many others [14, 35, 37, 38, 43, 53, 54, 106]. Some interesting technical problems with A/D conversion are described in Data Translation [31, 32] and Analog Devices [48–50] (by Kester) technical reports and

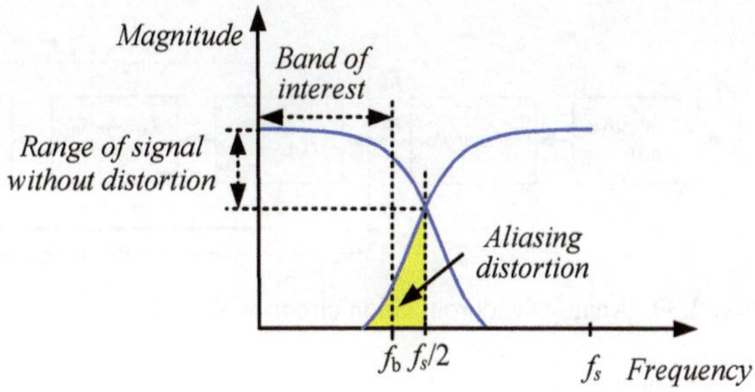

Fig. 2.28 Impact of aliasing on the signal dynamics

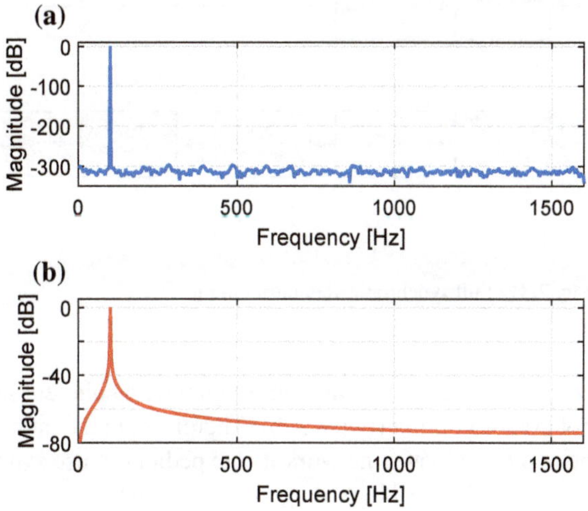

Fig. 2.29 Spectra of sinusoidal signals: **a** coherent sampling, **b** non-coherent sampling

books. Other problems about signal oversampling and signal sampling rate changes are discussed in Chap. 3.

2.6.1 Synchronization of Sampling Process

The properties of the majority of digital signal processing algorithms (e.g. DFT) to a large extent depend on whether the processed signal is sampled coherently [90]. Coherent sampling refers to a certain relationship between input frequency, f_{in}, sampling frequency f_s, N_{per}—integer number of signal periods in block of N signal samples

$$f_s = N_{per} \frac{f_{in}}{N} \quad . \tag{2.13}$$

With coherent sampling one is assured that the signal magnitude in an DFT is contained within one DFT bin, assuming single input frequency. For example, Fig. 2.29

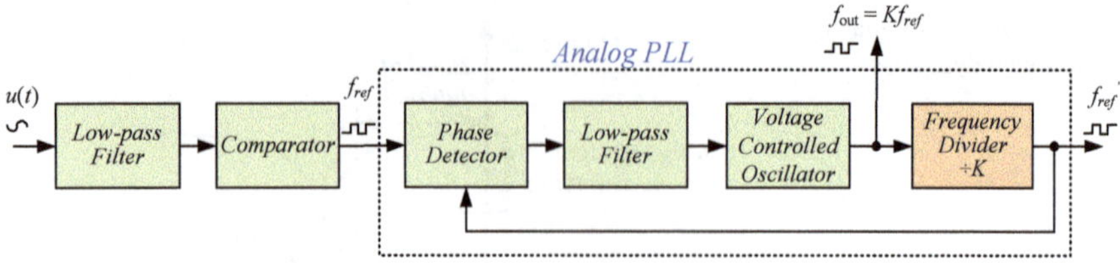

Fig. 2.30 Analog synchronization circuit with PLL

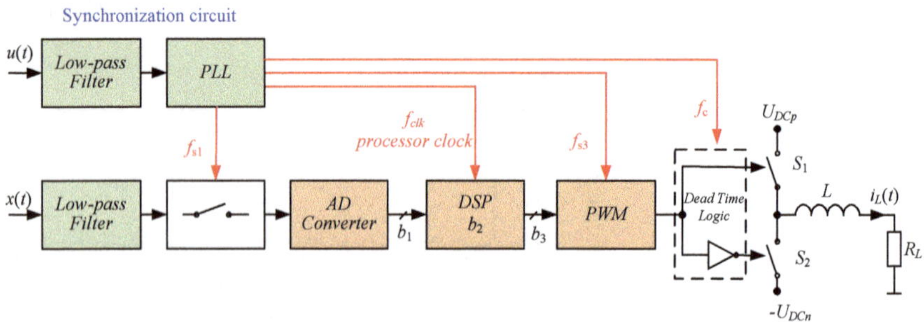

Fig. 2.31 Full synchronized control circuit

shows the spectrum of the same signal with coherent sampling (Fig. 2.29a) and non-coherent sampling (Fig. 2.29b). Therefore, in the author's opinion, for systems connected to the power network it is expedient to use synchronization.

In Fig. 2.30 a block diagram of an analog phase-locked-loop (PLL) circuit is depicted. Using this circuit it is possible to generate signal with frequency K times bigger than input frequency. A PLL circuit can track a reference frequency and it can generate a frequency that is a multiple of the input frequency. The phase of both signals are synchronized too. The PLL generates output signal with frequency

$$f_{out} = Kf_{ref} \quad , \tag{2.14}$$

where: f_{ref}—reference input frequency, K—integer frequency multiplication factor. While designing the analog synchronization circuit, the output of the power electronic device should be taken into consideration. In most cases it is a pulse width modulator (PWM) generating pulse controlled output switches (typically transistors). When the modulation frequency is independent of the reference system frequency (e.g., power line frequency), it will certainly result in generation of low frequency components. This is an effect of the beat frequency between the reference and modulation frequency. Hence, in the author's opinion, the input and output should be synchronized, which will minimize errors and eliminate unwanted components. The block diagram of such a solution is depicted in Fig. 2.31.

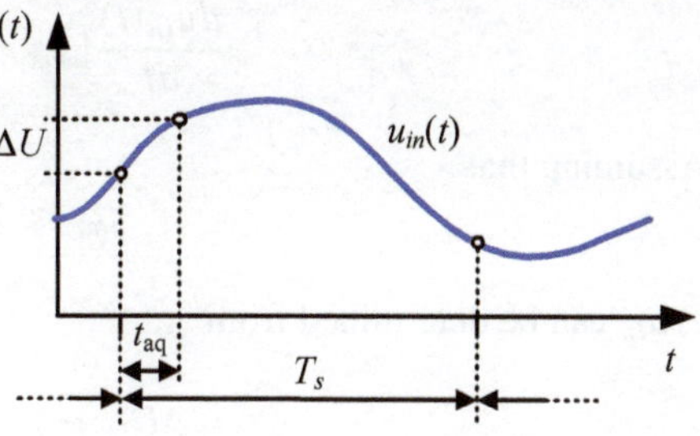

Fig. 2.32 Signal acquisition process

In the author's opinion, the same situation occurs during simulation tests, and if it is possible a coherent frequency of test signals should be used. A simple listing of the MATLAB program for coherent frequency calculation is shown in Listing 2.2.

Listing 2.2 Coherent frequency calculation

```
N=2048;        % length of signal block
fs=10000;      % sampling frequency
f=50;          % required frequency of the signal
f_koh=round(f/(fs/N))*fs/N; % nearest coherent frequency
```

2.6.2 *Maximum Signal Frequency Versus Signal Acquisition Time*

The signal acquisition time is the time required by the circuit to settle to its final value after it is paced in the hold mode. Signal acquisition time t_{aq} relates to A/D converters which use a sample–and–hold (or track–and–hold) circuit on the input to acquire and hold (to a specified tolerance) the analog input signal, see Brannon and Barlow [19]. For an A/D converter without a sample–and–hold circuit on the input, the signal acquisition time is equal to the converter conversion time t_c. The only exception concerns flash converters with well matched comparators. An illustration of a sampling process is shown in Fig. 2.32, assuming that the amplitude of the input signal in the acquisition process does not change more than half of the LSB of the A/D converter. Assuming maximum signal change during sampling process

$$\Delta U \leq 0.5\Delta = 0.5\frac{A_p}{2^{b-1}} = \frac{A_p}{2^b} \quad . \tag{2.15}$$

The analog input sinusoidal signal with amplitude A_p and frequency f

$$u_{in}(t) = A_p sin(2\pi f t) \quad . \tag{2.16}$$

The maximum speed of the signal change is determined by the equation

$$\frac{du_{in}(t)}{dt}|_{max} = 2\pi A_p f \quad . \tag{2.17}$$

Assuming that

$$t_{aq} \ll 1/f \quad . \tag{2.18}$$

ΔU_{in} can be determined from

$$\Delta U_{in} = 2\pi A_p f t_{aq} \quad . \tag{2.19}$$

As a result of a straightforward algebraic manipulation we obtain an equation describing the maximum signal frequency

$$\Delta U_{in} \leq \Delta U \quad , \tag{2.20}$$

$$2\pi A_p f t_{aq} \leq \frac{A_p}{2^b} \quad , \tag{2.21}$$

$$f \leq \frac{1}{2\pi 2^b t_{aq}} \quad , \tag{2.22}$$

2.6.2.1 Example—Acquisition Time

With maximum signal frequency for a number of bits $b = 16$, acquisition time $t_{aq} = 10\,\text{ns}$, and determined from the above inequality the maximum signal frequency is $f < 242.8\,\text{Hz}$. As explained, this is one of the most important factors in the A/D conversion.

Using the A/D converter with a very small value of acquisition time meeting the requirements of Eq. 2.22 is very difficult and expensive. Therefore systems are used where the acquisition time is much greater. Of course, in a single channel system, there will be differences in the signal sampling moment, and often the requirements of Eq. 2.22 can be omitted. However, in such a case, in multi-channel systems it is vital to provide the same value of acquisition time for all channels.

2.6.3 Errors in Multichannel System

In multichannel systems it is very important that the input signals are simultaneously sampled to reduce amplitude and phase errors. Multichannel A/D converters with a sampling circuit have three very common architectures, which are depicted in a two channel version in Fig. 2.33. Among the three presented solutions the two use simultaneous sampling (Fig. 2.33a, b) and the other uses sequential sampling

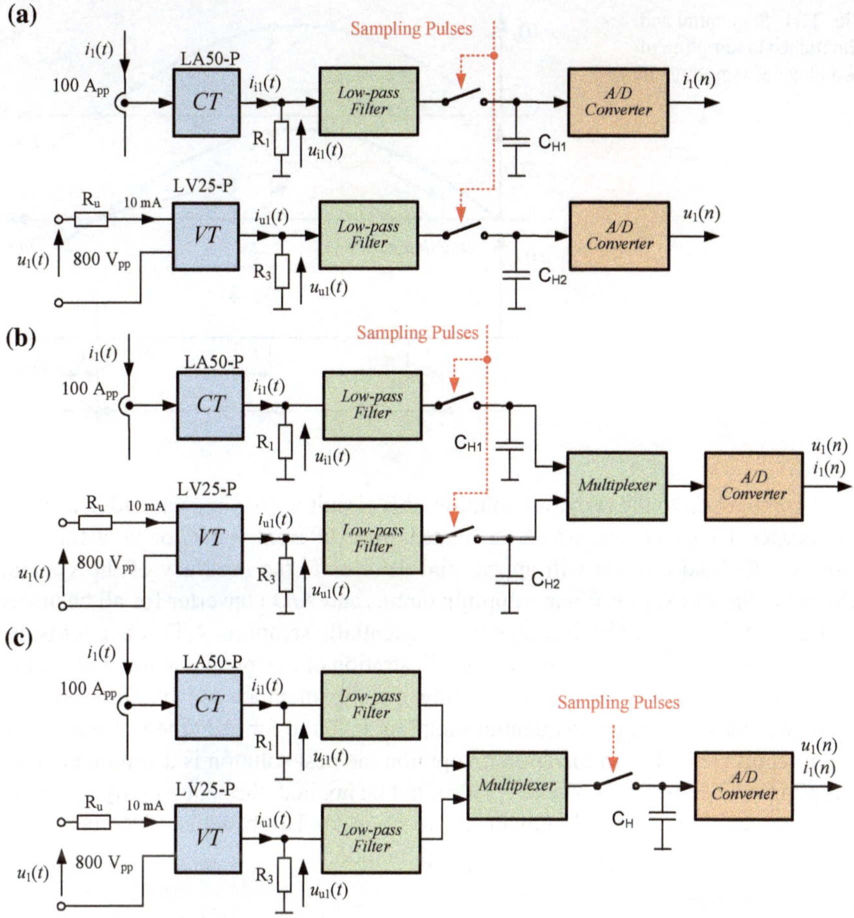

Fig. 2.33 Two-channel sampling circuits: **a** simultaneous sampling with A/D converter in each channel, **b** simultaneous sampling with single A/D, **c** sequential sampling

(Fig. 2.33c). A two-channel sequentially sampling analog-to-digital converter is depicted in Fig. 2.33a. In this circuit current $i_1(t)$ is converted to a current signal $i_{i1}(t)$ using galvanically isolated current transducer CT, and in the same way a voltage signal using galvanically isolated voltage transducer VT is processed. Then the signals pass through anti-aliasing low-pass filters and are simultaneously sampled. After this the A/D converters process and convert them to digital form. This solution is the fastest and the most comfortable for control circuit designers, however, it is also the most expensive. An alternative solution is a system with a single A/D converter with a few simultaneous sample-and-hold circuits. A block diagram of such a solution for the two channels is shown in Fig. 2.33b.

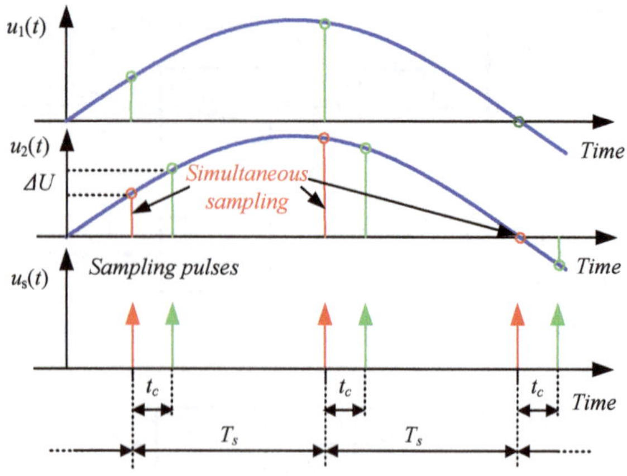

Fig. 2.34 Sequential and simultaneous sampling of two-channel sampling circuit

In comparison to the previous solution, this circuit is the simplest and hence can be cheaper. However, the errors associated with different capacitor hold times in sample–and–hold circuits will appear and deteriorate the accuracy of the system. The third circuit uses only one sampling circuit and A/D converter for all channels (Fig. 2.33c). The main disadvantage of a sequentially sampling A/D converter is the time error between channel samples. An illustration of this phenomenon is shown in Fig. 2.34, where two sample signals in time misalignment are shown. Discussion of simultaneous sampling vs. sequential sampling is found for example in Data Translation reports [31, 32]. In the author's opinion the best solution is a simultaneously sampling A/D converter, however, if it cannot be applied, the sequentially sampling A/D converter with time alignment has to be used. The benefits of simultaneous sampling compared to sequential sampling include:

- less jitter error,
- higher bandwidth of the system,
- less channel-to-channel crosstalk,
- less settling time.

2.6.4 Amplitude and Phase Errors of Sequential Sampling A/D Conversion

Due to the fact that typical A/D converters built-in microprocessor or separate IC allow only sequential sampling. Therefore errors in such a solution will be considered. Two sinusoidal input signals with the same frequency f and A_p amplitude

$$u_1(t) = A_p sin(2\pi f t) \quad , \quad u_2(t) = A_p sin(2\pi f (t + t_c)) \quad . \tag{2.23}$$

Difference of signals for sequential sampling with time difference t_c (Fig. 2.34)

$$\Delta U(t) = u_1(t) - u_2(t) = 2A_p cos\left(\frac{2\pi f t + 2\pi f(t+t_c)}{2}\right) sin\left(\frac{2\pi f t - 2\pi f(t+t_c)}{2}\right) \quad (2.24)$$
$$= 2A_p cos(2\pi f t + \pi f t_c) sin(-\pi f t_c) \quad .$$

Maximum of $\Delta U(t)$ is when derivative is equal zero

$$\frac{du(t)}{dt} = 4A_p \pi f sin(2\pi f t + \pi f t_c) sin(\pi f t_c) = 0 \quad . \quad (2.25)$$

So the derivative is zero when

$$2\pi f t + \pi f t_c = 0 \rightarrow t = -0.5t_c \quad . \quad (2.26)$$

Maximum value of signal error can be calculated by the formula

$$\begin{aligned} \Delta U_{max} &= \Delta U(t)|_{t=-0.5t_c} = 2A_p \cos\left(\frac{2\pi f(-0.5t_c) + 2\pi f t_c}{2}\right) \sin(-\pi f t_c) \\ &= 2A_p \cos(0) \sin(-\pi f t_c) \\ &= 2A_p \sin(-\pi f t_c) \quad . \end{aligned} \quad (2.27)$$

However, phase error can be determined from the equation

$$\Delta\phi = \frac{t_1}{T}360 - \frac{t_1 - t_c}{T}360 = 360 t_c f \quad . \quad (2.28)$$

2.6.4.1 Example—Sequential Sampling

With maximum signal frequency for sequential sampling for an A/D conversion time $t_c = 5\,\mu s$, signal amplitude $A_p = 1$ and signal frequency $f = 50\,Hz$ it is possible to determine from the above equation: maximum signal error is $\Delta U = 1.57\,mV$, phase error $\Delta\phi = 0.09$, and results for the 50th harmonics ($f = 2500\,Hz$): $\Delta U = 39.26\,mV$ and $\Delta\phi = 4.5$. The result will be worse in multichannel systems with sequential sampling, where the A/D conversion time t_c for the last channel will be multiplied by the number of channels.

For b-bit system signal error should be less than

$$\begin{aligned} |\Delta U_{max}| &\le 0.5\Delta \quad , \\ 2A_p sin(\pi f t_c) &\le \frac{A_p}{2^b} \quad , \end{aligned} \quad (2.29)$$

Assuming

$$t_c \ll 1/f \quad . \quad (2.30)$$

Finally

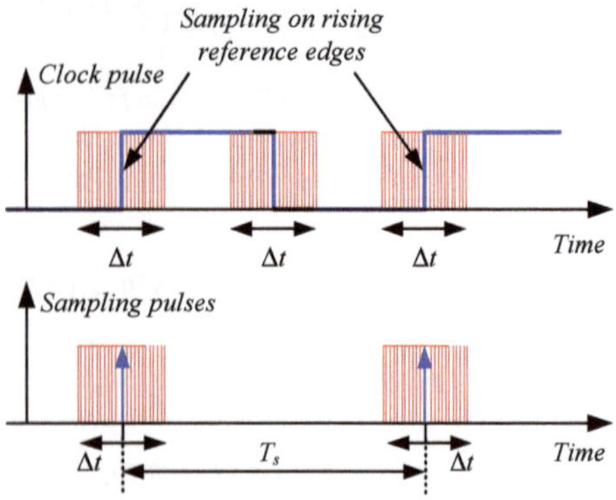

Fig. 2.35 Sampling clock with jitter

$$t_c \leq \frac{1}{\pi f 2^{b+1}} \quad . \tag{2.31}$$

The result is similar to the one obtained in Sect. 2.6.2.

Another source of errors in multi-channel systems are the channel crosstalk and channel-to-channel offset. Channel-to-channel offset is the difference in the characteristics of analog input channels which causes measurement error, as if small voltage were added to or subtracted from the input signal. Channel crosstalk is the leakage of signals between analog input channels in a data acquisition system. Channel crosstalk has the potential to increase uncorrelated noise in the A/D conversions, reducing the signal to noise ratio (*SNR*), while coupled signals can create spurs similar to harmonic terms, reducing spurious free dynamic range (*SFDR*) and total harmonic distortion (*THD*).

2.6.5 Sampling Clock Jitter

An important factor in the sampling process is the sampling clock in the A/D converter. Due to hardware error and noise the sampling moments in real A/D converters are uncertain. Problems of sampling signal uncertainty are described by many authors [12, 18, 19, 62, 79, 90]. Variation in the sampling time is known as aperture uncertainty, or jitter, and will result in an error voltage that is proportional to the magnitude of the jitter and the input signal slew rate. In other words, the greater the input frequency and amplitude, the more susceptibility to jitter in the clock source. Figure 2.35 shows a sampling pulse clock with jitter.

Figure 2.36 shows how jitter generates a signal error.

$$\Delta U = \Delta t \frac{\mathrm{d}u(t)}{\mathrm{d}t} \quad . \tag{2.32}$$

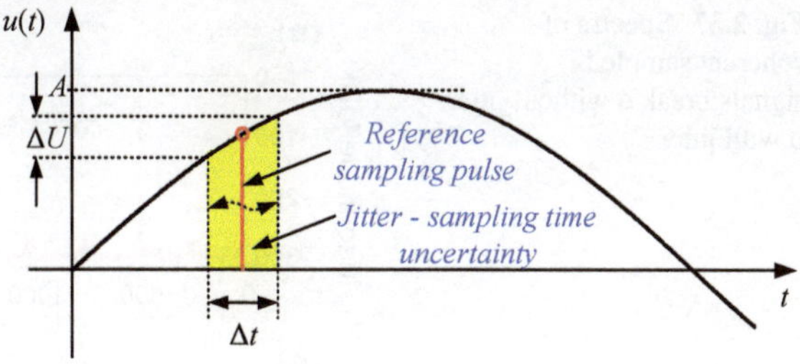

Fig. 2.36 Sampling time uncertainty—jitter

The maximum value of voltage error for sinewave of frequency f and amplitude A is at zero crossing

$$\Delta U_{max} = \Delta t \frac{du(t)}{dt}|_{max} = 2\pi f A \Delta t \quad . \tag{2.33}$$

This error cannot be corrected later because it is already attached to the sampling sequence that is being processed for digitization and will impact the overall performance of the A/D converter and finally of the control system, as shown in Eq. 2.33. Assuming that

$$\Delta U_{max} < 0.5\Delta \quad , \quad \Delta U_{max} < \frac{A}{2^b} \quad . \tag{2.34}$$

then

$$\Delta t < \frac{1}{2\pi f 2^b} \quad . \tag{2.35}$$

Finally

$$SNR_{jitter} = -20 \log(2\pi f \Delta t_{rms}) \quad , \tag{2.36}$$

where: t_{rms}—root means square of time jitter.

2.6.5.1 Example—Jitter

This case illustrates deterioration of signal quality by jitter. Assuming the following parameters: sampling frequency $f_s = 6400$ Hz, signal frequency $f = 500$ Hz, jitter value $\Delta t_{rms} = 1.57\,\mu$s, signal-to-noise ratio SNR the value calculated from the formula 2.36 is equal to $SNR = 46.18$ dB. For such assumed values a simulation in the MATLAB environment was also made. In Fig. 2.37 the results of the simulation are shown. The spectrum presented in Fig. 2.37a is for sinusoidal signal, which was coherently sampled without jitter, the $SNR = 248.93$ dB of this signal is only limited by the MATLAB arithmetic accuracy. The spectrum of signal sampled with jitter is shown in Fig. 2.37b. In this case the value of signal-to-noise ratio calculated on the basis of simulation is equal to $SNR = 46.17$ dB and it is almost equal to the value

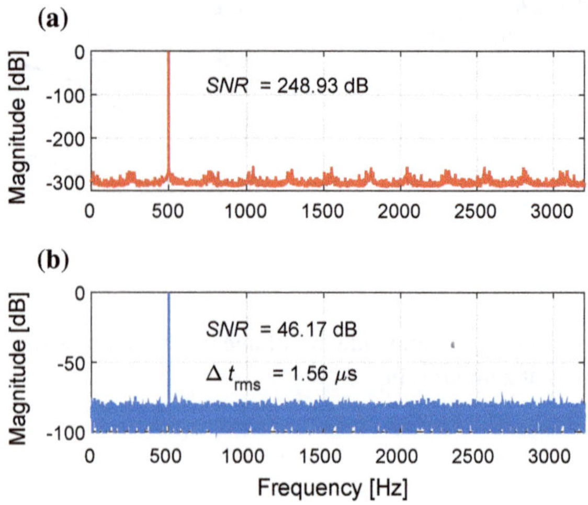

Fig. 2.37 Spectra of coherent sampled signals:break **a** without jitter, **b** with jitter

calculated by formula 2.36. The MATLAB program for calculation this example is described in Listing 2.3.

Listing 2.3 The jitter calculation

```
clear all; close all;
N=2^14;   fs=6400; Ts=1/fs; font=10;
f=500; fkoh=round(f/(fs/N))*fs/N; %koherent signal frequency
t=(0:N-1)*Ts; A_jitter=0.01*Ts;
tjitter=randn(1,N)*A_jitter; % normal (or Gaussian) distribution
A_jitter_rms=(sum(tjitter.^2)/N)^0.5; % RMS value of jiitter
SNR_calc=-20*log10(A_jitter_rms*2*pi*fkoh); % SNR from equation
wej_jitter=1.0*sin(2*pi*fkoh*(t+tjitter)); % signal with jitter
wej=1.0*sin(2*pi*fkoh*t); % signal without jitter
w_wej=fft(wej)/N*2; w_wej_jitter=fft(wej_jitter)/N*2;
w_wej_dB=20*log10(abs(w_wej)+eps);
w_wej_jitter_dB=20*log10(abs(w_wej_jitter)+eps);
f=(0:N-1)*fs/N;
%% SNR from spectrum
n=round(fkoh/(fs/N)); % find number of signal bin
SNR=20*log10((sum(abs(w_wej([1:n, n+2:(N/2-1)]).^2)))^0.5);
C = num2str(-SNR,'%3.2f');
SNR_jitter=-20*log10((sum(abs(w_wej_jitter([1:n, n+2:(N/2-1)]).^2)))^0.5);
A = num2str(SNR_jitter,'%3.2f'); B = num2str(SNR_calc,'%3.2f');
figure('Name','Spectra','NumberTitle','off')
subplot(211),plot(f,w_wej_dB,'r','linewidth',2); grid on;
set(gca,'fontsize',font); title('(a)'); ylabel('Magnitude␣[dB]'),
axis([0 fs/2 -320 0]);
text(700,-75,['\it␣SNR␣\rm␣=␣',C,'␣dB'],'FontSize',font);
subplot(212),plot(f,w_wej_jitter_dB,'linewidth',2); grid on;
set(gca,'fontsize',10);   title('(b)'); ylabel('Magnitude␣[dB]'),
xlabel('Frequency␣[Hz]'); axis([0 fs/2 -100 0]);
text(700,-15,['\it␣SNR␣\rm␣=␣',A,'␣dB'],'FontSize',font);
text(700,-35,['\it␣SNR_{calc}␣\rm␣=␣',B,'␣dB'],'FontSize',font);
```

```
D = num2str(A_jitter_rms*1e6,'%3.2f');
text(700,-60,['\it \Delta t\rm_{rms}  = ',D,' \mus'],'FontSize',font);
print('jitter_matlab.pdf','-dpdf');
```

2.7 Signal Quantization

In the process of A/D conversion the amplitude resolution of digital signal is limited to digital representation. In most cases the b-bit fixed-point system is used, which eliminates the excess of digits by discarding them or by rounding off the number. An illustration of the sampling and quantization process of a sinusoidal analog signal is shown in Fig. 2.38.

In Fig. 2.39 spectra of quantized analog sinusoidal signals are shown, for $f = 700\,\text{Hz}$, $f_s = 12800\,\text{Hz}$, $b = 7$ bit (Fig. 2.39a) and $b = 14$ bit (Fig. 2.39b).

The digital signal is a sequence of numbers in which each number is represented by a finite number of digits

$$x(n) \longleftrightarrow x(nT_s), \quad -\infty < n < \infty \quad , \tag{2.37}$$

where: n—number of samples, $x(n)$ discrete signal obtained by sampling of analog signal $x(t)$ every T_s period of time. The amplitude of the signal corresponding to the least significant bit (LSB) is determined by the equation

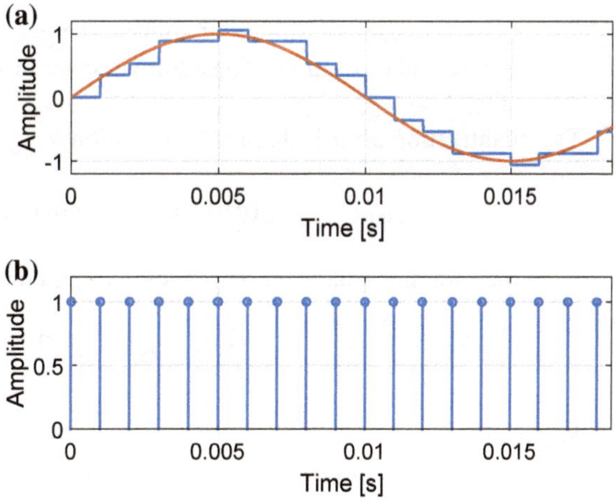

Fig. 2.38 Analog sinusoidal signal sampling and quantization process: **a** input signal and sampled signal, **b** sampling pulses

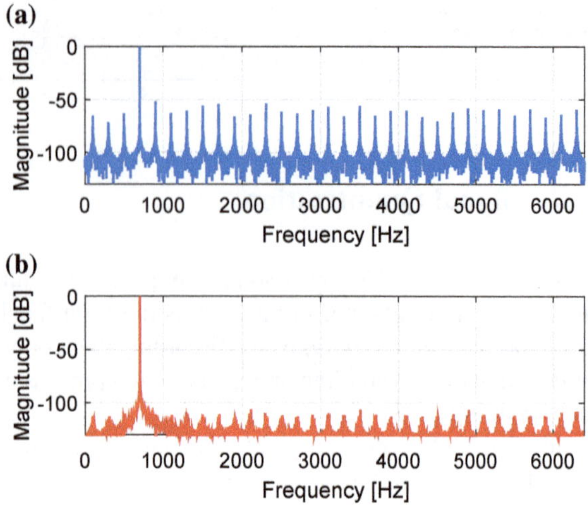

Fig. 2.39 Spectra of quantized analog sinusoidal signals: **a** for $b = 7$ bit, **b** for $b = 14$ bit

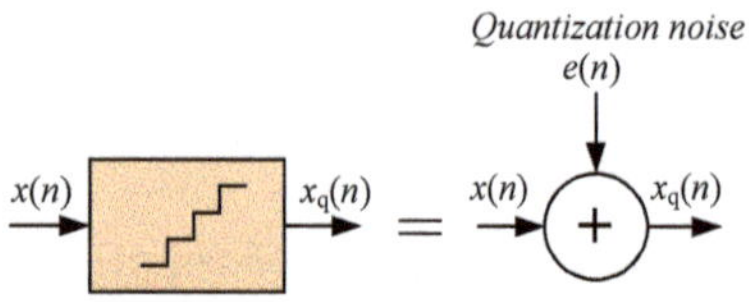

Fig. 2.40 The additive linear model of quantization process

$$\Delta = \frac{A_p}{2^{b-1}} \quad , \tag{2.38}$$

where: A_p—maximum amplitude of converted signal. The Δ is also called quantization step size and resolution. The additive linear model of quantization process is shown in Fig. 2.40.

The quantization error is defined by equation

$$e_q(n) = x_q(n) - x(n) \quad , \tag{2.39}$$

$e_q(n)$ in the rounding quantization process is limited to the range of $-\Delta/2$ to $\Delta/2$, that is

$$-\frac{\Delta}{2} \leq e_q(n) \leq= \frac{\Delta}{2} \quad , \tag{2.40}$$

Signal quantization adds noise to the signal, which deteriorates the signal dynamic range. For a sinusoidal signal and quantization noise, which are uniformly distributed, in fixed-point b-bit system, the noise power can be calculated by

$$P_n = \frac{\Delta^2}{12} \quad . \tag{2.41}$$

When a full-scale sine-wave is used as the input signal, SNR can be written as

$$SNR = 10\log\left(\frac{P_x}{P_n}\right) = 10\log\left(\frac{\frac{A_p{}^2}{2}}{\frac{\Delta^2}{12}}\right) = 10\log\left(\frac{3}{2}2^{2b}\right) \cong 1.76 + 6.02b \quad . \tag{2.42}$$

2.7.1 Dynamic Range of Signal

The dynamic range of a signal processing system can be defined as the ratio of the maximum sustainable signal level without overflow (or other distortion) to minimum signal level

$$DR = 20\log\left(\frac{|X_{max}|}{|X_{min}|}\right) \quad . \tag{2.43}$$

where: $|X_{max}|$—maximum amplitude of the signal (in digital systems typically the most significant bit, MSB, is used), $|X_{min}|$—minimum amplitude of the signal (least significant bit, LSB, in digital systems).

2.7.2 Signal Headroom

The SNR value from Eq. (2.42) is only possible when the signal amplitude is equal to A_p; in practice it is not possible to work with such high amplitude. It is therefore necessary to leave an adequate margin to exceed the value of the signal and additional space for the signal pulse components called signal headroom. Figure 2.41 shows an illustration of this phenomenon, where A_{p1} is nominal amplitude of the input signal, and A_{p2} is an extended amplitude of the input signal.

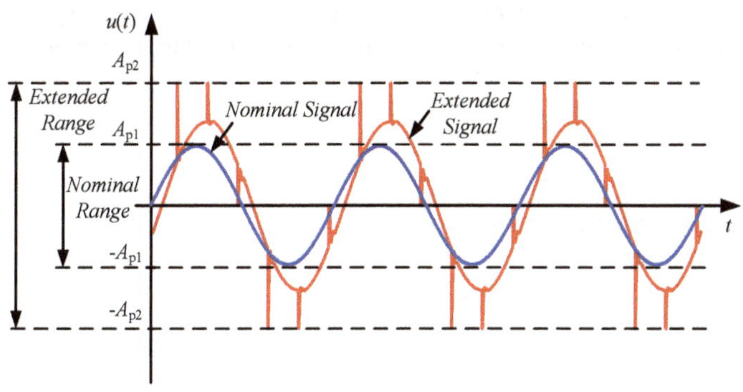

Fig. 2.41 Signal headroom

Therefore, in real systems, the value of SNR will be lower than is apparent from the number of A/D converter bits and can be calculated by the formula

$$\begin{aligned} SNR &= 10\log\left(\frac{\frac{A_{p1}{}^2}{2}}{\frac{\Delta^2}{12}}\right) = 10\log\left(\frac{\frac{A_{p1}{}^2}{2}}{\frac{\frac{A_{p2}^2}{2^{2b-1}}}{12}}\right) \\ &= 10\log\left(\left(\frac{A_{p1}}{A_{p2}}\right)^2\frac{3}{2}2^{2b}\right) \\ &\cong 1.76 + 6.02b + 20\log\left(\frac{A_{p1}}{A_{p2}}\right) . \end{aligned} \tag{2.44}$$

2.7.2.1 Example—Signal Headroom

A 12-bit A/D converter is used for measuring load current in a shunt switching compensator (active power filter) circuit for maximal value of load current equal to 100 A. Theoretically for a typical 12-bit A/D converter it is possible to achieve $SNR \approx 72$ dB. Practically, with 30% signal headroom we can achieve $SNR \approx 69$ dB. The situation can be even worse if the value of the load current is less. For example, if the load current is equal to 20 A then $SNR \approx 58$ dB.

2.7.3 *Noise Shaping Technique*

Figure 2.42 shows the spectra for the different methods of D/A conversion. The spectrum of the classical method of D/A conversion is shown in Fig. 2.42a. In the assumed model of quantization noise, spectral density is constant throughout the band 0 to f_b. By using oversampling, the noise power in band 0 to f_b can be determined from the expression

$$P_{nb} = P_n\frac{2f_b}{f_s} , \tag{2.45}$$

where: P_n—noise power in band 0 to $f_s/2$.

Therefore, for D/A conversion with oversampling, the expression for the signal-to-noise ratio (2.42) can be modified to form

$$SNR = 1.76 + 6.02b + 10\log\frac{f_s}{2f_b} . \tag{2.46}$$

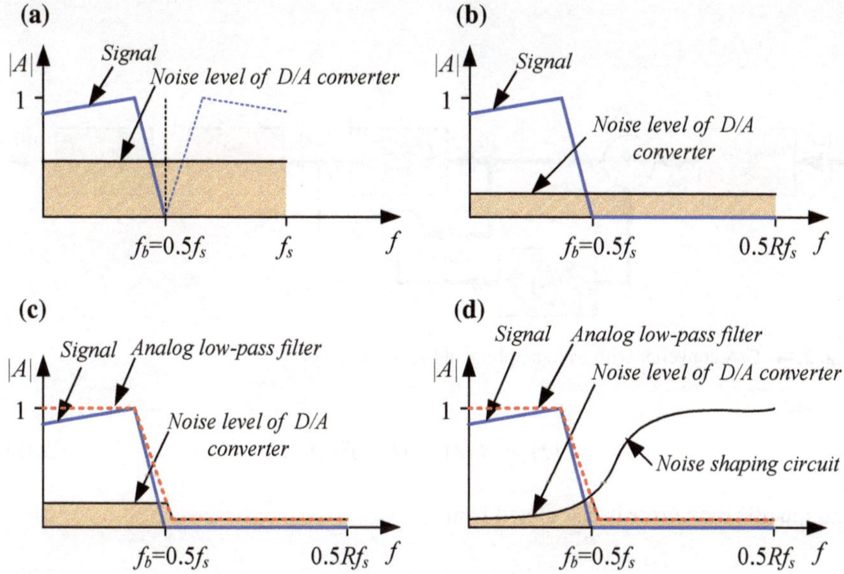

Fig. 2.42 Spectra of D/A conversions: **a** classical method, **b** method with oversampling, **c** method with oversampling and analog filter, **d** method with oversampling, analog filter and noise shaping circuit

Doubling the sampling rate increases the signal to noise ratio by 3 dB.

Out-of-band noise can be suppressed by the output analog low-pass filter as shown in Fig. 2.42c. Further increasing the signal to noise ratio can be achieved by the use of the digital noise shaping circuit. With this solution, the noise is moved outside the band of interest. The spectrum of such a solution is shown in Fig. 2.42d.

The noise shaping circuit works by putting the quantization error in a feedback loop. Different circuit architectures can be used for spectral shaping of the quantization noise, i.e., for moving it away from the band of interest toward higher frequencies [21]. Noise shaping circuits with feedback were first presented by Cutler [28] in 1954, and their detailed analysis was done by Spang and Schultheiss [91]. A block diagram of a quantization noise shaping circuit using a linear quantizer model with feedback [91, 103] is shown in Fig. 2.43. Output signal can be calculated as

Fig. 2.43 Noise shaping circuit

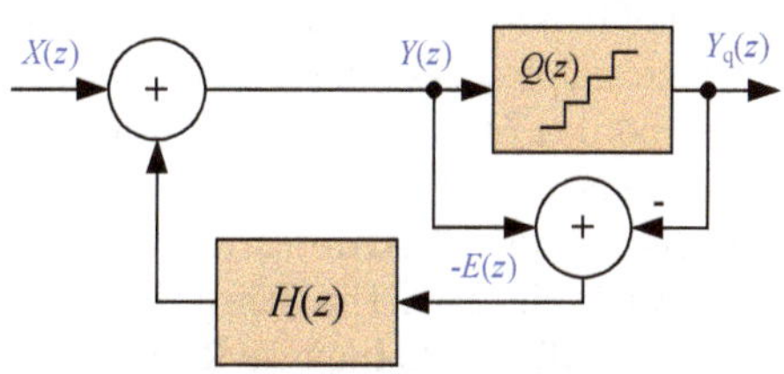

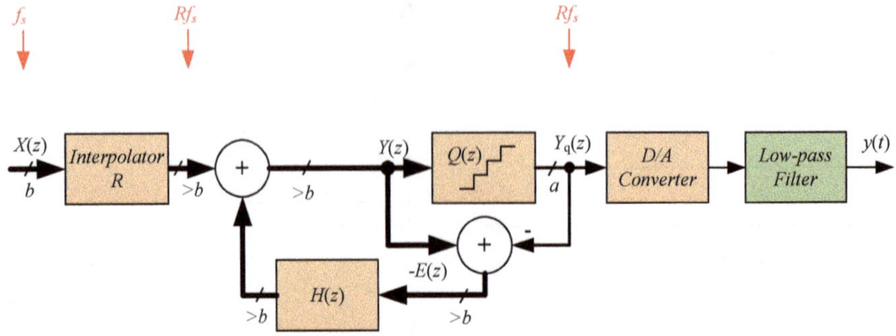

Fig. 2.44 D/A converter with oversampling and noise shaping

$$Y(z) = X(z) - H(z)E(z) \quad , \tag{2.47}$$

and quantization error is calculated from

$$-E(z) = Y(z) - Y_q(z) \quad . \tag{2.48}$$

Finally the output quantized signal is calculated by formula

$$Y_q(z) = X(z) - \overbrace{(1 + H(z))}^{H_n(z)} E(z) = X(z)H_s(z) + E(z)H_n(z) \quad , \tag{2.49}$$

where: $H_s(z) = 1$—transfer function for signal, $H_n(z) = 1 - H(z)$—transfer function for noise. A properly designed circuit with noise shaping has flat frequency response $H_s(z)$ in the signal frequency. On the other hand, $H_n(z)$ should have high attenuation in the frequency band of interest and a low attenuation in the rest of the band. For a low oversampling ratio R, an efficient way to increase the signal to noise ratio is by using a second-order loop filter.

A block diagram of the processing of b-bit digital signal $X(z)$ of the sampling rate f_s by means of a-bit D/A converter to the analog signal is shown in Fig. 2.44. Digital input signal $X(z)$ with a resolution of b-bit is interpolated by a factor of R, and produces a signal with a resolution of b-bit or more. For example, for the SHARC digital signal processor has is 32/40-bits resolution. Then the resolution of the signal is reduced to a-bits and the difference of signals $Y(z)$ and $Y_q(z)$ is transformed by a system of noise shaping transfer function $H(z)$.

In the simplest case, $H(z) = z^{-1}$ is only delay, to the input signal $X(z)$ is added to this portion of the signal, which has not been processed in the previous cycle. A block diagram of such a system, with the b-bit input samples the D/A converter converts only the oldest a-bits, and the remaining bits are added to the next sample, as shown in Fig. 2.45. The transfer function of the noise $H_n(z)$ in the simplest case can be the FIR (**F**inite **I**mpulse **R**esponse) filter and the N-th order is defined by the equation

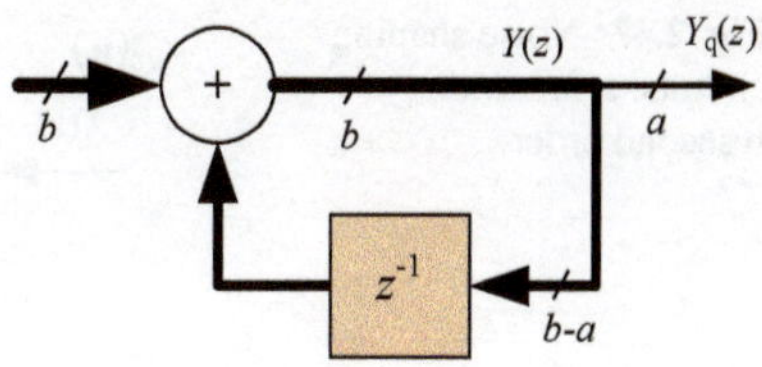

Fig. 2.45 Simplest noise shaping circuit

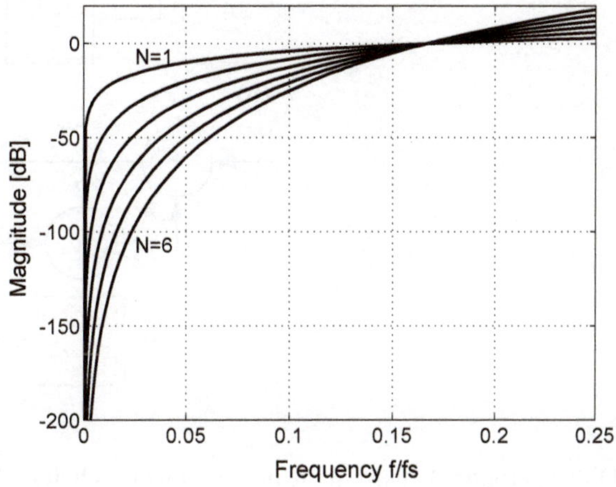

Fig. 2.46 Frequency characteristics of noise shaping circuits

$$H_n(z) = (1 - z^{-1})^N \quad . \tag{2.50}$$

The frequency characteristics of noise attenuation of the noise shaping circuits described by the above Eq. (2.50) for the orders $N = 1$–6 is shown in Fig. 2.46. Figure 2.47 shows the block diagrams of noise shaping circuits of the first and second order. In order to prevent overflow of numbers in circuits amplitude limiters are introduced.

The presented noise shaping method is one of several techniques, another example is the commonly used the delta-sigma modulator [22, 23, 64, 71, 81]. The author has applied the noise shaping circuit for high quality digital class-D audio amplifier [85, 88].

The noise shaping technique for D/A conversion shown in this section can be also used for A/D conversion.

2.7.4 Dither

The resolution of A/D and D/A conversion can be increased by adding to the input signal a low-level noise signal; this signal is called a dither [64, 68, 71, 81, 85]. It is an intentionally applied form of noise signal used to randomize quantization error.

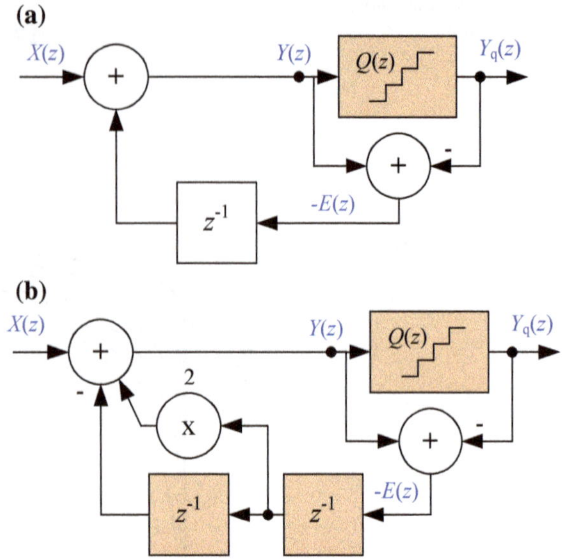

Fig. 2.47 Noise shaping circuits: **a** first order, **b** second order

Dither is routinely used in the processing of both digital audio and digital video data, and is often one of the last stages of audio production of compact discs. Figure 2.48a shows circuit with analog dither added before the A/D converter, but after the SH circuit. Similarly, a digital dither signal may be used to improve the D/A conversion as shown in Fig. 2.48. It can also be successfully used in power electronics systems. The amplitude of this signal must be small in order to avoid reduction of the dynamic range of the signal. The author's research shows that it should have an amplitude in the range from 0.5 to 2 LSB [85]. Figure 2.49 depicts a simple digital circuit where dither $D(z)$ signal is added to input signal $X(z)$ before its quantization. Application of a dither signal can also improve the performance of the noise shaping circuit as shown in Fig. 2.50a. A pseudorandom signal is added outside the loop feedback, so that it will not be followed by the tracking loop. Simulation studies carried out by the author [85] shows that the optimum pseudorandom signal amplitude in terms of obtaining the greatest signal to noise ratio *SINAD* is about 0.3 LSB. The amplitude of the random signal to be added is so small due to the fact that noise shaping circuit has a feedback loop that reduces the size of the quantization step.

Adding to the input signal pseudorandom signal degrades *SNR*, therefore it would be beneficial to change the amplitude depending on the amplitude of the input signal. For input signals with large amplitudes, the amplitude of the random signal is reduced. The block diagram of such a circuit is shown in Fig. 2.50b. The amplitude of the random signal $d(nT)$ can be modulated depending on the input module [63] according to the equation

$$d_m(nT_s) = (1 - \sqrt{|x(nT_s)|})^2 d(nT_s) \quad . \tag{2.51}$$

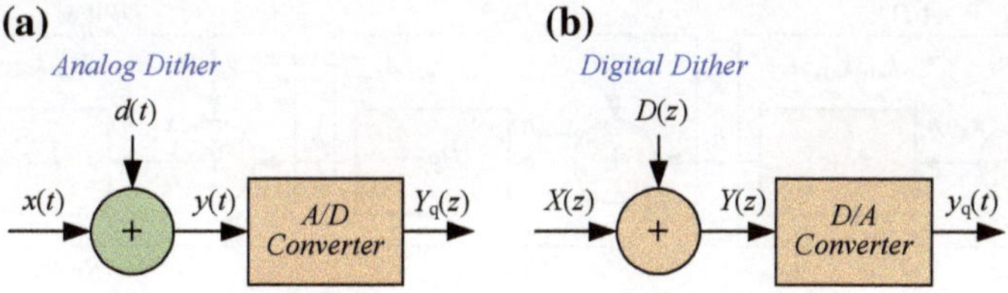

Fig. 2.48 Dither circuits: **a** the A/D conversion, **b** the D/A conversion

Fig. 2.49 Digital dither circuit

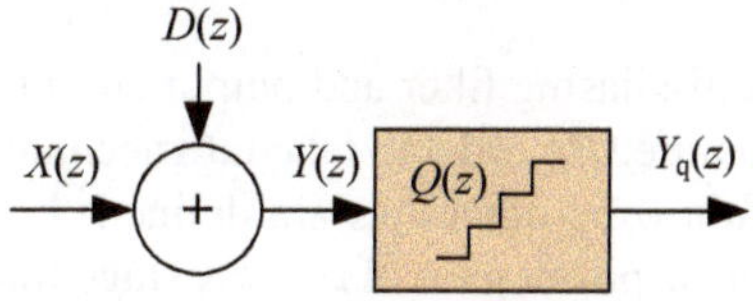

Fig. 2.50 Dither circuits with noise shaping circuit: **a** with constant level of the dither, **b** with dynamic level of the dither

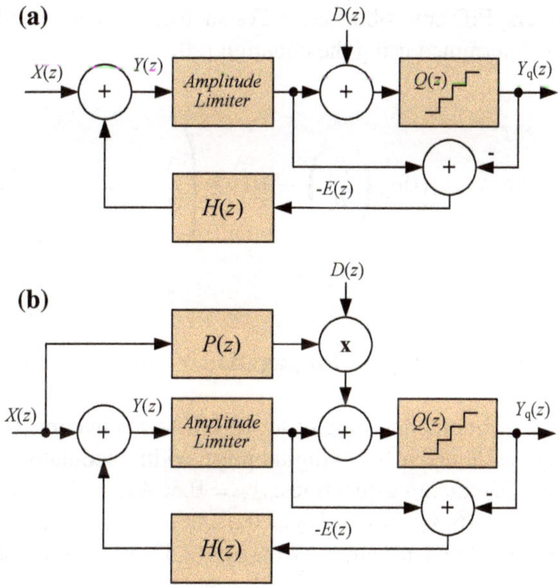

2.7.5 *Propagation of Quantization Noise*

An important issue is the resultant signal to noise ratio after passing through a circuit with different signal resolutions in subsequent steps. In analog systems, the Friis noise formula [34] is used for such analysis, and is independent of the signal level. In the case of quantization noise, it depends on the number of bits and the signal range. Figure 2.51 presents an example of a typical open-loop control system. A circuit consists of an A/D converter, DSP circuit, and an output DPWM modulator. Focusing only on the quantization noise analysis, the influence of the input

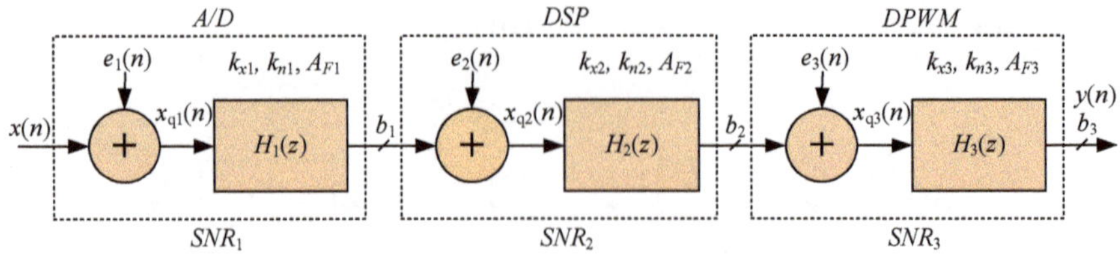

Fig. 2.51 Quantization model of the open-loop digital control circuit

anti-aliasing filter and output circuit, as well as of some other parameters, has been omitted. In order to take into account noise and signal propagation, at each stage the following quantities are defined: k_{xk}—k^{th} stage signal power gain, k_{nk}—k^{th} stage noise power gain, A_{Fk}—k_{th} stage full-scale signal amplitude, and b_k—k^{th} stage signal resolution. This approach allows for the analysis of digital circuits (such as digital filters, PID controllers etc.). For such circuits, the resultant signal-to-noise ratio can be determined using the equation [90]

$$SNR = 10\log\left(\frac{P_x}{P_n}\right) = 10\log\left(\frac{3}{2}\cdot\frac{k_{x1}k_{x2}k_{x3}A_x^2}{k_{n1}k_{n2}k_{n3}\frac{A_{F1}^2}{2^{2b_1}} + k_{n2}k_{n3}\frac{A_{F2}^2}{2^{2b_2}} + k_{n3}\frac{A_{F3}^2}{2^{2b_3}}}\right) \tag{2.52}$$

2.7.5.1 Example—Propagation of Quantization Noise

For a circuit consisting of an 18-bit A/D converter, 16-bit fixed point digital signal processor and a 10-bit digital pulse width modulator shown in Fig. 2.52 and assuming the remaining quantities: $A_x = 0.5$, $A_{F1} = 1$, $b_1 = 18$, $k_{x1} = 1$, $k_{n1} = 1$, $A_{F2} = 1$, $b_2 = 16$, $k_{x2} = 0.7$, $k_{n2} = 0.5$, $A_{F3} = 1$, $b_3 = 10$, $k_{x3} = 1$, $k_{n3} = 1$, the resultant SNR value determined by the Eq. (2.52) is equal to 54.4 dB.

2.7.6 *Effective Number of Bits*

An interesting parameter of the digital signal is the effective number of bits (*ENOB*). This parameter takes into account all errors made during the conversion. This parameter is especially important today when there is easy access to 18-bit A/D converters and by incompetent application it is easy to reduce the effective number of bits to 10–12. Major sources of errors during the A/D conversion are digital transmission and clock signals of high-speed processors. Among the most important are: quantization noise, jitter, A/D converter noise, aliasing, integral and differential nonlinearity,

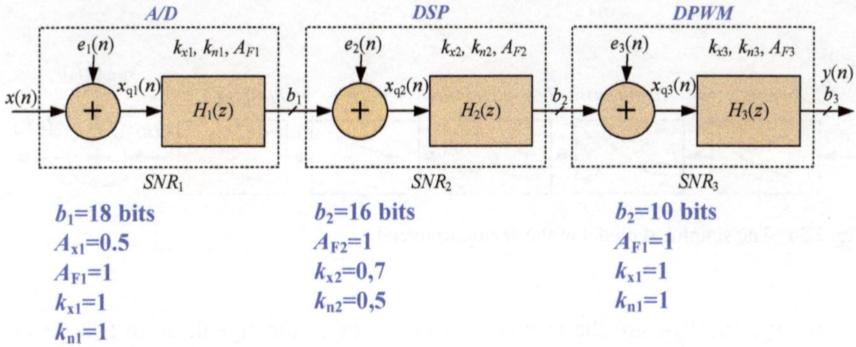

Fig. 2.52 Block diagram of an example of the digital circuit

Fig. 2.53 Effective number of bits

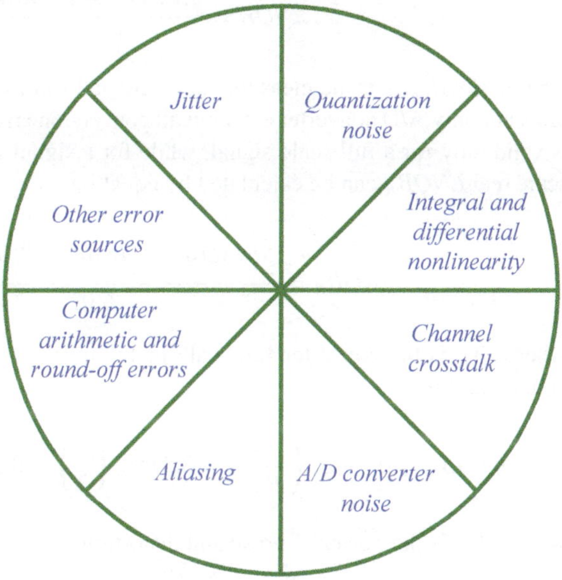

channel crosstalk, channel-to-channel crosstalk, and other error sources (Fig. 2.53). The simplified analog front end used for effective number of bit calculation is shown in Fig. 2.54. *SINAD* is the ratio of the wanted signal (the fundamental) to the sum of all distortion and noise products, after the DC term is removed. SINAD is a measure of the quality of a signal, defined as:

$$SINAD = 10 \log \left(\frac{P_x}{P_d + P_n} \right) , \qquad (2.53)$$

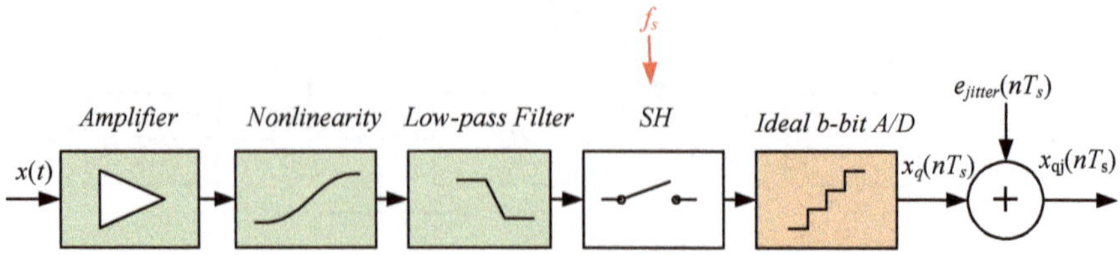

Fig. 2.54 The simplified model of the analog front end

where: P_x, P_n, P_d—are the average power values of the signal, noise and distortion components respectively. The effective number of bits can be described by the equation

$$ENOB = \frac{SINAD_M - 1.76\,\text{dB}}{6.02}\ . \tag{2.54}$$

where: $SINAD_M$—is the measured or calculated value of $SINAD$. The $SINAD_M$ contains not only A/D converter error but all conversion errors (Fig. 2.54). Equation 2.54 is valid only for a full scale signal, while for a signal with amplitude less than full scale real $ENOB_R$ can be calculated by equation

$$ENOB_R = \frac{SINAD_M - 1.76\,\text{dB} - 20\log\left(\frac{A_F}{A_{in}}\right)}{6.02}\ . \tag{2.55}$$

where: A_F is the converter full scale input signal amplitude, and A_{in} is the input signal amplitude.

$$ENOB_R = 0.5\log_2\left(\frac{P_s}{P_{nd}}\right) - 0.5\log_2\left(\frac{2}{3}\right) - 0.5\log_2\left(\frac{A_F}{A_{in}}\right)\ , \tag{2.56}$$

where: P_{nd} is the power of noise and distortion.

2.8 A/D Converters Suitable for Power Electronics Control Circuits

Currently the number of types of integrated circuits manufactured for A/D converters exceeds a few hundred, so the designer can be confused when choosing the best one for his application. A subjective choice of the A/D converter will be presented in this section. During the selection process the designer should consider following features:

- sampling speed,
- accuracy and resolution,

- input voltage range,
- interface—serial or parallel,
- sampling synchronization,
- power consumption and supply voltage,
- conversion delay time,
- standalone IC or integrated with microprocessor,
- cost and availability.

Given the above sources of error, in the author's opinion the A/D converter used for power electronics systems control circuits must meet the following requirements:

- simultaneous sampling for multichannel solution—multiple SH circuits or multiple A/D converters,
- a number of bits greater than 12,
- minimum delay introduced by the A/D converter, the best solution is A/D converter with successive approximation (SA), most popular and cheap A/D converter with delta sigma modulator has to be carefully considered,
- the ability to synchronize the sampling time by an external signal,
- if possible, coherent synchronous sampling should be used.

Despite the large number of available A/D converters, these requirements are only met by a small number of commercial integrated circuits and the choice is very limited. Analog Devices and Texas Instruments can be considered leading companies in this area.

2.8.1 A/D Converter with Successive Approximation

An A/D converter with successive approximation (SA) is best solution for power electronics, because of its short response time. Of course, there are known faster A/D converters, such as flash, pipeline etc., but typically their resolution is less than 12-bit. A block diagram of an A/D converter with successive approximation is shown in Fig. 2.55. In this A/D converter the processed sampled input signal is compared with a signal from the D/A converter by analog comparator. The D/A converter is controlled by the successive approximation register, which sequentially switches on the individual bits, starting from the most significant bit (MSB) to the least significant bit (LSB). During this process, a decision is made about whether to leave the bit switched on or turn it off. Hence the number of cycles is equal to the number of bits. Maximum response time is equal

$$t_{max} = T_s + t_{cov} + t_{tran} \quad \text{and} \quad T_s > t_{cov} + t_{tran} \quad , \tag{2.57}$$

where: t_{cov}—A/D converter conversion time, t_{tran}—data from A/D converter to microprocessor transmission time.

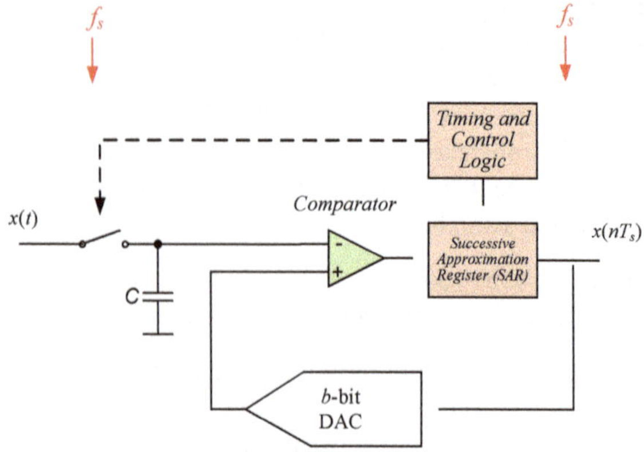

Fig. 2.55 Block diagram of an A/D converter with successive approximation

2.8.2 *A/D Converter with Delta Sigma Modulator*

In recent years, the A/D converter with delta sigma modulator (DSM) is the most common used A/D converter, due to its simplicity of implementation and low price [64, 71]. In this converter the oversampling technique is used which allows an increase in the resolution from 1-bit to 24-bit. An additional advantage is the absence of SH and lower requirements for the antialiasing filter. A block diagram of A/D converter with delta sigma modulator is shown in Fig. 2.56. A one bit A/D converter consists of: integrator, D flip-flop, comparator and a 1-bit D/A converter. It produces a bit stream with sampling speed equal to Rf_s. Then the signal is processed by a low-pass filter and its sampling speed is reduced to f_s. This filter is responsible for the signal delay. Typically FIR filters, with an order range from one hundred to several hundred are used. The delay introduced by the filter FIR is approximately equal to the number of samples which is equal to half the filter order

$$t_{delay} = 0.5\frac{T_s}{R}N \quad , \tag{2.58}$$

where: N—FIR filter order, T_s/R—conversion period. The output data are fully settled after N conversion periods. Therefore application of this kind of converter should be carefully considered. The value of R is typically ranging from 64 to 2048 and it is most often a power of two.

2.8.3 *Selected Simultaneous Sampling A/D Converters*

In this section some selected simultaneous sampling A/D converters useful for power electronics circuit are considered.

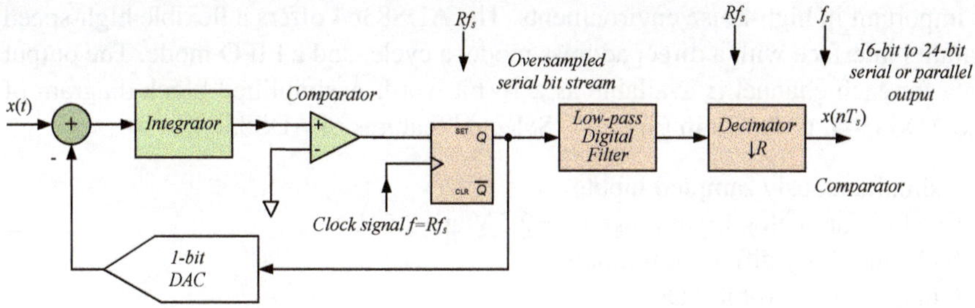

Fig. 2.56 Block diagram of an A/D converter with delta sigma modulator

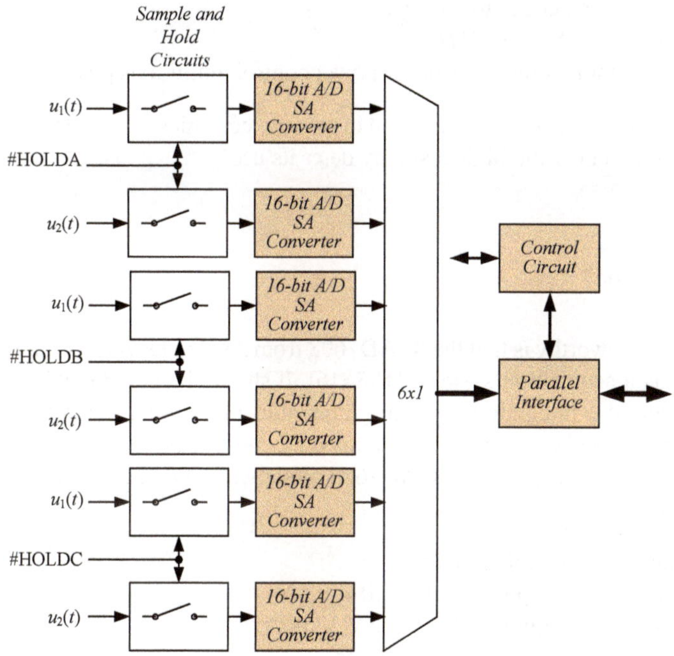

Fig. 2.57 Simplified block diagram of the AD8364: 6-channel, 16-Bit A/D converter

2.8.4 ADS8364

A typical A/D converter with simultaneous sampling suitable for power electronics applications is IC ADS8364 from Texas Instruments. The ADS8364 includes six, 16-bit, 250 KHz A/D converters with 6 fully differential input channels grouped into two pairs for high-speed simultaneous signal acquisition [94]. The A/D converters work using a successive approximation (SA) algorithm. Inputs to the SH amplifiers are fully differential and are kept differential with respect to the input of the A/D converter. This provides excellent common-mode rejection of 80 dB at 50 KHz, which is important in high-noise environments. The ADS8364 offers a flexible high-speed parallel interface with a direct address mode, a cycle, and a FIFO mode. The output data for each channel is available as a 16-bit word. A simplified block diagram of the ADS8364, is shown in Fig. 2.57. Selected features of ADS8364:

- 8 simultaneously sampled inputs,
- true bipolar analog input range of ±2.5 V at +2.5 V,
- 6-channel fully differential inputs,
- 6 independent 16-bit ADC,
- 4 μs total throughput per channel,
- on-chip accurate reference and reference buffer,
- testing no missing codes to 14-bits,
- 83.2 dB *SNR*, 82.5 dB *SINAD*,
- applications: motor control, 3-phase power control, multi-axis positioning system.

For the designer of a control circuit used in power electronics, it is a very comfortable arrangement and only the high cost may deter its use.

2.8.5 AD7608

Particularly noteworthy is that the IC AD7608 from Analog Devices, is an 8-channel, 18-bit SA data acquisition system (DAS) [6]. It should be noted that in one single chip there are also integrated eight programmable antialiasing second order filters. Analog signals can be simultaneous sampled by eight track-and-hold (TH) circuits. A simplified block diagram of the AD7608, is shown in Fig. 2.58. Selected features of AD7608:

- 8 simultaneously sampled inputs,
- true bipolar analog input ranges: ±10 V, ±5 V,
- single 5 V analog supply and 2.3 V–5 V VDRIVE,
- fully integrated data acquisition solution,
- analog input clamp protection,
- input buffer with 1 MΩ analog input impedance,
- second-order programmable antialiasing analog filter,
- on-chip accurate reference and reference buffer,
- 18-bit SA A/D converter with 200 kSPS on all channels,
- oversampling capability with digital filter,
- 98 dB *SNR*, −107 dB *THD*,
- parallel or serial interface.

Similarly to ADS8364, for the designer of a control circuit used in power electronics, it is a very comfortable solution and only the high cost may deter its use.

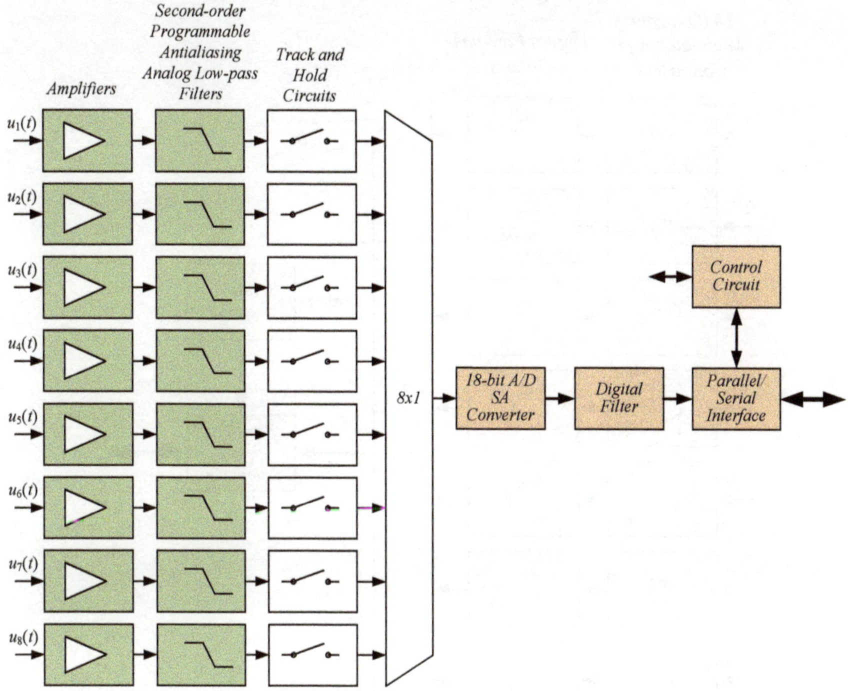

Fig. 2.58 Simplified block diagram of AD7608: 8-channel DAS with 18-Bit, bipolar, with simultaneous sampling TH

2.8.6 ADS1278

ADS1278 is an octal channel 24-bit, DSM A/D converter with data rates up to 144 kSPS, allowing simultaneous sampling of eight channels [99]. The A/D converter offers the highest possible resolution of A/D converters. A simplified block diagram of the ADS1278, is shown in Fig. 2.59. The A/D converter consists of eight advanced, 6-order chopper-stabilized, delta-sigma modulators followed by low-ripple, linear phase FIR filters. Oversampling ratio R is equal 64 or 128. After the step change on the input occurs, the output data changes very little prior to 30 conversion periods. The output data are fully settled after 76 or 78 periods depending on the converter mode. For High-Speed mode, the maximum clock f_{clk} input frequency is 37 MHz and output signal sampling rate f_{data} is equal to

$$f_{data} = \frac{f_{clk}}{4R} = 144531.25 \text{ SPS} \quad . \tag{2.59}$$

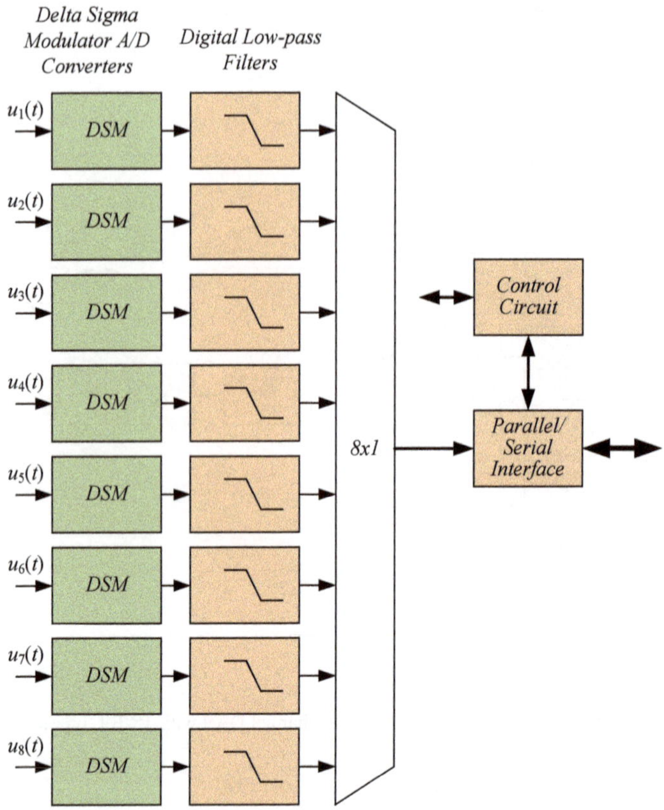

Fig. 2.59 Simplified block diagram of the ADS1278 8-channel 24-bit, DSM A/D converter

2.8.7 ADS8568

Another converter, the ADS8568 A/D converter from Texas Instruments is suitable for a power electronics circuit. The ADS8568 contain eight low-power, 16-bit, successive approximation register (SA) A/D converters with true bipolar inputs [100]. This architecture is designed on the charge redistribution principle, which inherently includes a SH function. A simplified block diagram of the ADS8568 is depicted in Fig. 2.60. The devices support a selectable parallel or serial interface with daisy-chain capability. The programmable reference allows handling of analog input signals with amplitudes up to ±12 V. Selected features of the ADS8568 [100]:

- 8 simultaneously sampled inputs,
- true bipolar analog input ranges: ±10 V (±4 * VREF), ±5 V (±2 * VREF),
- reference voltage 0.5–2.5 V or 0.5–3.0 V,
- 510 kSPS (parallel interface) or 400 kSPS (serial interface),
- fully integrated data acquisition solution,

Fig. 2.60 Simplified block diagram of the ADS8568 8-channel 16-bit, SA A/D converter

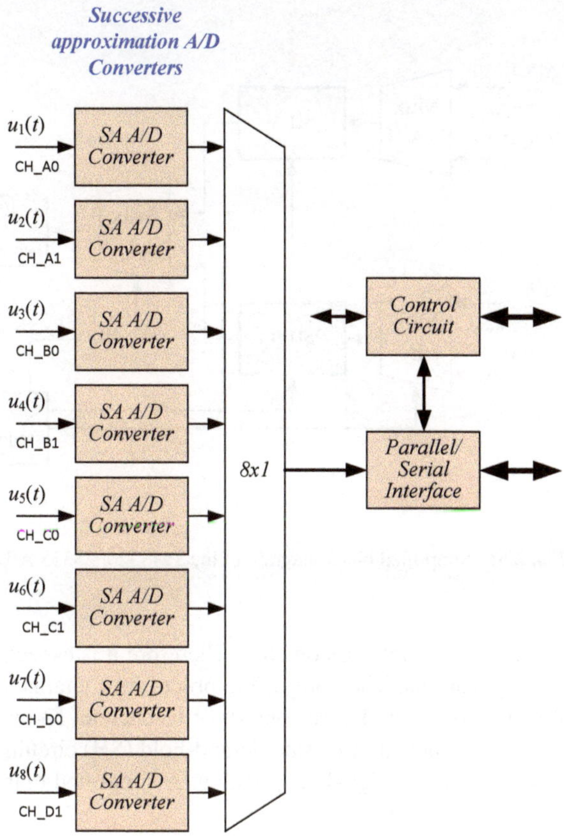

- analog input clamp protection,
- 91.5 dB *SNR*, −94 dB *D*,
- parallel or serial interface.

Unfortunately, the ADS8568 requires four separate supplies: an analog supply for the A/D converter (AVDD), the buffer I/O supply for the digital interface (DVDD), and the high-voltage supplies driving the analog input circuitry (HVDD and HVSS). It is best when the A/D converter needs only one supply voltage.

2.8.8 A/D Conventer of TMS320F28335

The rapidly growing market of microprocessors for power electronics circuits, have caused manufacturers to create integrated circuits which can fully meet the needs of the control system. An example of such a system is the digital signal controller (DSC), from Texas Instruments [95, 98]. It is a complete system with many useful

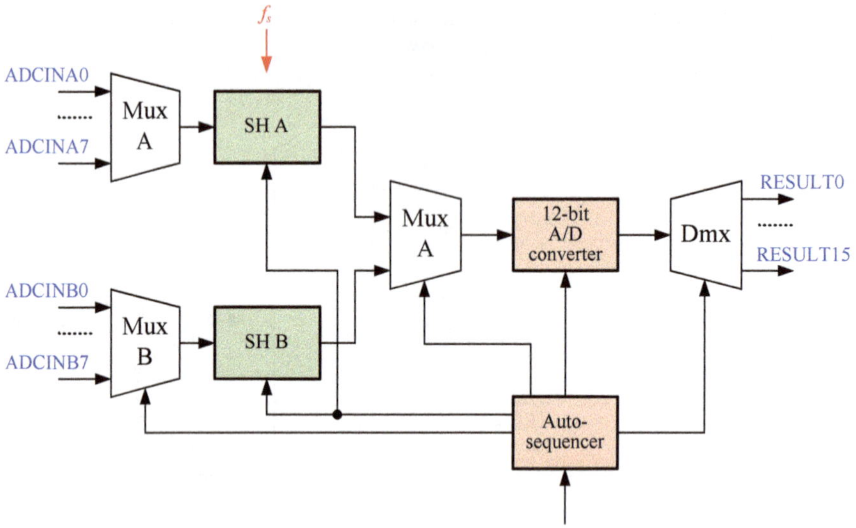

Fig. 2.61 Simplified block diagram of the TMS320F28335 A/D converter

features in a single silicon chip. Therefore it is especially good for power electronics applications. The core of the processor contains an IEEE-754 single-precision floating-point unit. It also consists of 16-channel 12-bit SA A/D converter, with 80 ns conversion rate and two sample-and-hold (SH) circuits. Therefore the simultaneous sampling of two signals is possible. A simplified diagram of this A/D converter is shown in Fig. 2.61.

On the input of each sample and hold circuit is located an 8-channel analog multiplexer that allows sequential converting of 8-pairs of signal (sampled simultaneously). The voltage input range is equal to 0–3 V. The converter input voltage U_in can be determined from the equation

$$U_{in} = \frac{D(U_{ref+} - U_{ref-})}{2^b - 1} + U_{ref-} \quad , \tag{2.60}$$

where: D—converter digital output, U_{ref}—reference voltage, b—number of bits. For $U_{ref+} = 3V$, $U_{ref-} = 0V$ and $b = 12$

$$U_{in} = \frac{3D}{4095} \quad . \tag{2.61}$$

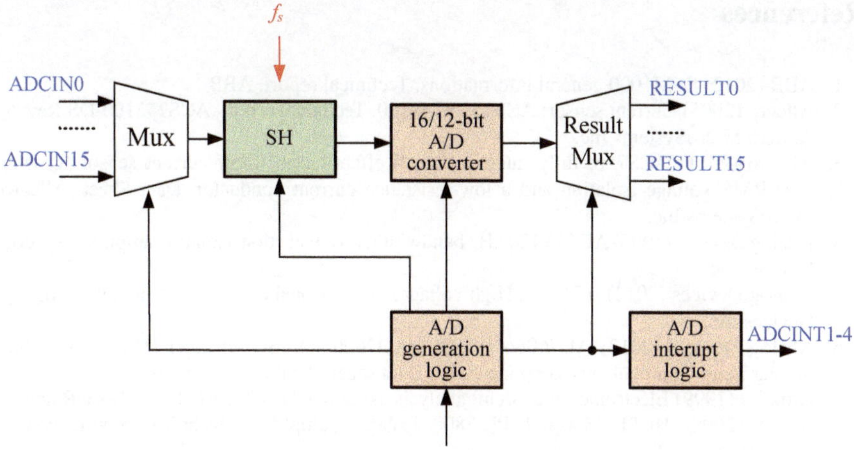

Fig. 2.62 Simplified block diagram of one of the TMS320F2837xD A/D converters

2.8.9 A/D Converters of TMS320F2837xD

Texas Instruments is permanently engaged in the design of a wide range of microcontrollers intended for use in power electronics systems. One of the most advanced is the TMS320F2837xD microcontroller family [101, 102]. The F2837xD includes four independent high-performance A/D converters, allowing the device to efficiently manage multiple analog signals for enhanced overall system throughput. Each A/D converter has a single SH circuit, and using multiple ADC modules enables it to perform simultaneous sampling or independent operation. The A/D converter is implemented using a successive approximation and it has the configurable resolution of either 16-bits or 12-bits. The Fig. 2.62 shows a simplified diagram of one of the four A/D converters. The A/D converters have many possible modes of operation [101, 102]. In the opinion of the author, the most important is the possibility to sample four analog signals. It is only a pity that it is not possible to simultaneously sample more analog signals, e.g., eight.

2.9 Conclusions

The chapter shows the most common sources of errors during conversion of analog signal to its digital form. This process is very important for the quality of the entire digital control system. However, in practical control systems the price is one of the most important limiting factors and the designer is forced to compromise solution. The discussion in this chapter gives better understanding of the selection of control system parameters. For extended studies of the problems discussed in this chapter may be used [15, 16, 20, 24, 26, 30, 33, 36, 44, 45, 52, 59, 61, 69–73, 86, 87, 104, 105, 108, 109].

References

1. ABB (2012) ESM1000 general informations. Technical report, ABB
2. Allegro (2005) Current sensor: ACS752SCA-100. Technical report, ACS752100-DS Rev. 6, Allegro MicroSystems Inc.
3. Allegro (2011) ACS756, fully integrated, hall effect-based linear current sensor IC with 3 kVRMS voltage isolation and a low-resistance current conductor. Data Sheet, Allegro MicroSystems Inc.
4. Analog Devices (1996) AD215 120 kHz bandwidth, low distortion, isolation amplifier, Analog Devices Inc.
5. Analog Devices (2012) AD8210, High voltage, bidirectional current shunt monitor, Analog Devices Inc.
6. Analog Devices (2012) AD7606/AD7606-6/AD7606-4 8-/6-/4-channel DAS with 16-bit, bipolar input, simultaneous sampling ADC. Data sheet, Analog Devices Inc.
7. Attia JO (1999) Electronics and circuit analysis using MATLAB. CRC Press, Boca Raton
8. Avago (2008) HCPL-7800A/HCPL-7800 isolation amplifier. Technical report, AV02-1436EN, Avago Technologies
9. Avago (2011) HCPL-7860/HCPL-786J optically isolated sigma-delta (S-D) modulator. Technical report, AV02-0409EN, Avago Technologies
10. Avago (2013) Optoisolation products, application block diagrams. Reference Guide, AV00-0271EN, Avago Technologies
11. Avago (2014) Optocouplers. Designer's guide, AV02-4387EN, Avago Technologies
12. Azeredo-Leme C (2011) Clock jitter effects on sampling: a tutorial. IEEE Circuits Syst Mag 3:26–37
13. Baggini A (ed) (2008) Handbook of power quality. Wiley, New York
14. Baird RT, Fiez TS (1995) Linearity enhancement of multibit A/D and D/A converters using data weighted averaging. IEEE Trans Circuits Syst II Analog Digital Sig Process 42(12):753–762
15. Bateman A, Paterson-Stephens I (2002) The DSP handbook: algorithms, applications and design techniques. Prentice Hall, New York
16. Baxandall PF (1959) Transistor sine-wave LC oscillators. In: ICTASD—IRE, pp 748–759
17. Bossche AV, Valchev VC (2005) Inductors and transformers for power electronics. CRC Press, Boca Raton
18. Brannon B (2004) Sampled systems and the effects of clock phase noise and jitter. Application note AN-756. Technical report, Analog Devices Inc.
19. Brannon B, Barlow A (2006) Aperture uncertainty and ADC system performance. Application note AN-501. Technical report, Analog Devices Inc.
20. Bruun G (1978) Z-transform DFT filters and FFT's. IEEE Trans Acoust Speech Signal Process 26(1):56–63
21. Candy J, Temes G (eds) (1992) Oversampling delta-sigma data converters. Theory, design, and simulation. IEEE Press
22. Carley RL, Schreier R, Temes GC (1997) Delta-sigma ADCs with multibit internal converters. In: Norsworthy SR, Schreier R, Temes GC (eds) Delta-sigma data converters. Theory, design, and simulation. IEEE Press
23. Cataltepe T, Kramer AR, Larson LE, Temes GC, Walden RH (1992) Digitaly corrected multi-bit $\Sigma\Delta$ data converters. In: Candy JC, Temes GC (eds) IEEE proceedings of oversampling delta-sigma data converters theory, design, and simulation, ISCAS'89, May 1989. IEEE Press
24. Chen WK (ed) (1995) The circuits and filters handbook. IEEE Press, Boca Raton
25. ChenYoung (2011) Closed loop precise Hall current sensor CYHCS-SH. Technical report, ChenYoung
26. Crochiere RE, Rabiner LR (1983) Multirate digital signal processing. Prentice Hall Inc., Englewood Cliffs
27. Cummings J, Doogue MC, Friedrich AP (2007) Recent trends in hall effect current sensing (Rev. 1). AN295045. Technical report, Allegro MicroSystems Inc.

28. Cutler C (1960) Transmission system employing quantization. US Patent 2927962
29. Dabrowski A (ed) (1997) Digital signal processing using digital signal processors. Wydawnictwo Politechniki Poznanskiej, Poznan (in polish)
30. Dabrowski A, Sozanski K (1998) Implementation of multirate modified wave digital filters using digital signal processors. In: XXI Krajowa Konferencja Teoria Obwodow i Uklady Elektroniczne, KKTUIE'98, Poznan
31. Data Translation (2008) The battle for data fidelity: understanding the SFDR spec. Technical report, Data Translation
32. Data Translation (2009) Benefits of simultaneous data acquisition modules. Technical report, Data Translation
33. Farhang-Boroujeny B, Lee Y, Ko C (1996) Sliding transforms for efficient implementation of transform domain adaptive filters. Signal Process 52(1):83–96
34. Friis HT (1944) Noise figures of radio receivers. Proc IRE 32(7):419–422
35. Galton I (1997) Spectral shaping of circuit errors in digital-to-analog converters. IEEE Trans Circuits Syst II Analog Digit Signal Process 44(10):789–797
36. Goertzel G (1958) An algorithm for the evaluation of finite trigonometric series. Am Math Monthly 65:34–35
37. Goldberg JM, Sandler MB (1994) New high accuracy pulse width modulation based digital-to-analogue convertor/power amplifier. IEE Proc Circuits Devices Syst 141(4):315–324
38. Gwee BH, Chang JS, Adrian V (2007) A micropower low-distortion digital class-d amplifier based on an algorithmic pulsewidth modulator. IEEE Trans Circuits Syst I Regul Pap 52(10):2007–2022
39. Hartley RVL (1928) Transmission of information. Bell Syst Tech J 7:535–563
40. Hartmann M, Biela J, Ertl H, Kolar JW (2009) Wideband current transducer for measuring ac signals with limited DC offset. IEEE Trans Power Electron 24(7):1776–1787
41. Holmes DG, Lipo TA (2003) Pulse width modulation for power converters: principles and practice. Institute of Electrical and Electronics Engineers, Inc.
42. Honeywell (2008) Current sensors line guide. Technical report, Honeywell International Inc.
43. IEEE (2011) IEEE standard for terminology and test methods for analog-to-digital converters. IEEE Std. 1241-2010. Technical report, IEEE
44. Jacobsen E, Lyons R (2003) The sliding DFT. IEEE Signal Process Mag 20(2)
45. Jacobsen E, Lyons R (2004) An update to the sliding DFT. IEEE Signal Process Mag 21:110–111
46. Kazimierkowski M, Malesani L (1998) Current control techniques for three-phase voltage-source converters: a survey. IEEE Trans Ind Electron 45(5):691–703
47. Kazmierkowski MP, Kishnan R, Blaabjerg F (2002) Control in power electronics. Academic Press, San Diego
48. Kester W (2004) Analog-digital conversion. Analog Devices Inc., Norwood
49. Kester W (2005) The data conversion handbook. Newnes, New York
50. Kester W (2009) Understand SINAD, ENOB, SNR, THD, THD+N, and SFDR so you don't get lost in the noise floor. Technical report, Analog Devices Inc.
51. Kotelnikov AV (1933) On the capacity of the 'ether' and of cables in electrical communication. In: Proceedings of the first All-Union Conference on the technological reconstruction of the communications sector and low-current engineering, Moscow
52. Kurosu A, Miyase S, Tomiyama S, Takebe T (2003) A technique to truncate IIR filter impulse response and its application to real-time implementation of linear-phase IIR filters. IEEE Trans Signal Process 51(5):1284–1292
53. Larson LE, Cataltepe T, Temes G (1992) Multibit oversampled—A/D converter with digital error correction. In: Candy JC, Temes GC (eds) Oversampling delta-sigma data converters. Theory, design, and simulation, IEEE Electronics Letters, 24 August 1988. IEEE Press
54. Leung BH, Sutarja S (1992) Multibit—A/D converter incorporating a novel class of dynamic element matching techniques. IEEE Trans Circuits Syst II Analog Digital Signal Process 39(1):35–51
55. LEM (2004) Isolated current and voltage transducer, 3rd edn. LEM Components

56. LEM (2012) Current transducer LA 55-P. LEM Components
57. LEM (2012) Current transducer LA 205-S. Technical report, LEM Components
58. Lyons R (2004) Understanding digital signal processing, 2nd edn. Prentice Hall, Upper Saddle River
59. Lyons R, Bell A (2004) The Swiss army knife of digital networks. IEEE Signal Process Mag 21(3):90–100
60. Mitra S (2006) Digital signal processing: a computer-based approach. McGraw-Hill, New York
61. Mohan N, Undeland TM, Robbins WP (1995) Power electronics, converters, applications and design. Wiley, New York
62. Mota M (2010) Understanding clock jitter effects on data converter performance and how to minimize them. Technical report, Synopsis Inc.
63. Norsworthy SR (1997) Quantization errors and dithering in modulators. In: Norsworthy SR, Schreier R, Temes GC (eds.) Delta-sigma data converters. Theory, design, and simulation. IEEE Press
64. Norsworthy SR, Schreier R, Temes GC (eds) (1997) Delta-sigma data converters, theory, design, and simulation, IEEE Press
65. Nyquist H (1924) Certain factors affecting telegraph speed. Bell Syst Tech J 3:324–346
66. Nyquist H (1928) Certain topics in telegraph transmission theory. AIEE Trans 47:617–644
67. Oppenheim AV, Schafer RW (1999) Discrete-time signal processing. Prentice Hall, New Jersey
68. Orfanidis SJ (2010) Introduction to signal processing. Prentice Hall, Inc., Upper Saddle River
69. Oshana R (2005) DSP Software development techniques for embedded and real-time systems. Newnes, Boston
70. Owen M (2007) Practical signal processing. Cambridge University Press, Cambridge
71. Plassche R (2003) CMOS integrated analog-to-digital and digital-to-analog converters. Springer, New York
72. Powell SR, Chau PM (1991) A technique for realizing linear phase IIR filters. IEEE Trans Signal Process 39(11):2425–2435
73. Press WH, Teukolsky SA, Vetterling WT, Flannery BP (2007) Numerical recipes: the art of scientific computing, 3rd edn. Cambridge University Press, Cambridge
74. Proakis JG, Manolakis DM (1996) Digital signal processing, principles, algorithms, and application. Prentice Hall Inc., Englewood Cliffs
75. Rabiner LR, Gold B (1975) Theory and application of digital signal processing. Prentice Hall Inc., Englewood Cliffs
76. Ray WF, Davis RM (1993) Wide bandwidth Rogowski current transducer: part 1—the Rogowski coil. EPE J 3(2):116–122
77. Ray WF, Davis RM (1993) Wide bandwidth Rogowski current transducer: part 2—the integrator. EPE J 3(1):51–59
78. Ray WF, Davis RM (1997) Developments in Rogowski current transducer. In: Conference proceedings, EPE, Trondheim, vol 3, pp 308–312
79. Redmayne D, Trelewicz E, Smith A (2006) Understanding the effect of clock jitter on high speed ADCs. Design Note 1013. Technical report, Linear Technology, Inc.
80. Rollier S (2012) High accuracy, high technology: the perfect choice! ITB 300-S/IT 400-S/IT 700-S current transducers. Technical report, LEM Components
81. Schreier R, Temes GC (2004) Understanding delta-sigma data converters. Wiley-IEEE Press, New York
82. Shannon CE (1948) A mathematical theory of communication. Bell Syst Tech J 27: 379–423 and 623–656
83. Silicon Laboratories (2015) Si8920ISO-EVB, Si8920ISO-EVB user's guide, Silicon Laboratories
84. Silicon Laboratories (2016) Si8920 Isolated amplifier for current shunt measurement. Data Sheet, Silicon Laboratories

85. Sozański K (1999) Design and research of digital filters banks using digital signal processors. PhD thesis, Technical University of Poznan (in Polish)
86. Sozański K (2002) Implementation of modified wave digital filters using digital signal processors. In: Conference proceedings of 9th international conference on electronics, circuits and systems, pp 1015–1018
87. Sozański K (2012) Realization of a digital control algorithm. In: Benysek G, Pasko M (eds) Power theories for improved power quality. Springer, London, pp 117–168
88. Sozański K, Strzelecki R, Fedyczak Z (2001) Digital control circuit for class-D audio power amplifier. In: Conference proceedings of 2001 IEEE 32nd annual power electronics specialists conference, PESC'01, pp 1245–1250
89. Sozanski K (2015) Selected problems of digital signal processing in power electronic circuits. In: Conference proceedings, SENE'15, Lodz Poland
90. Sozanski K (2016) Signal-to-noise ratio in power electronic digital control circuits. In: Conference proceedings of signal processing, algorithms, architectures, arrangements and applications, SPA'16. Poznan University of Technology, pp 162–171
91. Spang H, Schulthessis P (1962) Reduction of quantizing noise by use of feedback. IRE Trans Commun Syst 10:373–380
92. Strzelecki R, Fedyczak Z, Sozanski K, Rusinski J (2000) Active power filter EFA1. Technical Report, Instytut Elektrotechniki Przemyslowej, Politechnika Zielonogorska (in Polish)
93. Texas Instruments (2005) ISO124 precision lowest-cost isolation amplifier. Data sheet, Texas Instruments Inc.
94. Texas Instruments (2006) ADS8364 250kSPS, 16-bit, 6-channel simultaneous sampling analog-to-digital converter, Data sheet. Texas Instruments Inc.
95. Texas Instruments (2008) TMS320F28335/28334/28332, TMS320F28235/28234/28232, digital signal controllers (DSCs). Data Manual, Texas Instruments Inc.
96. Texas Instruments (2009) ISO120, ISO121 precision low cost isolation amplifier. Technical report, iso121.pdf, Texas Instruments, Inc.
97. Texas Instruments (2010) INA270, INA271 voltage output, unidirectional measurement current-shunt monitor. Data sheet, Texas Instruments Inc.
98. Texas Instruments (2010) C2000 teaching materials, tutorials and applications. SSQC019, Texas Instruments Inc.
99. Texas Instruments (2011) ADS1274, ADS1278, quad/octal, simultaneous sampling, 24-bit analog-to-digital converters. Data sheet, Texas Instruments Inc.
100. Texas Instruments (2011) Simultaneous sampling analog-to-digital converters 12-, 14-, 16-bit, eight-channel. Data sheet, Texas Instruments Inc.
101. Texas Instruments (2016) TMS320F2837xD dual-core delfino microcontrollers. Data sheet, Texas Instruments Inc.
102. Texas Instruments (2016) The TMS320F2837xD architecture: achieving a new level of high performance. Technical Brief, Texas Instruments Inc.
103. Tewksbury S (1978) Oversampled, linear predictive and noise-shaping coders of order N > 1. IEEE Trans Circuits Syst 25(7):436–447
104. Trzynadlowski A (2010) Introduction to modern power electronics. Wiley, New York
105. Vaidyanathan PP (1992) Multirate systems and filter banks. Prentice Hall Inc., Englewood Cliffs
106. Verona J (2001) Power digital-to-analog conversion using sigma-delta and pulse width modulations, ECE1371 Analog Electronics II, ECE University of Toronto 2001, vol II, pp 1–14
107. Vishay (2011) Linear optocoupler, high gain stability, wide bandwidth. Data sheet, Vishay Semiconductor GmbH
108. Wanhammar L (ed) (1999) DSP integrated circuit. Academic Press, London
109. Zieliński T (2005) Digital signal processing: from theory to application. Wydawnictwo Komunikacji i Lacznosci, Warsaw (in Polish)
110. Zolzer U (ed) (2002) DAFX—digital audio effects. Wiley, New York
111. Zolzer U (ed) (2008) Digital audio signal processing. Wiley, New York
112. Zumbahlen H (ed) (2007) Basic linear design. Analog Devices Inc., Norwood

Chapter 3
Selected Methods of Signal Filtration and Separation and Their Implementation

3.1 Introduction

This chapter considers selected methods of digital signal filtration, separation and their implementation. Special attention is paid to digital filters and filter banks useful for the control circuits in power electronics. The author's efficient realizations of lattice wave digital filters (LWDF) and modified lattice wave digital filters (MLWDF) using digital signal processors [23] are presented. The author has carried out implementations of modified wave digital filters for modern digital signal processors for the first order and second order adapters [63, 64, 64]. For systems in which linear phase shift and FIR filters require too many arithmetic operations integrated IIR filters with linear phase shift are considered. The author presents his own solutions for these filters. They are particularly useful in audio systems, and they are also used to interpolate signals in the class-D amplifier. Considered too are multirate circuits and the influence of changing the signal sample rate on the quality of the signal. This chapter also presents the author's very useful implementations of an interpolator using bireciprocal LWDF. Many circuits are supplemented by the author's listings in MATLAB®, used for simulation of selected algorithms. The diagrams illustrating the characteristics of the presented circuits are the work of the author. One of the selected power electronics devices is the APF, for which filter banks are especially useful. The filter banks allow separation and selection of compensating harmonics. For this purpose, the author has chosen filter banks such as: strictly complementary [68], sliding DFT [65–67, 69], sliding Goertzel DFT, moving DFT and LWDF [64]. In the next part of this Chapter analysis of the features of selected digital signal processors (DSP) is presented.

K. Sozański, *Digital Signal Processing in Power Electronics Control Circuits*,
Power Systems, DOI 10.1007/978-1-4471-7332-8_3

3.2 Digital Filters

Digital filters are used to transform signal from one form to another, especially to eliminate specific frequencies in the signal. The word "filter" is derived from electrical engineering, and in the past a filter was primarily electrical. Therefore, mainstream filter theory was developed in electrical engineering. Digital filters are medium-less, being a linear combination of the input signal $x(n)$ and possibly the output signal $y(n)$ and include many of the operations for signal processing. The digital filter is realized using LTI discrete-time circuits (see in Chap. 1). There are two basic types of digital filter:

- recursive filters, also known as infinite impulse response (IIR) filters, filters with feedback,
- nonrecursive filters, filters without feedback, also called finite impulse response (FIR) filters.

The problems of digital filter design are widely described by many authors. The author can particularly recommend a few books ("bricks") on this topic, written by: Oppenheim et al. [52], Rabiner and Gold [61], Proakis and Manolakis [59], Hamming [39], Mitra [50], Zieliński [88, 89], Chen et al. [16], Vaidyanathan [82], Wanhammar [86], Pasko [57], Izydorczyk and Konopacki [40], Dabrowski [20, 21], Venezuela and Constantindes [83], Orfanidis [53, 54], Owen [56] for audio application Zolcer [90, 91] and many others. Among many of the digital filters described in the above publications, the author has chosen those which in his opinion are especially well suited for use in control circuits of power electronics devices.

3.2.1 Digital Filter Specifications

The filter design problem involves constructing the transfer function of a filter that meets the desired frequency response specifications. Typical response specifications for an ideal low-pass digital filter are depicted in Fig. 3.1. For an ideal digital filter with cutoff frequency f_{cr}, the passband is defined from 0 to f_{cr}, stopband from f_{cr} to $f_s/2$ and magnitude $D(\omega)$ for positive frequency

$$D(\omega) = \begin{cases} 1, \text{ if } & 0 \le \omega \le \omega_{cr} \\ 0, \text{ if } & \omega_{cr} < \omega \le \omega_s/2 \end{cases} . \tag{3.1}$$

Figure 3.1 also depicts the specification for an ordinary (not ideal) digital filter, and in such case: $0 \le f \le f_p$—passband, $f_p < f < f_z$—transitionband and $f_z \le f \le f_s/2$—stopband. Additionally, there are defined parameters: δ_p—ripple in the passband, δ_z—ripple in the stopband. With these defined parameters of the filter specification it is possible to find a desired transmittance for a digital circuit. In the same way specifications are defined for high-pass, passband and bandstop digital filters.

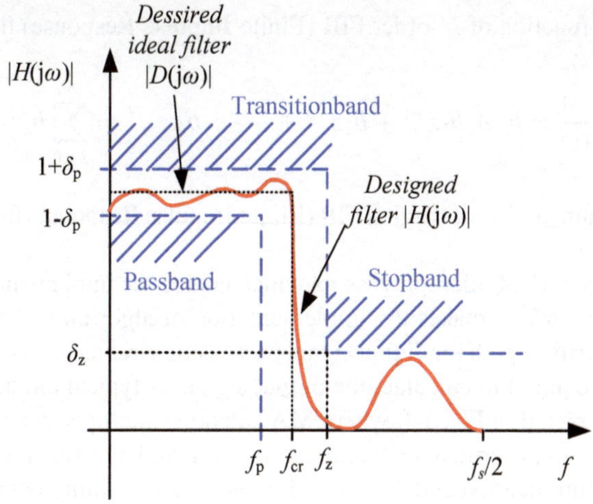

Fig. 3.1 Magnitude response specifications of low-pass digital filter

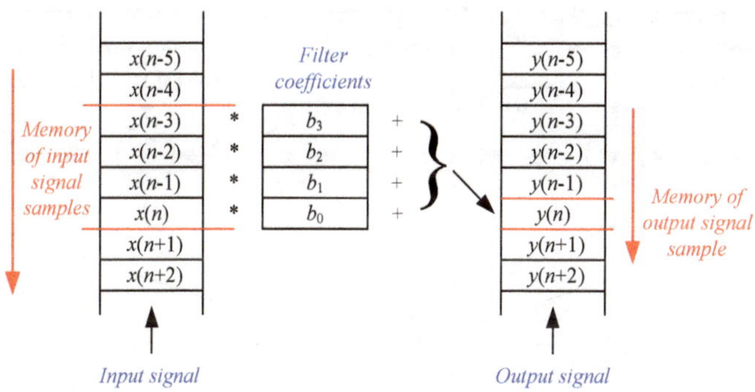

Fig. 3.2 The signal flow of 3-order nonrecursive digital filter

3.2.2 *Finite Impulse Response Digital Filters*

One of the simplest digital filters is a filter based on the idea of moving average, briefly described in Chap. 1. A signal flow of such a 3-order filter is shown in Fig. 3.2. This is a simple type of digital filter, which is defined by the linear formula

$$y(n) = b_0 x(n) + b_1 x(n-1) + b_2 x(n-2) + b_3 x(n-3) = \sum_{k=0}^{3} b_k x(n-k) \quad . \tag{3.2}$$

The coefficients b_k are the constants of the filter (for time-invariant circuit), $x(n-k)$ is the input data (input sample) and $y(n)$ the output.

The transfer function of N-order FIR (**F**inite **I**mpulse **R**esponse) filter

$$H(z) = \frac{Y(z)}{X(z)} = b_0 + b_1 z^{-1} + b_2 z^{-2} + \cdots + b_N z^{-N} = \sum_{k=0}^{N} b_k z^{-k} \quad . \tag{3.3}$$

The block diagram of N-order digital FIR (**F**inite **I**mpulse **R**esponse) filter is depicted in Fig. 3.3.

Today there are lot of microprocessors which enable the implementation of operations in parallel, which makes the implementation of algorithms determined by a critical path. A critical path for digital signal processing circuits is a list of all sequential operations required to calculate the output signal. A typical circuit of FIR filter realization is depicted in Fig. 3.4, where MAC denotes multiply-accumulate operation. In the diagram the critical path realization is marked by a red dotted line, and it contains one multiplication and $N + 1$ additions. The remaining operations can be performed in parallel to the operations from the critical path. The critical path is the

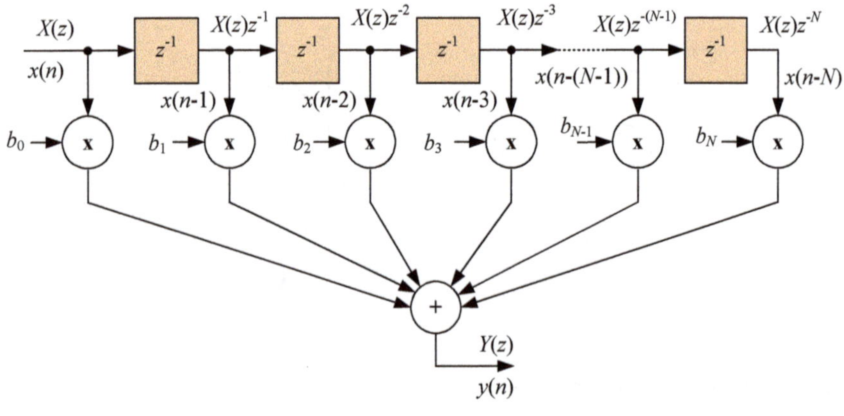

Fig. 3.3 The block diagram of N-order digital FIR filter

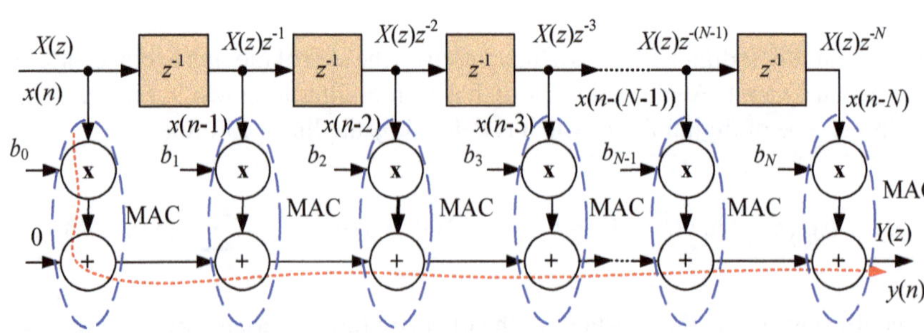

Fig. 3.4 Realization of N-order digital FIR filter

longest necessary path through a digital circuit when taking into consideration its interdependencies. It should be noted, however, that modern signal processors are designed for the implementation of FIR filters, so they can execute in a single cycle multiplication, accumulation operations and two transfer operations. Thus the critical path can contain the full implementation of the FIR filter. It should also be noted that the algorithm implements FIR filter can be easily divided into parts executed parallel.

FIR filters have many advantages such as: guaranteed stability, linear phase response, simplicity. Therefore they are used in many different fields. However the main drawback of such filters is the necessity of using high order (even several hundred), finally signal delay is equal to $N/2$ samples.

3.2.3 *Infinite Impulse Response Digital Filters*

In infinite impulse response (IIR) digital filters not only input samples $x(n)$ are used for computing the output signal $y(n)$, but also other samples of output. Third-order IIR filter is depicted in Fig. 3.5. The output signal is calculated as follows

$$\begin{aligned} y(n) = {} & b_0 x(n) + b_1 x(n-1) + b_2 x(n-2) + b_3 x(n-3) + \\ & - a_1 y(n-1) - a_2 y(n-2) - a_3 y(n-3) = \\ & \sum_{k=0}^{3} b_k x(n-k) - \sum_{k=1}^{3} a_k y(n-k) \quad . \end{aligned} \tag{3.4}$$

Generally IIR filter are digital linear time-invariant circuits (system) (LTI) and they can be described by the transfer function

Fig. 3.5 The signal flow of third-order IIR filter

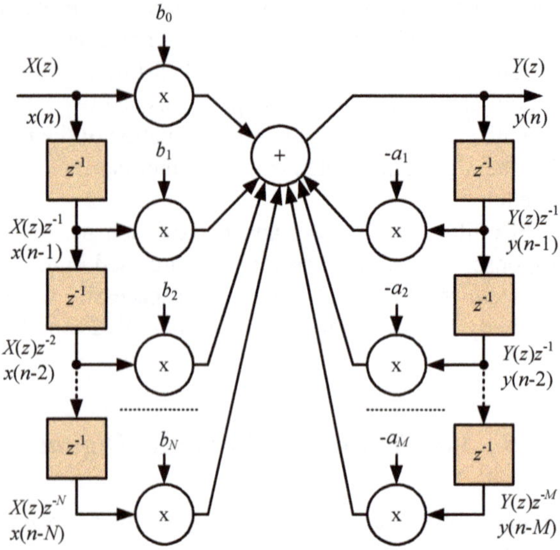

Fig. 3.6 The block diagram of a direct form digital LTI circuit

$$H(z) = \frac{Y(z)}{X(z)} = \frac{b_0 + b_1 z^{-1} + b_2 z^{-2} + \cdots + b_N z^{-N}}{1 + a_1 z^{-1} + a_2 z^{-2} + \cdots + a_M z^{-M}} \quad . \tag{3.5}$$

The order of such a circuit is determined by $\max(N, M)$. The block diagram of direct form of such a circuit realization is shown in Fig. 3.6. The output signal can be calculated as follows

$$\begin{aligned} Y(z) &= (b_0 X(z) + b_1 X(z) z^{-1} + b_2 X(z) z^{-2} + \cdots + b_N X(z) z^{n-N}) + \\ &\quad - (a_1 Y(z) z^{-1} + a_2 Y(z) z^{-2} + \cdots + a_M Y(z) z^{-M}) = \\ &= \sum_{k=0}^{N} b_k X(z) z^{-k} - \sum_{k=1}^{M} a_k Y(z) z^{-k} \quad , \end{aligned} \tag{3.6}$$

or using the equivalent difference equation

$$\begin{aligned} y(n) &= (b_0 x(n) + b_1 x(n-1) + b_2 x(n-2) + \cdots + b_N x(n-N)) + \\ &\quad -(a_1 y(n-1) + a_2 y(n-2) + \cdots + a_M y(n-N)) = \\ &= \sum_{k=0}^{N} b_k x(n-k) - \sum_{k=1}^{M} a_k y(n-k) \quad . \end{aligned} \tag{3.7}$$

Notation $x(n)$ is a simplified form of the full form $x(nT_s)$, but it is commonly used for the sake of simplicity. However, note that n represents the number of samples and that for uniform sampling systems, the distance between two samples is equal to the sampling period T_s. The filter structure shown in Fig. 3.6 is numerically inefficient, especially for higher orders, therefore the structure should be transferred to the $K + 1$ cascade connection of the second-order section

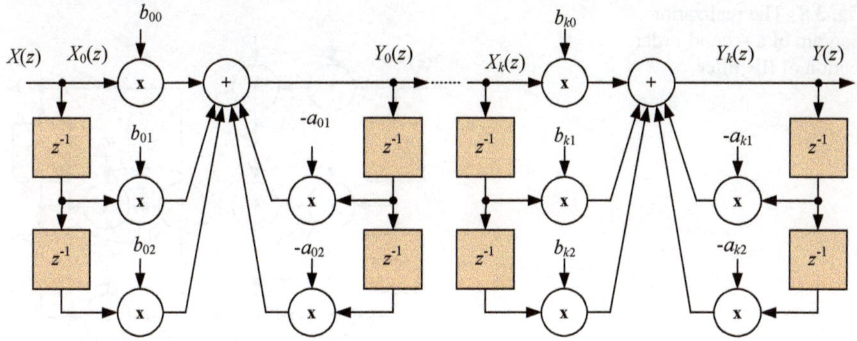

Fig. 3.7 The block diagram of N-order cascade digital IIR filter

$$H(z) = \frac{\sum_{k=0}^{N} b_k z^{-k}}{1 + \sum_{k=1}^{N} a_k z^{-k}} = \prod_{k=0}^{K} \frac{b_{k0} + b_{k1} z^{-1} + b_{k2} z^{-2}}{a_{k0} + a_{k1} z^{-1} + a_{k2} z^{-2}} \quad , \tag{3.8}$$

where: $K = N/2$ for even value of N and $K = (N+1)/2$ for odd value of N.

The block diagram of N-order digital cascade IIR filter is depicted in Fig. 3.7.

Due to the occurrence of feedback in the IIR filters, they are not always stable and, therefore, their stability must always be tested. Instability of IIR filters can also result from the limited precision of arithmetic, so the implementation of such a filter must be carefully checked. Table 3.1 presents a comparison of FIR and IIR filters.

A realization block diagram of second-order IIR filter with marked critical path is depicted in Fig. 3.8, where MAC denotes multiply-accumulate operation. The critical path consists of one multiplication, five additions and one delay. Similar to the FIR filter the remaining operations can be performed in parallel to the operations from the critical path. However, using digital signal processors it is possible to per-

Table 3.1 Comparison of digital filters

Feature	FIR	IIR
Typical order N	20–500	1–8
Stability	Always	Should be considered
Phase response	Linear	Nonlinear
Delay	N-samples	Moderate
Group delay	$N/2$-samples	Moderate
Length of buffer	N	$2N$
Limit cycles	No	Possible
Implementation	Very easy	Moderate
Parallel realization	Very easy	Possible

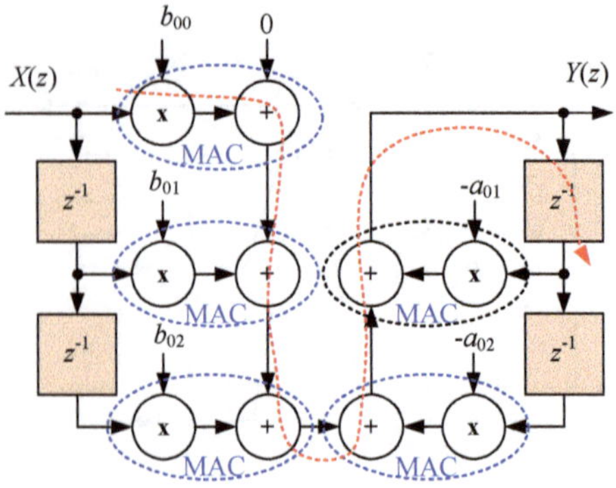

Fig. 3.8 The realization diagram of a second-order section of IIR filter

form multiplication and accumulation (MAC) and two transfer operations in a single machine cycle.

3.2.4 *Design of Digital IIR Filters*

Fortunately, nowadays many tools are available for the design and implementation of digital filters, e.g. **Signal Processing Toolbox** and **DSP System Toolbox**, (especially functions **fdatool** and **fvtool**) in the MATLAB, **QEDesign** from Momentum Data System etc. Hence designers have an easier task, but it does not absolve them from the necessity of understanding the phenomena occurring in digital filters. In the process of designing digital IIR filters one of the best design methods is based on an analog filter prototype. Figure 3.9 illustrates the conversion of a second-order analog circuit to a second-order digital circuit. The block diagram of a IIR filter design process is depicted in Fig. 3.10. There are several types of analog-to-digital transformations: backward difference approximation, forward difference approxima-

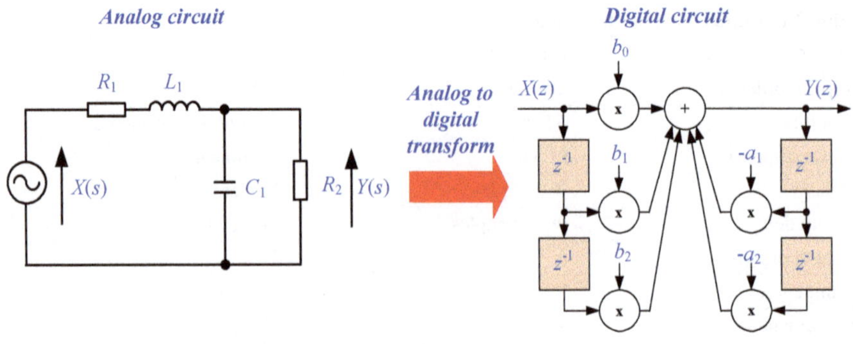

Fig. 3.9 A conversion of second-order analog circuit to second-order digital one

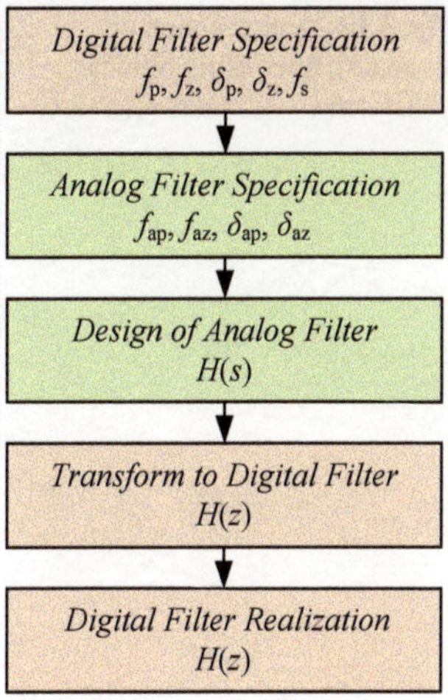

Fig. 3.10 IIR filter design

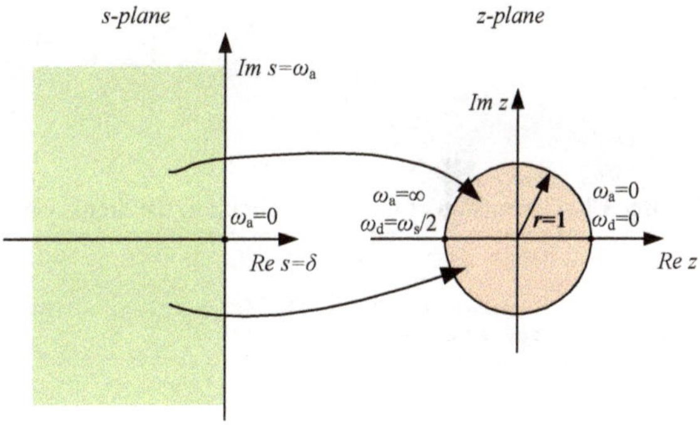

Fig. 3.11 The bilinear transform

tion, impulse invariant method, bilinear transform (also known as Tustin's method), matched bilinear, matched z-transformation or pole–zero matching method. They are widely described in the literature [48, 50, 52, 59, 74]. From among those listed the most useful and versatile is bilinear transform. The bilinear transform allows a stable continuous system mapping to a stable discrete system, with stability in the discrete domain meaning that there are no system poles that lie outside the unit circle in the z-plane. This process is illustrated in Fig. 3.11.

For bilinear transform where the first order Pade approximation is used for z^{-1} instead of a first-order series approximation, then

$$z^{-1} = e^{-sT_s} \approx \frac{1 - \frac{T_s}{2}s}{1 + \frac{T_s}{2}s} \Rightarrow s = \frac{2}{T_s}\frac{1 - z^{-1}}{1 + z^{-1}} \quad . \tag{3.9}$$

where: T_s—is sampling period.

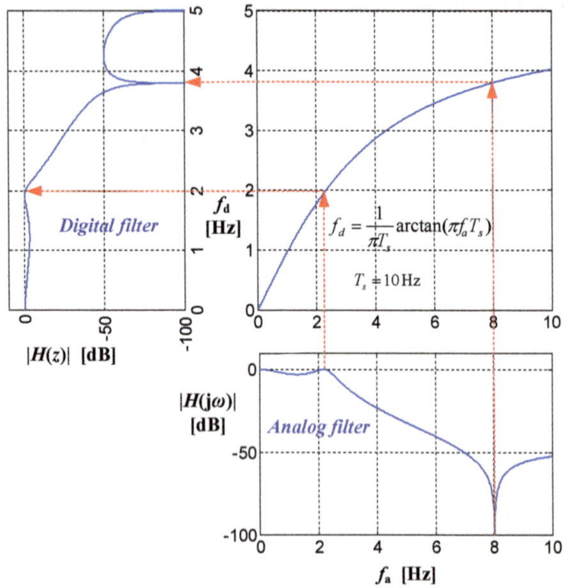

Fig. 3.12 A conversion of an analog filter using bilinear transform

Listing 3.1 Conversion of the analog circuit to the digital circuit

```
clear all; close all; l_width = 1.5; f_size = 10;
fs = 1.0;                   % sampling frequency
[z,p,k] = ellipap(8,7,50);  % elliptic analog low-pass filter prototype omega=1
[num,den] = zp2tf(z,p,k);   % convert to transfer function form
w = logspace(-1,1,2^14);    % generate logarithmically spaced vectors
figure('Name','Analog␣prototype','NumberTitle','off');
h = freqs(num,den,w); fa=w/(2*pi);
figure('Name','Frequency␣response␣of␣analog␣prototype','NumberTitle','off');
subplot(2,1,1), plot(fa,20*log10(abs(h)),'r','linewidth',l_width),
set(gca,'FontSize',f_size), grid on, axis([0 0.5 -100 5])
title('(a)'), ylabel('Magnitude␣[dB]');
subplot(2,1,2), plot(fa,180/pi*angle(h),'b','linewidth',l_width),
set(gca,'FontSize',f_size),grid on, set(gca,'Xlim',[0 0.5]);
xlabel 'Frequency␣(Hz)', ylabel('Phase␣[deg]'); title('(b)'),
print('analog_prototype.pdf','-dpdf');
[numd,dend] = bilinear(num,den,fs);    % analog to digital conversion
N=2^14; imp=[1 zeros(1,N-1)];
resp=filter(numd,dend,imp);
spect=fft(resp); f=(0:N-1)/(fs*N);
figure('Name','Frequency␣response␣of␣digital␣filter','NumberTitle','off');
subplot(2,1,1), plot(f,20*log10(abs(spect)),'r','linewidth',l_width),
set(gca,'FontSize',f_size), grid on, axis([0 fs/2 -100 5]);
title('(c)'), ylabel('Magnitude␣[dB]');
subplot(2,1,2), plot(f,180/pi*angle(spect),'b','linewidth',l_width),
set(gca,'FontSize',f_size),grid on, title('(d)'), set(gca,'Xlim',[0 fs/2]);
xlabel 'Frequency␣(f/f_{s})', ylabel('Phase␣[deg]');
print('digital_filter.pdf','-dpdf');
```

The bilinear transform maps the left-half of the s-plane to the entire unit circle in the z-plane. In this sense, it is a good approximation, although it still does not preserve the frequency response characteristics of the original z-transform. The nonlinear

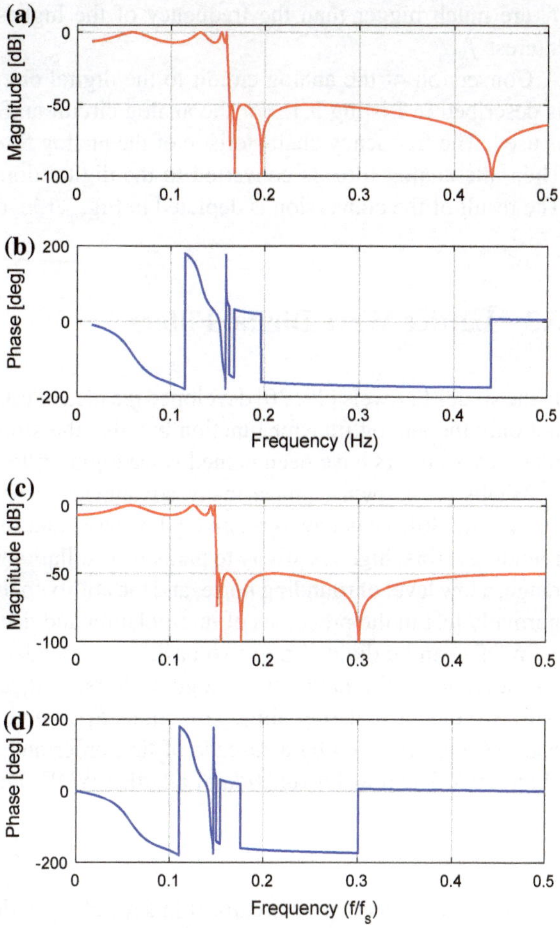

Fig. 3.13 The frequency characteristic of the filters: **a, b** analog low-pass filter prototype, **c, d** digital filter

relation between analog frequency ω_a and digital frequency ω_d is determined by the formula

$$\omega_d = \frac{2}{T_s} \arctan\left(\omega_a \frac{T_s}{2}\right) . \tag{3.10}$$

During the design process special attention should be paid to the non-linear transfer of analog to digital frequency. This is especially important for higher frequency. So correction of frequency should be made. An illustration of the conversion of an analog filter to a digital one using bilinear transform is depicted in Fig. 3.12.

The frequency response of an analog circuit is spanned between zero and frequency and it should be compressed from zero to $f_s/2$ frequency for the digital domain. Therefore, the characteristics of analog and digital circuits are different, especially near $f_s/2$ frequency. This may be acceptable if the sampling frequencies

f_s are much bigger than the frequency of the highest component of the band of interest f_b.

Conversion of the analog circuit to the digital circuit using MATLAB program is described in Listing 3.1. For the analog circuit an 8-order elliptic low-pass filter is used. The frequency characteristic of the analog filter is depicted in Fig. 3.13a, b. Then, the analog filter is converted to the digital domain using bilinear transform The result of the conversion is depicted in Fig. 3.13c, d.

3.3 Lattice Wave Digital Filters

In the sixties Fettweis [29, 30] developed the idea of transferring to the digital domain not only the analog transfer function but also the structure of the passive analogue filter. These filters have been named wave digital filters (WDF).

WDFs are known to have many advantageous properties [19, 21, 30–32, 34, 35, 45, 46, 86]. They have a relatively low passband sensitivity to coefficients, small rounding errors, high resistivity to parasitic oscillations (limit cycles), great dynamic range, a low level of rounding noise, and the ability to recover effective pseudopower, normally lost in the processes of interpolation and decimation, etc.

WDFs can be divided into two basic types, leader and lattice. Especially worth considering are the lattice wave digital filters. Lattice wave digital filters are built with two blocks realizing allpass functions $S_1(z)$ and $S_2(z)$. Typically blocks $S_1(z)$ and $S_2(z)$ are realized by a cascade of first order and second order allpass sections (Fig. 3.14). The transfer function of a lattice WDF can be written as

$$H(z) = 0.5(S_1(z) + S_2(z)) \quad . \tag{3.11}$$

These allpass filters can be realized in several ways described in [30, 32, 37]. One approach that yields parallel and modular filter algorithms is to use cascaded first and second-order sections. A detailed block diagram of N-order lattice WDF is shown in Fig. 3.15. The lattice WDF consists of one first and a few second order allpass sections.

The first and the second-order allpass sections are here realized using symmetric two-port adaptors. Wave digital filters were first proposed in the seventies when multiplication was a quite expensive operation. For this reason they were designed with

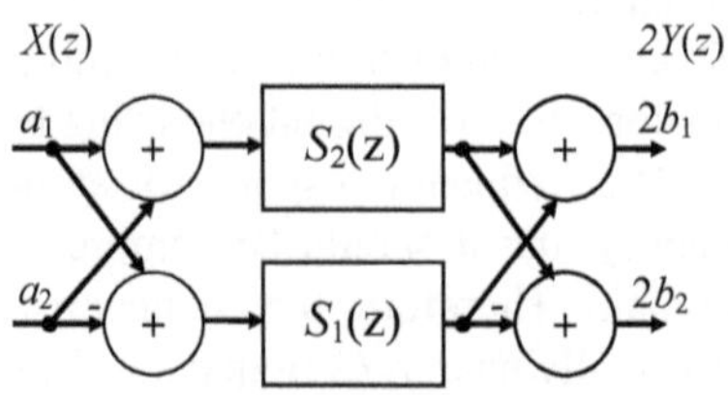

Fig. 3.14 Simplified block diagram of lattice wave digital filter

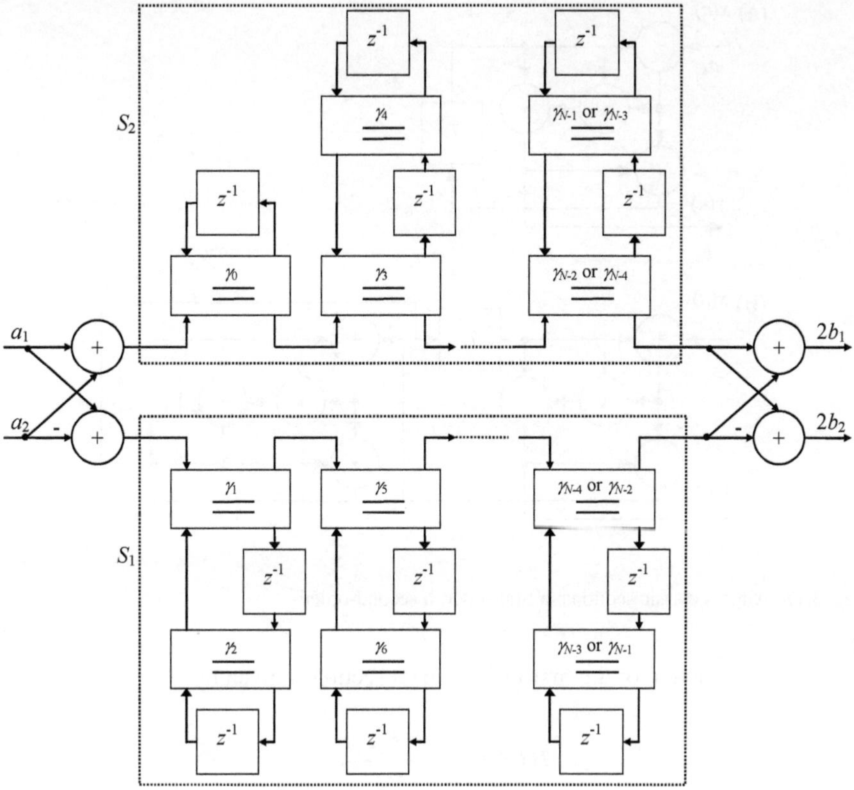

Fig. 3.15 N-order lattice WDF

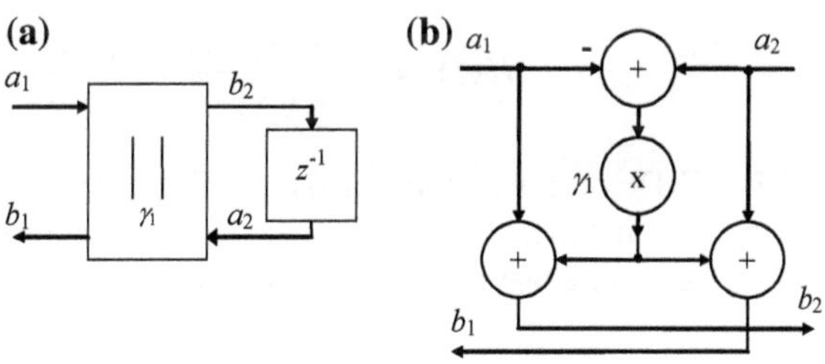

Fig. 3.16 Block diagram of classical first-order allpass section: **a** allpass filter, **b** two port adaptor

the minimum number of multipliers. A typical classical two port adaptor is depicted in Fig. 3.16b. It requires a single multiplier and three adders. Typical classical two-port adaptors used for building allpass sections are depicted in Fig. 3.17. Reflection signals b_1 and b_2 for first-order allpass section (Fig. 3.17a) can be calculated by

$$\begin{cases} b_1 = -\gamma_1 a_1 + (1+\gamma_1)a_2 \\ b_2 = (1-\gamma_1)a_1 + \gamma_1 + a_2 \end{cases} . \tag{3.12}$$

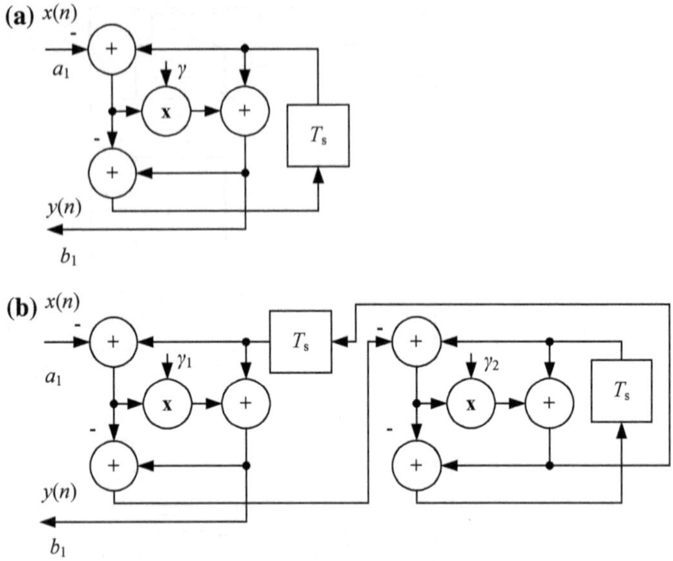

Fig. 3.17 Allpass classic sections: **a** first-order, **b** second-order

The transfer function of the first-order allpass section is given by

$$H(z) = \frac{-\gamma + z^{-1}}{1 - \gamma z^{-1}} \ . \tag{3.13}$$

The second-order allpass classical filter is shown in Fig. 3.17b. The transfer function of such a filter is given by

$$H(z) = \frac{-\gamma_1 + (\gamma_1\gamma_2 - \gamma_2)z^{-1} + z^{-2}}{1 + (\gamma_1\gamma_2 - \gamma_2)z^{-1} - \gamma_1 z^{-2}} \ . \tag{3.14}$$

The LWDF design methods and algorithms are very well described by Gazsi [37]. Using these methods the Delft University of Technology has prepared a very useful tool for designing wave digital filters, (L)WDF Toolbox for the MATLAB [7, 8].

3.3.1 Comparison of Classical IIR Filter and Lattice Wave Digital Filter

Typically, IIR digital filters are implemented by dividing them into second-order sections (Fig. 3.7). In order to compare the implementation of IIR filters, two 3-order Butterworth filters were designed with the parameters: crossover frequency $f_{cr} = 50$ Hz, sampling frequency $f_s = 10000$ Hz. A classical IIR filter was designed using the

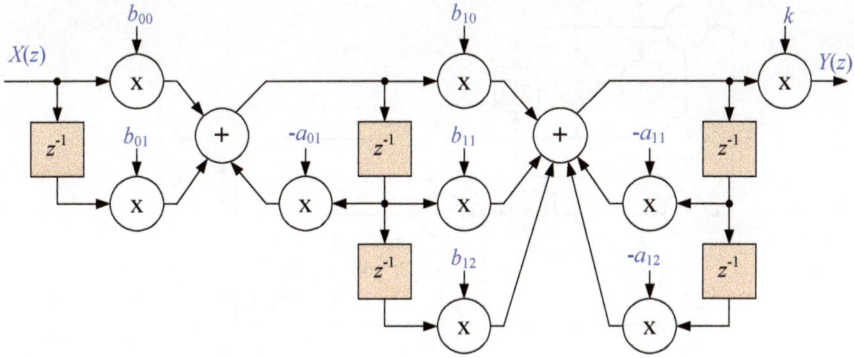

Fig. 3.18 3-order IIR filter

Table 3.2 3-order Butterworth digital filter coefficients

Section	$n = 0$	$n = 1$
b_{n0}	1.000000000	1.000000000
b_{n1}	0.999990222	2.000009777
b_{n2}	0	1.000009778
a_{n1}	–0.969067417	–1.968103311
a_{n2}	0	0.969074930
k	$3.756838019 \times 10^{-6}$	

standard MATLAB tools. Figure 3.18 shows the scheme of such a filter, the filter is implemented using two SOS sections. The value of the filter coefficients are shown in Table 3.2. Filter coefficients appear to be useful for fixed-point implementation. However, note the small differences between the values of b_{00} and b_{01}, as well as the small value of the scaling factor k.

Using the (L)WDF Toolbox for MATLAB [7, 8, 26] an LWDF with the same parameters was also designed. A block diagram of the implementation of such a filter is shown in Fig. 3.19. The values of coefficients are shown in Table 3.3. As can be seen, the values of filter coefficients are aligned and are well suited for fixed-point implementation. The advantages of such implementation are revealed also in the floating-point implementation.

3.3.2 *Realization of LWDF*

Two classic first-order allpass sections are shown in Fig. 3.20a, b. They require a single multiplication, three additions and one delay. For both allpass sections critical paths are marked. These two versions have different lengths of critical paths. The critical path of the first-order allpass section in Fig. 3.20a consists of a single

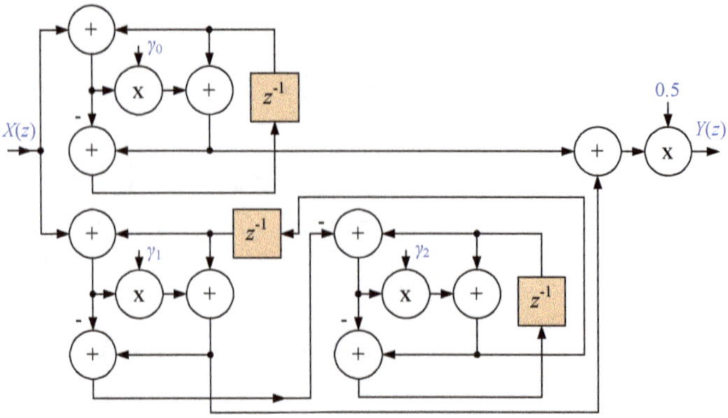

Fig. 3.19 3-order LWDF

Table 3.3 3-order Butterworth LWDF coefficients

γ	Value
γ_0	0.969069850
γ_1	–0.969077362
γ_2	0.999506639

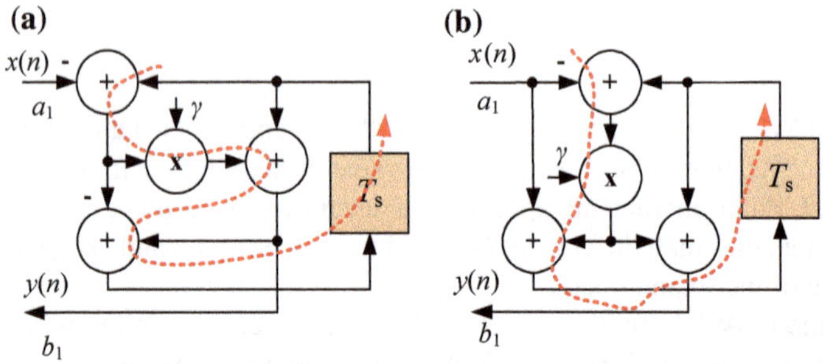

Fig. 3.20 Critical paths of first-order allpass classic sections: **a**, **b** two realizations of a two-port adaptor

multiplication, two additions and one delay, while that in Fig. 3.20b consists of a single multiplication, three additions and one delay. So the first version is better for implementation in a digital signal processor with a parallel instruction set. Figure 3.21 shows the author's realization of classical first-order sections using SHARC DSP. For this realization five machine cycles of SHARC DSP are needed.

The second-order allpass section in Fig. 3.22 consists of two multiplications, six additions and two delays. The critical path of the second-order allpass section consists

(a)

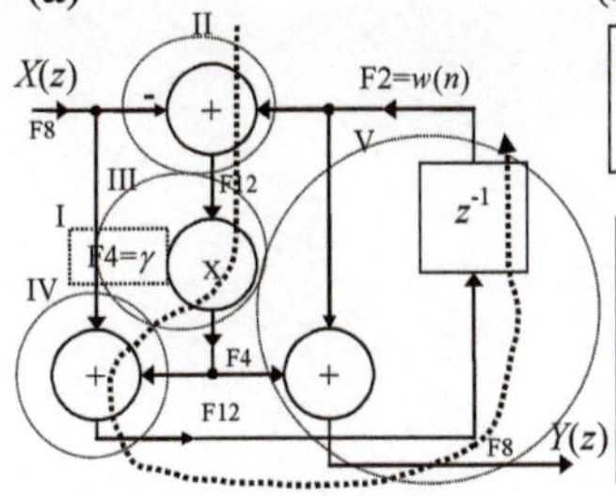

(b)

```
Initialization: input sample x(n) in register F8
   B0=adres_w; I0=0; M0=0;        { address w(n) →B0}
   B8=adres_g; I8=0; M8=0;        { address γ→B8}

I. Get from memory:F2=w, F4=γ                       F2=dm(I0,M0), F4=pm(I8,M8);
II. Calculate F12=w(n)-x(n)                         F12=F8-F2;
III. Calculate γ(w(n)-x(n)):                        F4=F12*F4;
IV. Calculate w(n+1)=F12=x(n)+γ(w(n)-x(n)):         F12=F4+F8;
V. Calculate y(n)=F8=x(n)+γ(w(n)-x(n)) and store in memory
   w(n+1): F8=F2+F4, dm(I0,M0)=F12;
```

Fig. 3.21 Realization of classical first-order sections using SHARC DSP: **a** block diagram, **b** corresponding assembler program

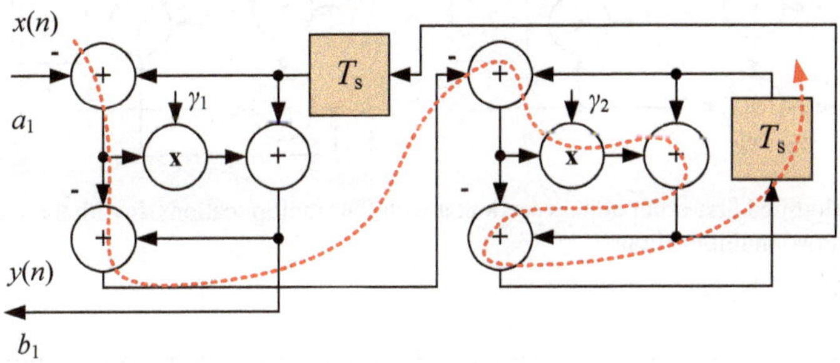

Fig. 3.22 Critical path of second-order allpass classic section

of a single multiplication, five additions and one delay. Looking at the block diagrams of the first and second order sections it can be seen that they are not well suited for modern digital signal processors. Considerations of LWDFs effective implementation can be found in the works of Fettweis [33], the author [64, 73], author and Dabrowski [23], Vesterbacka [84, 85], Wanhammar [86].

Using Eq. (3.12) describing the two-port adaptor, the block diagram can be converted to the form shown in Fig. 3.23a, so that the adaptor is obtained for the realization needed to perform four multiplications and two additions. By using the identity transformation of the circuit from Fig. 3.23a modifications are made to give an adapter with three multiplications and two additions, shown in Fig. 3.23b. The adaptor (Fig. 3.23b) requires the placement of an additional coefficient $1 - \gamma$ and is particularly suitable for the implementation of floating point arithmetic. Further transformation of the structure in search of an adaptor with compensated addition and multiplication values led to an adapter for the system as in Fig. 3.23c with such goals.

Wave digital filters were proposed by Fettweis in the seventies, i.e., [30, 32, 34], when multiplication was a very expensive operation. That is why they were originally designed to minimize the number of multipliers. This is still advantageous if the filter is implemented in a simple digital hardware structure, such as μC, μP,

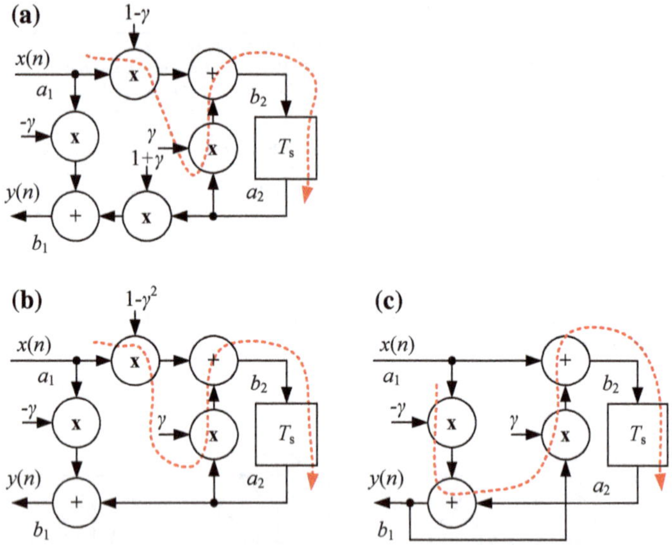

Fig. 3.23 Modified first-order allpass section: **a** with four multiplications, **b** with three multiplications, **c** with two multiplications

FPGA, ASIC, etc., but can be undesirable for realizations with modern digital signal processors (DSP's). In typical DSP's a single computing cycle consists of a multiplication combined with accumulation and to/from memory data moving operations. As a result, a single addition also requires one computing cycle. This is a disadvantage for implementing WDF's by modern digital signal processors, especially with floating point arithmetic.

3.4 Modified Lattice Wave Digital Filters

Today, modern digital signal processors are designed to be able to calculate multiplication together with addition (or more) in a single operational cycle. As a result the classical two-port adaptor structure of Fig. 3.17 is ineffective for the DSP realization, especially for floating point arithmetic. This is why modified structures have been proposed. The idea of modified wave digital filters, i.e., those with equal numbers of additions and multiplications and with short critical paths, were proposed by Fettweis in [33]. This idea has been developed by the author with implementations of modified lattice wave digital filters (MLWDF) for digital signal processors [63, 64].

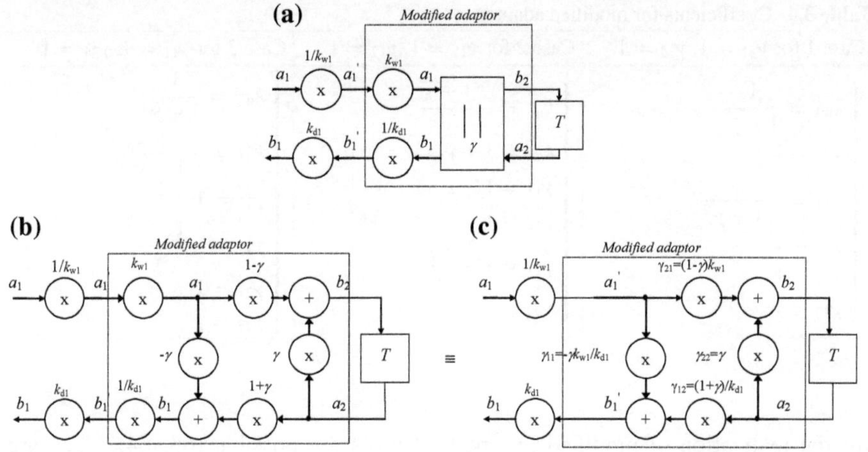

Fig. 3.24 Diagrams of first-order modified allpass sections: **a** idea, **b**, **c** realization

3.4.1 First-Order Sections

Figure 3.24a illustrates the idea behind modified wave digital filters by including two additional complementary multipliers in the first-order allpass section. Signals a_1 and b_1 are modified by coefficients k_{w1} and k_{d1}, and

$$a_1 = k_{w1}a_1' \quad , \quad b_1 = k_{w1}b_1' \quad . \tag{3.15}$$

Signals b_1' and b_2 of the modified two-port adaptor (Fig. 3.24c) may be expressed:

$$\begin{cases} b_1' = -\gamma_{11}a_1' + \gamma_{11}a_2 \\ b_2 = \gamma_{21}a_1 + \gamma_{22}a_2 \quad , \end{cases} \tag{3.16}$$

in which coefficients γ_{ij} are given by

$$\begin{cases} \gamma_{11} = -\gamma\dfrac{k_{w1}}{k_{d1}} \\ \gamma_{12} = \dfrac{1+\gamma}{k_{d1}} \\ \gamma_{21} = \dfrac{1-\gamma}{k_{w1}} \\ \gamma_{22} = \gamma \end{cases} \quad . \tag{3.17}$$

For the adaptor shown in Fig. 3.24c it is possible to select the value of coefficients k_{w1} and k_{d1} so that the values of the two γ coefficients are equal to one. Three cases

Table 3.4 Coefficients for modified adaptor

Case 1 for $\gamma_{21} = 1, \gamma_{12} = 1$	Case 2 for $\gamma_{11} = 1, \gamma_{12} = 1$	Case 3 for $\gamma_{11} = 1, \gamma_{21} = 1$
$\begin{cases} k_{w1} = \dfrac{1}{1-\gamma} \\ k_{d1} = 1+\gamma \end{cases}$	$\begin{cases} k_{w1} = -\dfrac{1+\gamma}{\gamma} \\ k_{d1} = 1+\gamma \end{cases}$	$\begin{cases} k_{w1} = \dfrac{1}{1-\gamma} \\ k_{d1} = -\dfrac{\gamma}{1-\gamma} \end{cases}$
$\begin{cases} \gamma_{11} = -\dfrac{\gamma}{1-\gamma^2} \\ \gamma_{12} = 1 \\ \gamma_{21} = 1 \\ \gamma_{22} = \gamma \end{cases}$	$\begin{cases} \gamma_{11} = 1 \\ \gamma_{12} = 1 \\ \gamma_{21} = -\dfrac{1-\gamma^2}{\gamma} \\ \gamma_{22} = \gamma \end{cases}$	$\begin{cases} \gamma_{11} = 1 \\ \gamma_{12} = -\dfrac{1-\gamma^2}{\gamma} \\ \gamma_{21} = 1 \\ \gamma_{22} = \gamma \end{cases}$

for the realization of modified two-port adaptors are possible [63, 64]. They are described by the equations listed in Table 3.4 and they are depicted in Fig. 3.25b–d.

Every realization needs five operations: two multiplications, two additions and one delay. In cases 1 and 3, the critical path consists of only two arithmetic operations and one delay. Realization of modified first-order sections using SHARC DSP is shown in Fig. 3.26.

Using this first-order section it is possible to build a branch of the lattice wave digital filter, and the realization of an N-order branch with modified first-order sections is depicted in Fig. 3.27 [63, 64]. The resulting value of the overall branch coefficient can be calculated as

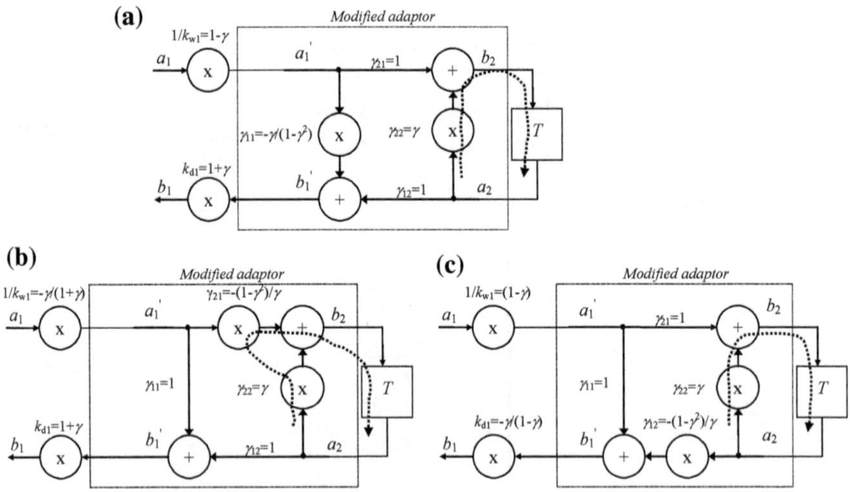

Fig. 3.25 Diagrams of first-order modified allpass sections: **a** case 1, **b** case 2, **c** case 3

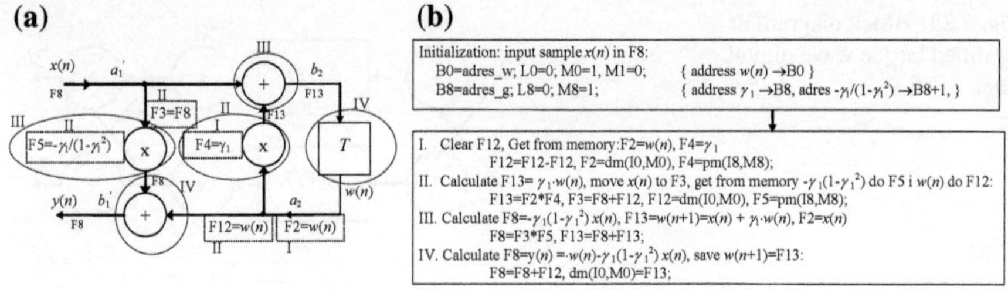

Fig. 3.26 Realization of modified first-order sections using SHARC DSP: **a** block diagram, **b** corresponding assembler program

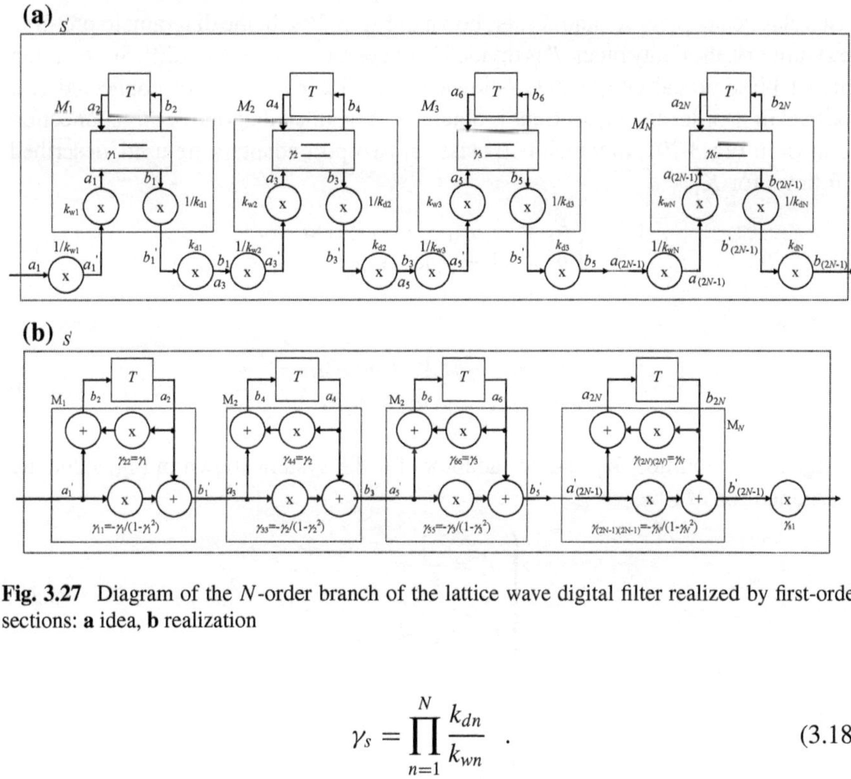

Fig. 3.27 Diagram of the N-order branch of the lattice wave digital filter realized by first-order sections: **a** idea, **b** realization

$$\gamma_s = \prod_{n=1}^{N} \frac{k_{dn}}{k_{wn}} \quad . \tag{3.18}$$

A block diagram of modified lattice wave digital filter is shown in Fig. 3.28

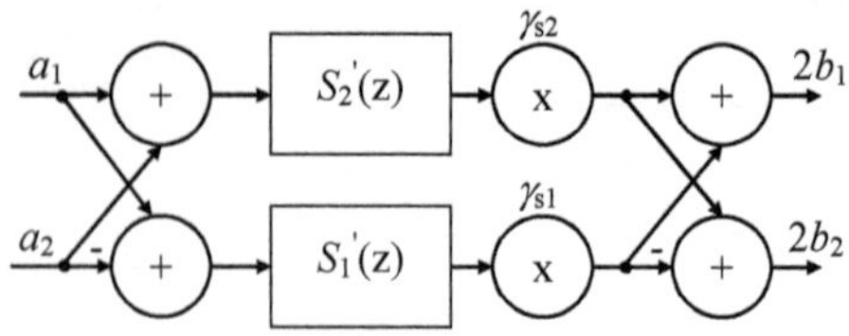

Fig. 3.28 Block diagram of modified lattice wave digital filter

3.4.2 Second-Order Sections

A second-order all-pass section is the next circuit which is necessary to build an MLWDF. A classical scheme of second-order allpass filter, consisting of a connection of two classic adapters K_1 and K_2, is shown in Fig. 3.29a. In the diagram, to preserve the symmetry, the delay block T is divided between two delays $T/2$ [29]. Substituting pairs of bireciprocal coefficients (shown in Fig. 3.29b) system of connected two modified first-order allpass sections is obtained. A detailed diagram of this connection is shown in Fig. 3.29c, in which two classical two-port adapters are used, described as follows for K_1

$$\begin{cases} b_1 = -\gamma_1 a_1 + (1 + \gamma_{11}) a_2 \\ b_2 = (1 - \gamma_1) a_1 + \gamma_1 a_2 \end{cases}, \tag{3.19}$$

and for K_2

$$\begin{cases} b_3 = -\gamma_2 a_3 + (1 + \gamma_2) a_4 \\ b_4 = (1 - \gamma_2) a_3 + \gamma_2 a_4 \end{cases}. \tag{3.20}$$

Signals of modified M_1 and M_2 adaptors for the system shown in Fig. 3.29d are determined for $M1$ by

$$\begin{cases} a_1' = \dfrac{1}{k_{w1}} a_1 \\ b_1' = \dfrac{1}{k_{w1}} b_1 \end{cases}, \tag{3.21}$$

and for M_2

$$\begin{cases} a_3' = \dfrac{1}{k_{w3}} a_1 \\ b_3' = \dfrac{1}{k_{d2}} b_3 \\ a_4' = \dfrac{1}{k_{w1}} a_4 \\ b_4' = \dfrac{1}{k_{d1}} b_4 \end{cases}. \tag{3.22}$$

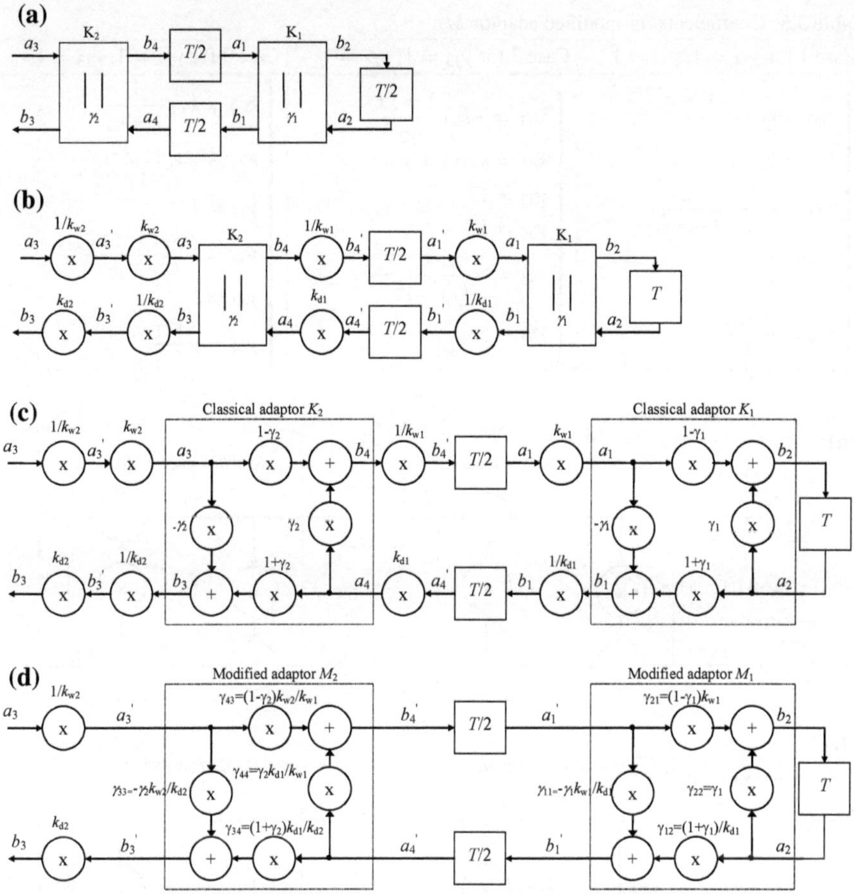

Fig. 3.29 Block diagrams of second-order allpass sections: **a** classical, **b–d** modified

Substituting according to the Eqs. (3.19) and (3.20), the Eqs. (3.21) and (3.22) give the equations describing the modified second-order allpass sections [63, 64], and adaptor M_1 is defined by the relationship

$$\begin{cases} b_1' = \overbrace{-\gamma_1 \frac{k_{w1}}{k_{d1}}}^{\gamma_{11}} a_1' + \overbrace{(1+\gamma_1)\frac{1}{k_{d1}}}^{\gamma_{12}} a_2 \\ b_2 = \overbrace{(1-\gamma_1)k_{w1}}^{\gamma_{21}} a_1' + \overbrace{\gamma_1}^{\gamma_{22}} a_2 \end{cases}, \tag{3.23}$$

Table 3.5 Coefficients for modified adaptor M_2

Case 1 for $\gamma_{34} = 1, \gamma_{43} = 1$	Case 2 for $\gamma_{33} = 1, \gamma_{24} = 1$	Case 3 for $\gamma_{33} = 1, \gamma_{43} = 1$
$\begin{cases} k_{w2} = k_{w1}\dfrac{1}{1-\gamma_2} \\ k_{d2} = k_{d2}(1+\gamma_2) \end{cases}$ $\begin{cases} \gamma_{33} = -\dfrac{k_{w1}}{k_{d1}}\dfrac{\gamma_2}{1-\gamma_2^2} \\ \gamma_{34} = 1 \\ \gamma_{43} = 1 \\ \gamma_{44} = \gamma_2\dfrac{k_{d1}}{k_{w1}} \end{cases}$	$\begin{cases} k_{w1} = -k_{d1}\dfrac{1+\gamma_2}{\gamma_2} \\ k_{d1} = k_{d1}(1+\gamma_2) \end{cases}$ $\begin{cases} \gamma_{33} = 1 \\ \gamma_{34} = 1 \\ \gamma_{43} = -\dfrac{k_{d1}}{k_{w1}}\dfrac{1-\gamma_2^2}{\gamma_2} \\ \gamma_{44} = \gamma_2\dfrac{k_{d1}}{k_{w1}} \end{cases}$	$\begin{cases} k_{w2} = k_{w1}\dfrac{1}{1-\gamma_2} \\ k_{d1} = -k_{w1}\dfrac{\gamma_2}{1-\gamma_2} \end{cases}$ $\begin{cases} \gamma_{33} = 1 \\ \gamma_{34} = -\dfrac{k_{d1}}{k_{w1}}\dfrac{1-\gamma_2^2}{\gamma_2} \\ \gamma_{43} = 1 \\ \gamma_{44} = \gamma\dfrac{k_{d1}}{k_{w1}} \end{cases}$

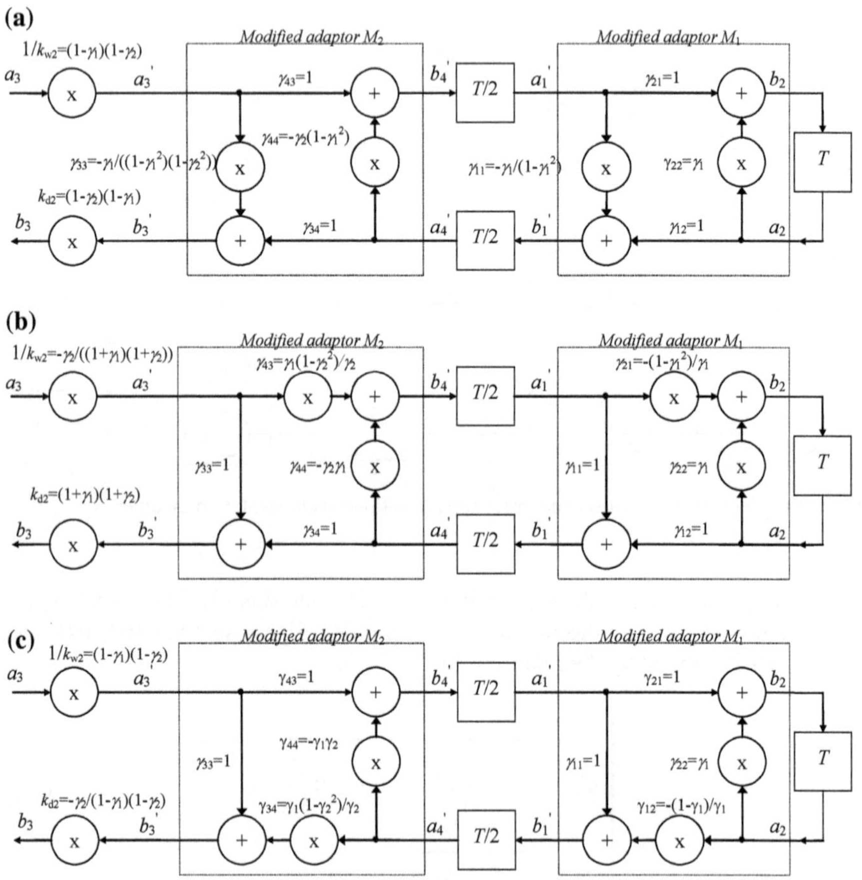

Fig. 3.30 Block diagrams of second-order allpass sections: **a** case 11, **b** case 22, **c** case 33

Table 3.6 Equations for modified second-order allpass sections

Case 11 for $\gamma_{21}=1,\gamma_{12}=1$ $\gamma_{34}=1,\gamma_{43}=1$	Case 22 for $\gamma_{11}=1,\gamma_{12}=1$ $\gamma_{33}=1,\gamma_{34}=1$	Case 33 for $\gamma_{11}=1,\gamma_{21}=1$ $\gamma_{33}=1,\gamma_{43}=1$
$\begin{cases} k_{w2}=\dfrac{1}{(1-\gamma_1)(1-\gamma_2)} \\ k_{d2}=(1-\gamma_1)(1-\gamma_2) \end{cases}$	$\begin{cases} k_{w1}=-\dfrac{(1-\gamma_1)(1-\gamma_2)}{\gamma_2} \\ k_{d1}=(1+\gamma_1)(1+\gamma_2) \end{cases}$	$\begin{cases} k_{w2}=\dfrac{1}{(1-\gamma_1)(1-\gamma_2)} \\ k_{d1}=-\dfrac{\gamma_2}{(1-\gamma_1)(1-\gamma_2)} \end{cases}$
$\begin{cases} b_1'=-\dfrac{\gamma_1}{1-\gamma_1^2}a_1'+a_2 \\ b_2=a_1'+\gamma_1 a_2 \\ b_3'=-\dfrac{\gamma_2^2}{(1-\gamma_1)(1-\gamma_2^2)}a_3'+a_4' \\ b_4'=a_3'-\gamma_2(1-\gamma_1^2)a_4' \end{cases}$	$\begin{cases} b_1'=a_1'+a_2 \\ b_2=\dfrac{1-\gamma_1^2}{\gamma_1^2}a_1'+\gamma_1 a_2 \\ b_3'=a_3'+a_4' \\ b_4'=\dfrac{\gamma_1(1-\gamma_2^2)}{\gamma_2}a_3'+\gamma_1\gamma_2 a_4' \end{cases}$	$\begin{cases} b_1'=a_1'+\dfrac{1-\gamma_1^2}{\gamma_1}a_2 \\ b_2=a_1'+\gamma_1 a_2 \\ b_3'=a_3'+\dfrac{\gamma_1(1-\gamma_2^2)}{\gamma_2}a_4' \\ b_4'=a_3'-\gamma_1\gamma_2 a_4' \end{cases}$

and adaptor M_2

$$\begin{cases} b_3' = \overbrace{-\gamma_2\dfrac{k_{w1}}{k_{d2}}}^{\gamma_{33}}a_3' + \overbrace{(1+\gamma_2)\dfrac{k_{d1}}{k_{d2}}}^{\gamma_{34}}a_4' \\ b_4' = \overbrace{(1-\gamma_2)\dfrac{k_{w2}}{k_{w1}}}^{\gamma_{43}}a_3' + \overbrace{\gamma_2\dfrac{k_{d1}}{k_{w1}}}^{\gamma_{44}}a_2 \end{cases} . \tag{3.24}$$

A detailed diagram of a second-order allpass section composed of modified two-port adaptors is shown in Fig. 3.29d. The rule to eliminate two multipliers in M_1 adaptor is the same as for first-order. This is possible for the three cases described in Table 3.4 and is shown in Fig. 3.24. However, for M_2 adapter for given k_{w1} and k_{d1} coefficients it is possible to calculate γ coefficients for the three cases as described in Table 3.5.

Using three choices for the adapter M_1 and three for adaptor M_2 nine feasibility allpass sections of the second-order are obtained. However, the author has realized three basic modified second-order allpass sections restricted to identical cases for M_1 and M_2. Figure 3.30 shows diagrams of modified second-order allpass sections. They need four multiplications, four additions and two delays. Equations for modified second-order allpass sections are described in Table 3.6.

3.5 Linear-Phase IIR Filters

In many applications it is desirable to use a linear phase shift filter, a condition which is satisfied by typical FIR filters. However, FIR filters require very large orders (hundreds), which results in a very large computational load. Therefore it

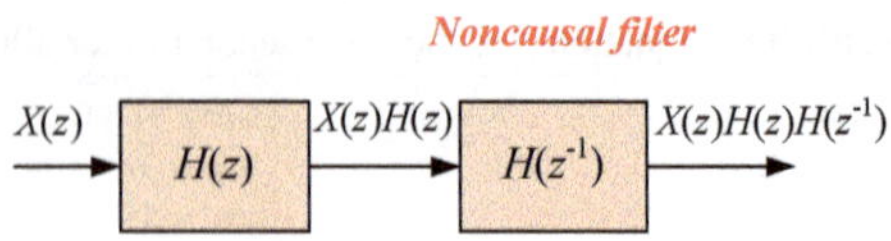

Fig. 3.31 The idea of linear-phase IIR filter

is desirable to use IIR filters with linear-phase shift. The aim of a linear-phase IIR filter (LF IIR) is to obtain higher computational efficiency than that offered by FIR filters at similar performance levels. The idea of a linear-phase IIR filter is shown in Fig. 3.31, it was presented by Rabiner and Gold [61] in the mid seventies. The filter consists of a causal filter $H(z)$ and a noncausal filter $H(z^{-1})$.

The whole filtration process realized by the circuit depicted in Fig. 3.31 can be described by equations, for the causal filter

$$Y_1(z) = X(z)H(z) \quad , \tag{3.25}$$

and after, for the non-causal filter

$$Y(z) = X(z)H(z)H(z^{-1}) = X(z)|H(e^{j\omega T})|^2 \quad . \tag{3.26}$$

The resultant amplitude characteristics of the filter is squared $|H(e^{j\omega T})|^2$ and should be taken in account in the filter design process.

A method which allows for a simple implementation of the IIR filter with linear-phase shift is a method with time reversal [18, 61] for the realization of a noncausal IIR filter $H(z^{-1})$. Figure 3.32 shows two versions of the filter with linear-phase shift: a filter in cascade connection (Fig. 3.32a) and a filter in parallel connection (Fig. 3.32b). In the cascade connection of filters (Fig. 3.32a) input signal is passed through the filter $H(z)$, then the order of the samples is reversed and again passed

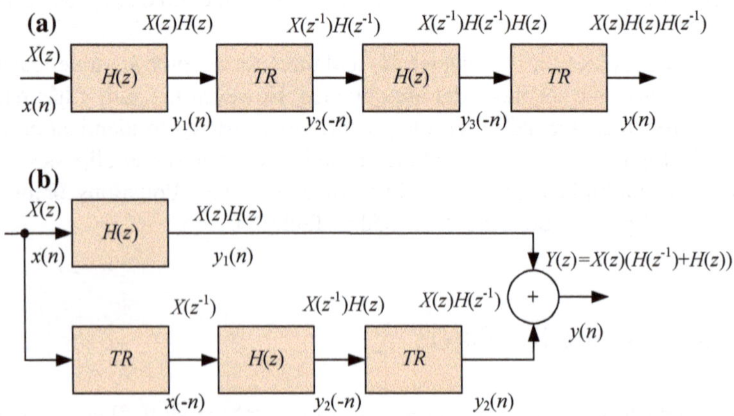

Fig. 3.32 Two linear-phase IIR filters: **a** cascade, **b** parallel

through the filter $H(z)$, and then the order of samples is once again reversed. The symbol TR determines time reversal. The TR circuit performs a reverse order of the signal samples. Time reversal can be described by

$$x(nT) \longrightarrow x(-nT) \quad , \tag{3.27}$$

and for z-transform

$$X(z) \longrightarrow X(z^{-1}) \quad . \tag{3.28}$$

Of course, methods with time reversal cannot be performed in real time, because it is impossible to reverse the flow of time (fortunately). The implementation of such a system is only possible if the signal is divided into blocks of samples. However, there is a problem with connecting blocks of signal samples in the output signal. The filter $H(z)$ transient effects generate amplitude distortion at the end of blocks of signal samples. To avoid this distortion, an overlap technique with additional N_{ov} samples can be used [70]. In Fig. 3.33 a block diagram of such a solution is presented. If transient effects are bothersome in a given application, discarding N_{ov} samples from the end of each signal block should be considered. The first $H(z)$ filter works continuously, while the second filter uses a block of samples which is reset before every block. Figure 3.34 shows a implementation diagram of such a linear-phase IIR filter.

The LF IIR filter delay (N_d) in samples is equal to

$$N_d = 2(N + N_{ov}) \quad . \tag{3.29}$$

By selecting the length of a block of signal samples N and length of overlap N_{ov} it is possible to choose output signal quality of the LF IIR filter. A good indicator of the output signal quality is the signal-to-noise and distortion ratio (*SINAD*).

In recent years there has been a significant interest in real-time implementation of IIR filters with a linear phase. Powell and Chau [58] have invented an efficient method for the design and implementation of real-time LF IIR filters using a suitable modification of a well-known time reversing technique. The block diagram of such

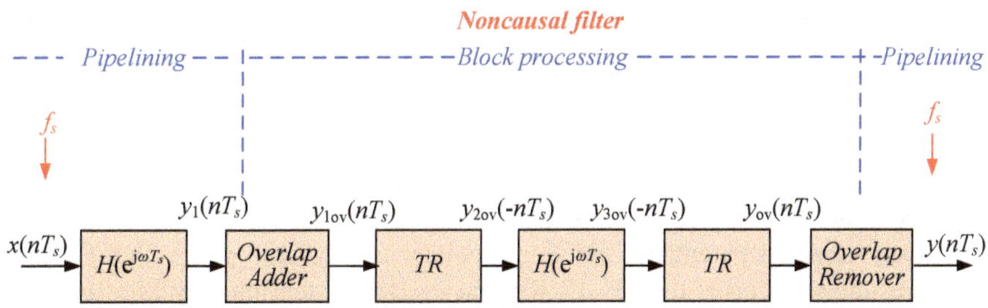

Fig. 3.33 The LF IIR filter with overlap

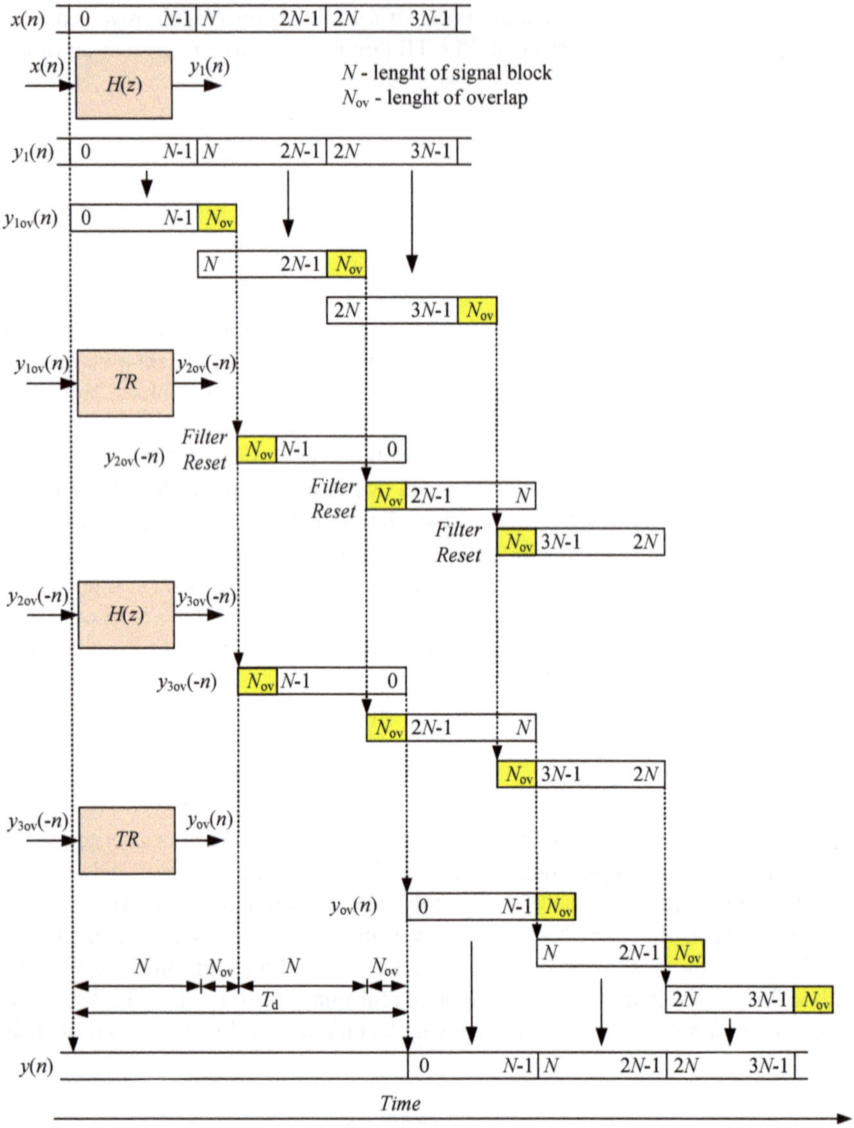

Fig. 3.34 Implementation of linear-phase IIR filter

an LF IIR filter is depicted in Fig. 3.35. This implementation has been modified by Willson and Orchard [87]. Kurosu et al. have reduced the LF IIR filter delay [44]. Also, Azizi has patented a signal interpolator using a zero-phase filter [9, 10]. An interesting solution of the LF IIR without overlap is presented by Mouffak and Belbachir [51].

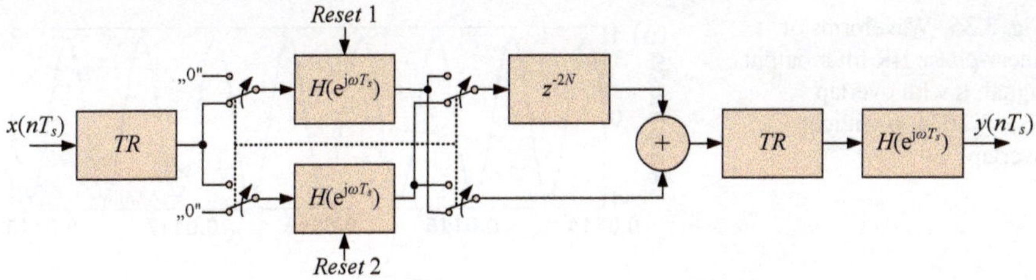

Fig. 3.35 The LF IIR filter with overlap by Powell and Chau

Table 3.7 Comparison of LF IIR filters realizations

Filter type	Filter dealy in samples, N_d	Quantity of MAC operation per sample, N_{MAC}
Powell and Chau	$3N$	$6L$
Proposed	$2(N + N_{ov})$	$2L(2 + N_{ov}/N)$

In Table 3.7 a comparison of the proposed L-order LF IIR filter with L-order Powell and Chau LF IIR filter [58] is presented. The proposed filter has less delay and fewer multiplications and additions (MAC) per input sample.

3.5.1 *Example of a Linear-Phase IIR Filter*

As an example the author has implemented an 8-order IIR elliptic filter with crossing frequency $f_{cr} = 20$ kHz and sampling frequency $f_s = 352.8$ kHz. The filter design parameter are presented in Table 3.8. The input sinusoidal signal with frequency $f_{syg} = 19.98$ kHz is divided into blocks of length $N = 2048$ samples. That makes delay of the output signal equal to $T_d = 2N/f_s = 11.69$ ms. The author's MATLAB program for the implementation of an LF IIR filter is shown in Listing 3.2. Figure 3.36b shows the results of connecting blocks of signal samples for the LF IIR filter without overlap. The graph shows visible signal amplitude distortion associated with transient effects of the filter when connecting the blocks of signal samples.

Table 3.8 LF IIR filter design parameters

Sampling rate, f_s	Pass-band frequency, f_p	Stop-band frequency, f_z	Pass-band ripple, A_p	Stop-band attenuation, A_z
352.8 kHz	20 kHz	24 kHz	0.1 dB	63 dB

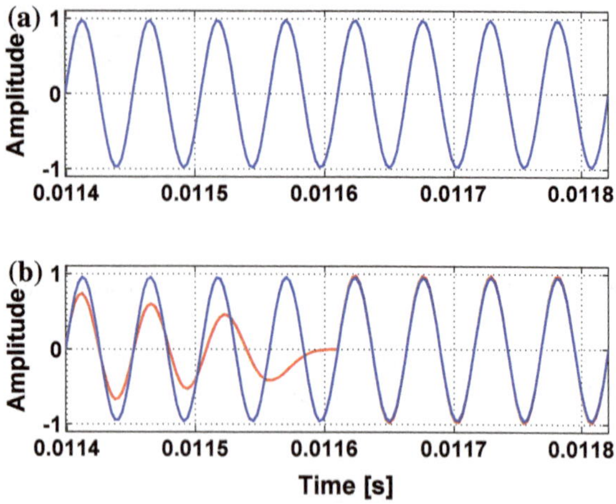

Fig. 3.36 Waveforms of linear-phase IIR filter output signal: **a** with overlap $N_{ov} = 1024$, **b** without overlap

Listing 3.2 LF IIR filter

```
clear all;
fs=352.8e3; % sampling frequency
fb=20e3; % end of band of interest
fsyg=19e3; % signal frequency
ib=6; % number of blocks
N=2048; % length of block
Ns=N*ib; % length of input signal
t=(0:Ns-1)/fs; % time vector
y=zeros(1,Ns); %space for y
% coherent frequency of input signal
fsyg_k=round(fsyg/(fs/N))*fs/N;
x=sin(2*pi*fsyg_k*t); % input signal
% ------ Filter design ---------------
Fg=2*fb/fs;
[b a]=ellip(8,0.1,63,Fg,'low');
% ----- Causal filtering -------------
y1=filter(b,a,x);
% Non-causal filtering, time reversing
N_ov=1024; % number of samples in overlap
y3=zeros(1,N+N_ov); %space for y3
for nb=1:ib-1
% time reversing and filtering
y3=filter(b,a,y1(nb*N+N_ov:-1:(nb-1)*N+1));
% time reversing
y3=y3(N+N_ov:-1:N_ov+1);
% output signal synthesis
y(1,(nb-1)*N+1:nb*N)=y3;
end
```

Using overlap as in Fig. 3.34, it is possible to reduce this distortion. In this particular case the overlap length used is equal to $N_{ov} = 1024$ samples. Figure 3.36a shows the results of connecting blocks of signal samples for the LF IIR filter with overlap. In this case the output signal delay is longer, $T_d = 2(N + N_{ov})/f_s = 17.41$ ms. Additionally, in Fig. 3.37 the difference between a reference signal and the filter output signal for a filter with overlap and without overlap is shown.

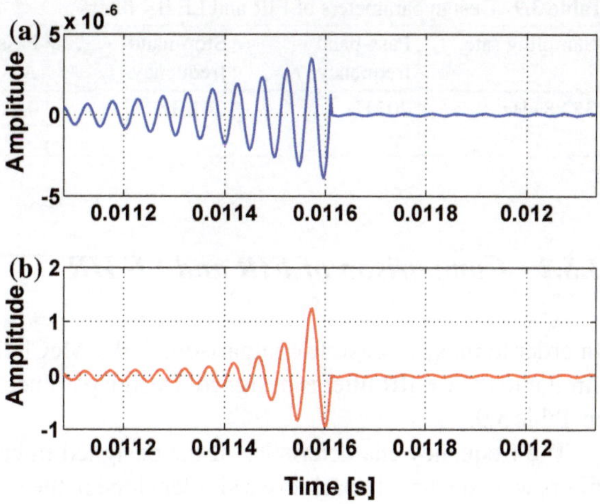

Fig. 3.37 Difference between reference signal and LF IIR filter output signal: **a** with overlap $N_{ov} = 1024$, **b** without overlap

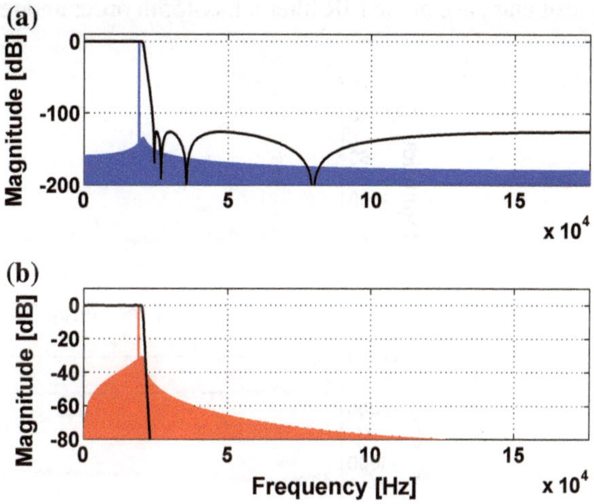

Fig. 3.38 Spectra of LF IIR filter output signal and IIR filter characteristic (*black*): **a** with overlap $N_{ov} = 1024$, **b** without overlap

The illustration of the dynamic range of the signal is presented in Fig. 3.38, showing spectra of output signals and characteristics of the filter for both versions: with and without overlap. The input sinusoidal signal has unity amplitude and frequency $f_{syg} = 19.98\,\text{kHz}$. The filter characteristic is a result of the responses of two IIR filters in cascade connection.

Table 3.9 Design parameters of FIR and LF IIR filters

Sampling rate, f_s	Pass-band frequency, f_p	Stop-band frequency, f_z	Pass-band ripple, A_p	Stop-band attenuation, A_z
352.8 kHz	20 kHz	24 kHz	0.1 dB	63 dB (for IIR) 140 dB (for FIR)

3.5.2 Comparison of FIR and LF IIR

In order to make a sensible comparison, Parks-McClellan FIR filter with parameters similar to the LF IIR filter is designed. Design parameters of the two filters are shown in Table 3.9.

The frequency characteristics of the designed filters are shown in Fig. 3.39. The filters were designed to achieve a similar slope in the transition band of the amplitude characteristic.

Results of the FIR filter and LF IIR filter design are shown in Table 3.10. In the particular case of the FIR filter it has 455th order and requires 456 multiplication and

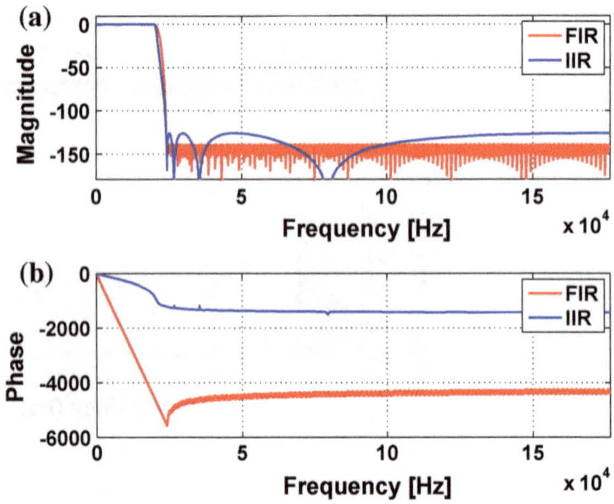

Fig. 3.39 Characteristics of FIR and IIR filters: **a** amplitude response, **b** phase response

Table 3.10 Comparsion of parameters of FIR and LF IIR filters

Filter type	Filter order, L	Filter delay in samples, N_d	MAC operations per sample, N_{MAC}
Parks-McClellan FIR	455	228	456
Elliptic LF IIR for $N = 2048$, $N_{ov} = 1024$	$2 * 8$	$4096 + 2048$	$32 + 16$

accumulation (MAC) operations per one input sample. Delay N_d for such a filter is equal to 228 samples. The solution for the LF IIR filter requires two 8th order elliptic filters and $32 + 16$ MAC operations per one input sample. The delay N_d is equal to 6144 samples.

It should be noted, however, that in the case using an LF IIR filter, zero phase shift of the output signal is obtained.

3.6 Multirate Circuits

The main reason for introducing multirate circuits (systems) is the necessity to improve the quality, whilst maintaining or reducing the cost of the system. The application of multirate circuits is necessary during the conversion of A/D and D/A signals, and when oversampling is used. Another reason for using the multirate circuits systems is the necessity to exchange data between systems, using different sampling rates. The process of reducing the sampling rate of a signal is commonly known as decimation, and the multirate circuit used for decimation is called decimator. The process of increasing signal sampling rate is called interpolation, and the circuit used for signal interpolation is called an interpolator. The interpolator and the decimator are the most common multirate circuits used for changing signal sampling rate. Multirate circuits are described in many publications, and the author can recommend a few books written by: Crochiere and Rabiner [17], Vaidyanathan [82], Flige [36], Proakis and Manolakis [59]. The exemplary block diagram of a digital multirate control circuit is depicted in Fig. 1.12. In this section circuits useful for power electronics control circuits are presented. In this case, the most important is the use of a signal interpolator for noise shaping circuits in the inverter output circuits.

3.6.1 Signal Interpolation

A signal interpolator made up of an upsampler and an anti-imaging low-pass filter for integer valued conversion factor R is depicted in Fig. 3.40a. The R is called oversampling ratio. The low-pass filter $H(z)$, also called the interpolation filter, removes the $R - 1$ unwanted images in the spectra of upsampled signal $w(kT_s/R)$. An illustration of interpolating process for $R = 3$ is depicted in Fig. 3.41. After the upsampling process, the out-of-band signal (unwanted images) is a potential source of interference for the input signal. The out-of-band signal (unwanted images) can dramatically decrease the signal dynamic ratio (*SINAD*). The anti-imaging filter must attenuate all unwanted images. The stopband cutoff frequency F_z must be selected to limit aliasing in the input signal frequency range.

Two types of stopband criteria can be used in practice, type 1, where aliasing is not allowed in the transition band (Fig. 3.42b), and type 2, where aliasing is allowed

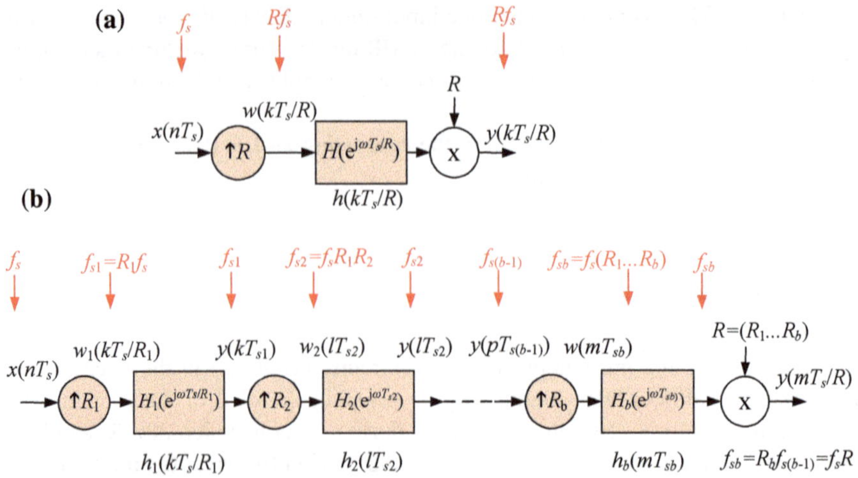

Fig. 3.40 Interpolator made up of upsampler and an anti-imaging filter: **a** single stage version, **b** multistage version

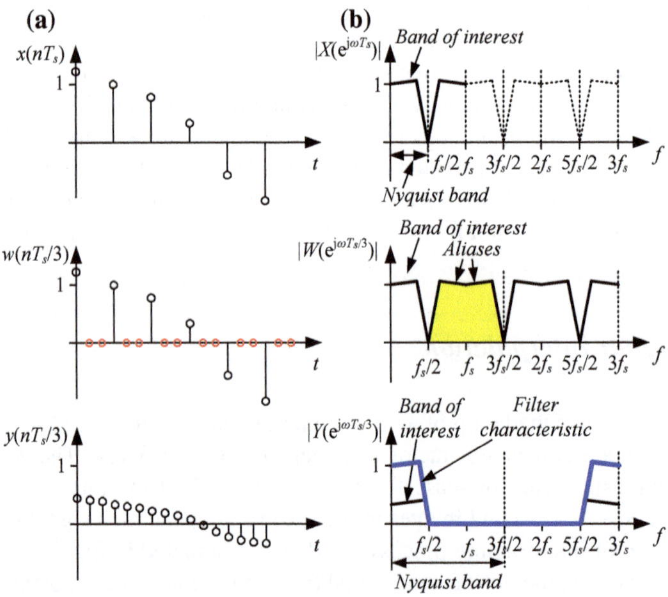

Fig. 3.41 Illustration of signal interpolation for $R = 3$: **a** waveform, **b** spectra

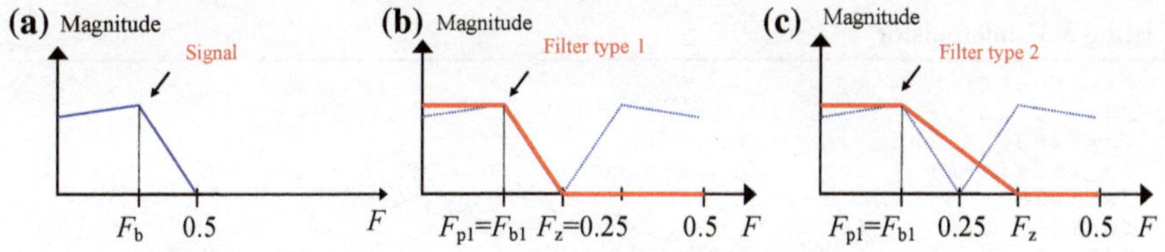

Fig. 3.42 Filter types: **a** input signal band width, **b** type 1 anti-imaging filter requirements with aliasing not allowed in transition band (for $R = 2$), **c** type 2 anti-imaging filter requirements with aliasing allowed in transition band (for $R = 2$)

in the transition band (Fig. 3.42c). The normalized stopband frequency F_z for filters of type 1 and type 2 is respectively equal to

$$F_z = \frac{F_s}{2R} \quad , \quad F_z = \frac{F_s}{R} - \frac{F_b}{R} \quad , \tag{3.30}$$

where: F_b—the normalized passband frequency of the input signal, F_s—the normalized sampling rate.

The multistage version of the interpolator is depicted in Fig. 3.40b. In this case, a design strategy is used, in which every stage attenuates its own unwanted images. Normalized passband signal frequency on the output of every stage is given by

$$F_{bk} = \frac{F_{b(k-1)}}{R_k} \quad , \tag{3.31}$$

where: R_k—interpolating ratio at stage k. For kth stage of the interpolator with a filter type 1 and type 2, the stopband frequency is given by

$$F_{zk} = \frac{F_{s(k-1)}}{2R_k} \quad , \quad F_{zk} = \frac{F_s}{R_k} - \frac{F_{b(k-1)}}{R_k} \quad . \tag{3.32}$$

respectively. Multistage interpolators are often used because of the lower number of required arithmetic operations.

The author's MATLAB program for the realization of a single stage signal interpolator for $R=4$ (depicted in Fig. 3.40a) is described in Listing 3.3.

Listing 3.3 Interpolator

```
clear all;
R=4; % interpolator factor
fs=3200; % sampling frequency
fs_int=fs*R; % sampling frequency after interpolation
fsig=200; % signal frequency
fb=500; % end of band of interest
N=2^10; % length of samples block
N_int=N*R; % length of samples block after interpolation
t=(0:N-1)/fs; % time vector
% coherent frequency of input signal
fsigk=round(fsig/(fs/N))*fs/N;
x=sin(2*pi*fsigk*t); % input signal
% ------ Upsampling ------------------------
w=zeros(1,N_int);
w(1,1:R:N_int)=x; % upsampled input signal
% ------ Filter design ----------------------
Fg=2*fb/fs_int;
[b a]=butter(2,Fg);
y=filter(b,a,w)*R; % filtering
```

Using the above program a simulation is made for a 200 Hz sinusoidal input signal with sampling rate $f_s = 3.2\,\text{kHz}$, which is interpolated for $R = 4$. A second-order IIR Butterworth filter is used as the interpolating filter for $f_{cr} = 500\,\text{Hz}$. Figure 3.43 shows the spectra of signals: $x(nT_s)$, $w(nT_s/4)$, $y(nT_s/4)$ (red) and interpolating filter amplitude response (blue). As shown in Fig. 3.43, the aliasing and image produced during the interpolation process can degrade the dynamic range of the signal. Therefore, the choice of low-pass filter parameters is particularly important for output signal quality.

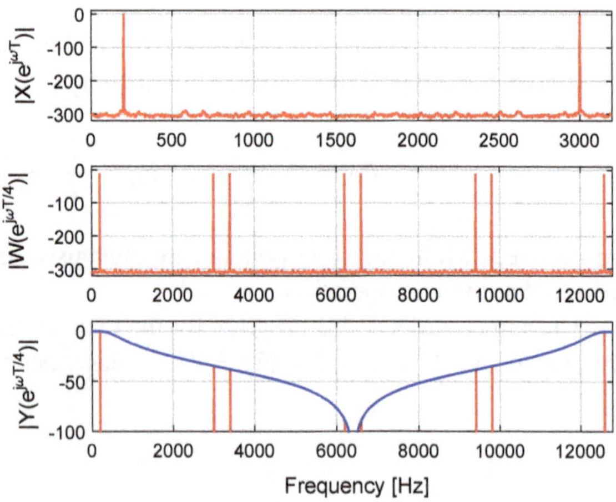

Fig. 3.43 Spectra of interpolation process for $R = 4$

3.6.2 Signal Decimation

Another multirate circuit is the signal decimator. It is used to reduce the sampling rate. During the signal decimation process the bandwidth of a signal must first be reduced by the low-pass filter before its sampling rate is reduced by a downsampler. A block diagram of a single stage decimator for an integer valued conversion factor M is depicted in Fig. 3.44a. A decimator consists of a low-pass filter and a downsampler. The bandwidth of signal should be reduced from $f_s/2$ to $f_s/(2M)$, otherwise aliasing components penetrate into usable bandwidth and deteriorate signal parameters (e.g. *SINAD*). Illustration of a signal decimation process for $M = 3$ is depicted in Fig. 3.45 As in the case of interpolation, in the decimation process two types of anti-imaging filters can be defined, such as shown in Fig. 3.42.

A multistage decimator is in depicted in Fig. 3.44b. Similar to the interpolator, a design strategy is used in which every stage attenuates its own aliasing components. A multistage version of the decimator requires typically fewer arithmetic operations than a single stage one, especially for high value or M. Therefore multistage decimators are more frequently used.

The author's MATLAB program for the realization of a single stage signal decimator for $M = 4$ (depicted in Fig. 3.44a) is described in Listing 3.4.

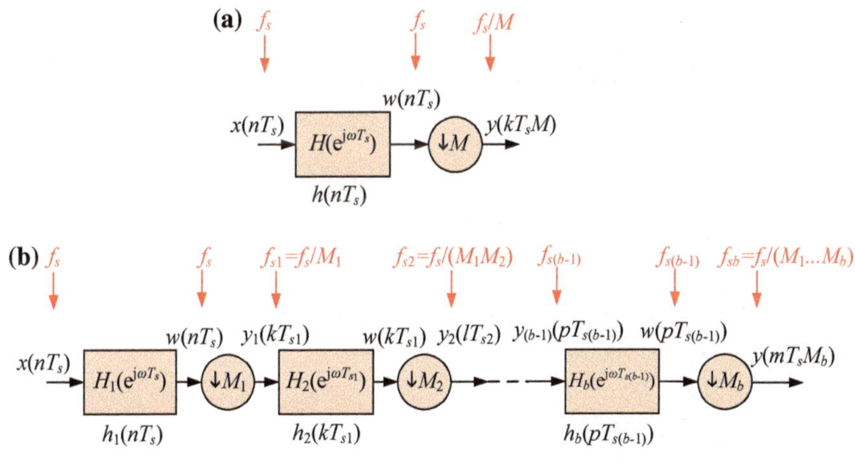

Fig. 3.44 Decimators: **a** single stage, **b** cascaded

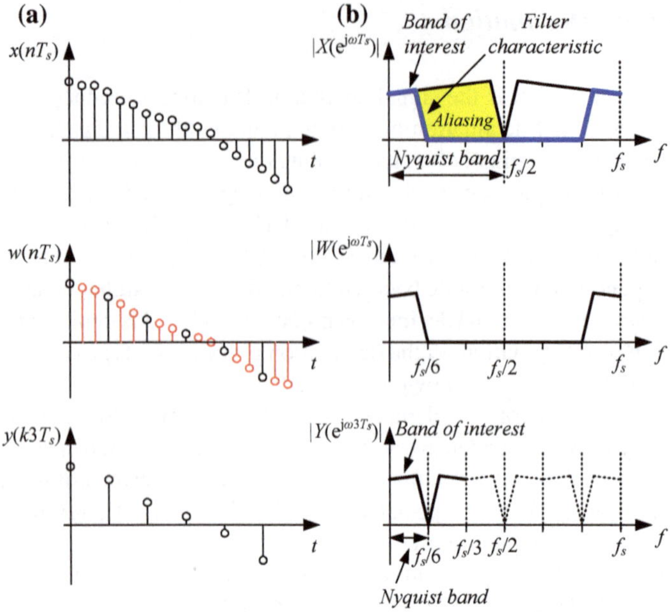

Fig. 3.45 Illustration of signal decimation process for $M = 3$: **a** waveform, **b** spectra

Listing 3.4 Decimator

```
clear all;
M=4; % decimation factor
fs=50*2^6; % sampling frequency
fsig=50; % signal frequency
fb=2.3*fsig; % end of band of interest
N=2^12; % length of block
t=(0:N-1)/fs; % time vector
% coherent frequency of input signal
fsigk=round(fsig/(fs/N))*fs/N;
% input signal
x=sin(2*pi*fsigk*t)+0.5*sin(21*pi*fsigk*t)+...
0.3*sin(41*pi*fsigk*t);
% ------ Filter design -----------------------
Fg=2*fb/fs;
[b a]=butter(3,Fg);
% ------ Decimation --------------------------
w=filter(b,a,x); % filtering
y=w(1,1:M:N); % downsampling
```

A simulation is made using the above program for harmonics input signal $x(t) = \sin(100\pi t) + 0.5\sin(1050\pi t) + 0.3\sin(2050\pi t)$ with sampling rate $f_s = 3.2\,\text{kHz}$, which is decimated for $M = 4$. As a decimating filter a third-order IIR Butterworth filter for $f_{cr} = 115\,\text{Hz}$ is used. Figure 3.46 shows the result of such a simulation, with the spectra of signals: $x(nT_s)$, $w(n4T_s)$, $y(n4T_s)$ (red) and decimating filter amplitude response (blue). As is in the case of interpolation, during the signal decimation the aliasing image can degrade the dynamic range of the signal. Therefore, the

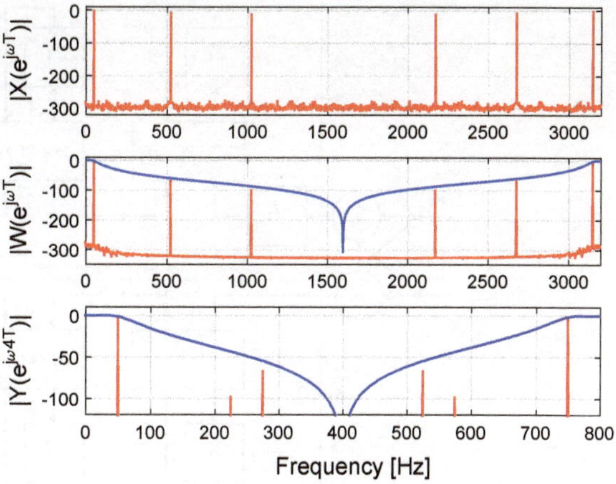

Fig. 3.46 Spectra of decimation process for $M = 4$

choice of low-pass decimator filter parameters is particularly important and affects the dynamics range of the signal.

3.6.3 *Multirate Circuits with Wave Digital Filters*

A special class of lattice wave digital filters, referred to as bireciprocal, are suitable for the realization of interpolators. The characteristic function $K(\psi)$ of bireciprocal filters satisfies the equation

$$K(\psi) = \frac{1}{K\left(\frac{1}{\psi}\right)}, \quad where \quad \psi = \frac{z-1}{z+1}. \tag{3.33}$$

For low-pass bireciprocal filters the passband is from 0 to $fs/4$, and for high-pass bireciprocal filters the passband is from $fs/4$ to $fs/2$. For this type of filter all even filter coefficients γ_k are equal to zero and the filter circuit can be simplified to the circuit shown in Fig. 3.47, so that the number of filter elements is halved. Bireciprocal lattice wave digital filters of this kind are very useful for building multirate circuits. The first-order allpass section of branch S_2 is replaced by a unit delay and the second-order allpass sections are replaced by two-port adaptors with double delay. Branches of bireciprocal filters work at two times lower speed. Therefore the output filter summing block can be replaced by the switch while reducing the speed of the filter to half. Figure 3.48a shows a signal interpolator for $R = 2$. In the same way it is possible to build a signal decimator for $M = 2$. The block diagram of the decimator is depicted in Fig. 3.48b. A cascaded version of such an interpolator with bireciprocal

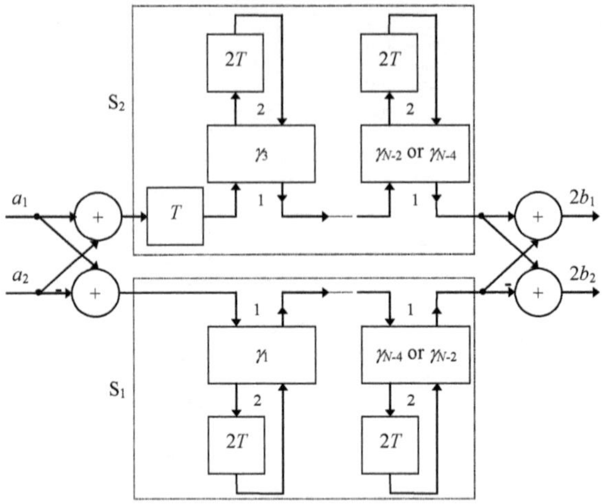

Fig. 3.47 Block diagram of bireciprocal lattice wave digital filter

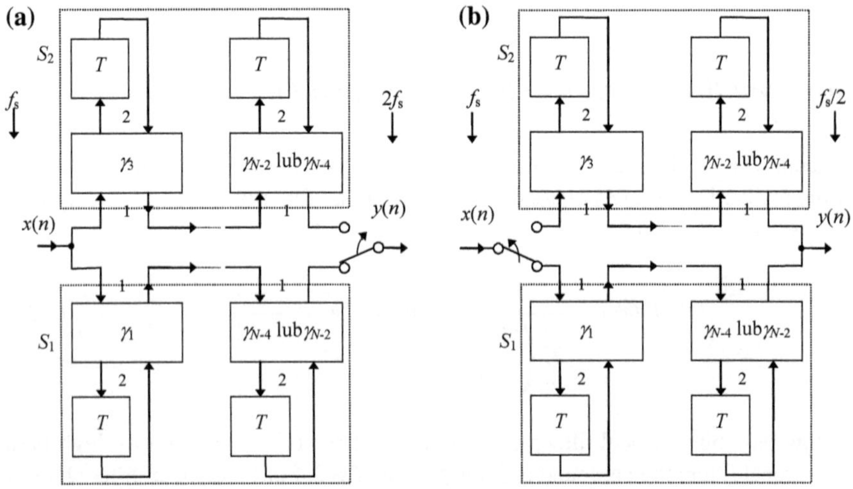

Fig. 3.48 Block diagram of multirate circuit using bireciprocal lattice wave digital filter: **a** signal interpolator, **b** signal decimator

lattice wave digital filters is shown in Fig. 3.49. As in the case of a cascade signal interpolator a cascade signal decimator may be created.

Of course, a modified LWDF can also be successfully used in multirate circuits. Examples of such applications made by the author are discussed in Chap. 6.

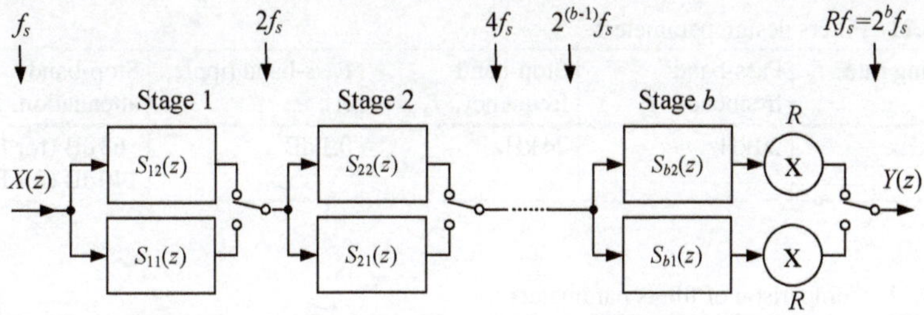

Fig. 3.49 Block diagram of cascade signal interpolator using bireciprocal lattice wave digital filter

3.6.4 Interpolators with Linear-Phase IIR Filters

In the considered signal interpolators low-pass filters are used to suppress the aliasing components. Filters introduce signal delay and in the case of IIR filters they have a nonlinear phase-response. Therefore, in order to obtain linear-phase response, the FIR filters should be used. An alternative may be the application of linear-phase IIR filters. An example of such a solution is presented by Azizi [9, 10]. Also, the author has designed an interpolator with a linear-phase IIR filter [70]. The block diagram of a signal interpolator with linear-phase IIR filter proposed by the author is depicted in Fig. 3.50. The first filter works in pipeline mode and the second in block mode. The number of MAC operations per one input sample for signal intergrator using L-order LF IIR filter can be calculated by the equation

$$N_{MAC} = 4RL\left(1 + \frac{N}{N_{ov}}\right) , \tag{3.34}$$

where: N—length of signal block, N_{ov}—length of signal overlap block, R—oversampling ratio.

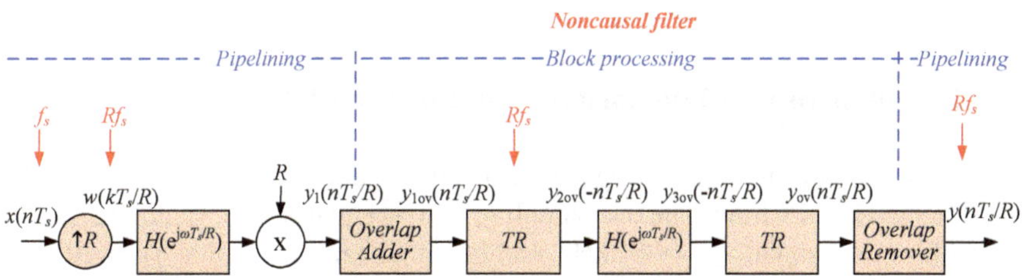

Fig. 3.50 Interpolator with linear-phase IIR filter

Table 3.11 Filters design parameters

Sampling rate, f_s	Pass-band frequency, f_p	Stop-band frequency, f_z	Pass-band ripple, A_p	Stop-band attenuation, A_z
352.8 kHz	20 kHz	24 kHz	0.1 dB	63 dB (for IIR) 140 dB (for FIR)

Table 3.12 Comparison of filters parameters

Filter type	Filter order, L	Filter delay in samples, N_d	MAC operations per sample, N_{MAC}
Parks-McClellan FIR	455	227.5	456
Elliptic LF IIR for $N = 2048$, $Nov = 1024$	$2 * 8$	$4096 + 2048$	$32 + 16$

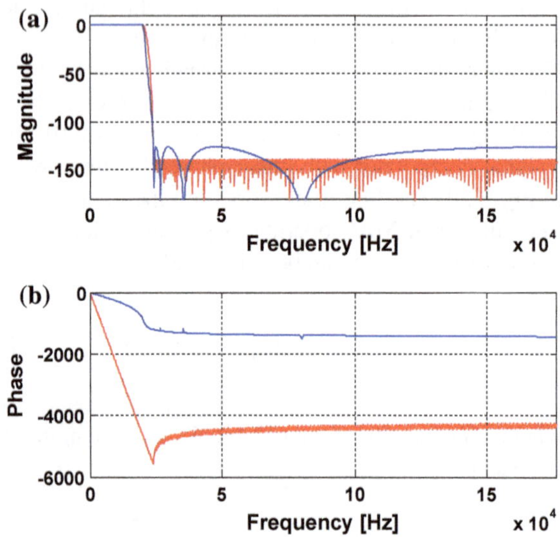

Fig. 3.51 FIR (*red*) and IIR (*blue*) filters characteristics: **a** amplitude response, **b** phase response

3.6.4.1 Comparison of Interpolators with FIR and LF IIR Filter

In order to compare the signal interpolators, an investigation was performed using the FIR filter and LF IIR filter. The oversampling ratio is equal $R = 8$, sampling ratio is typical for audio signal $f_s = 44.1$ kHz, end of signal band is equal to $f_b = 20$ kHz. The parameters of the designed filters are as shown in Tables 3.11 and 3.12. The frequency responses of the filters are shown in Fig. 3.51.

The FIR filter order is $L = 455$, So the interpolator requires $R(L + 1) = 3648$ multiplications and additions per one input sample, which is a huge amount of MAC operations. It is possible to decrease the quantity of arithmetic calculation by elimination of the multiplication and addition operations for zero value samples. In the

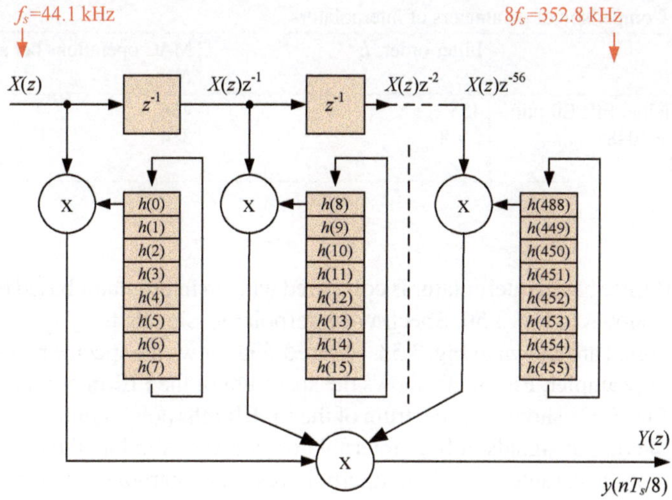

Fig. 3.52 Block diagram of FIR based signal interpolator for $R = 8$ and $L = 455$

Fig. 3.53 Spectra of interpolators signals: **a** upsampled input signal, **b** with FIR filter, **c** with LF IIR filter

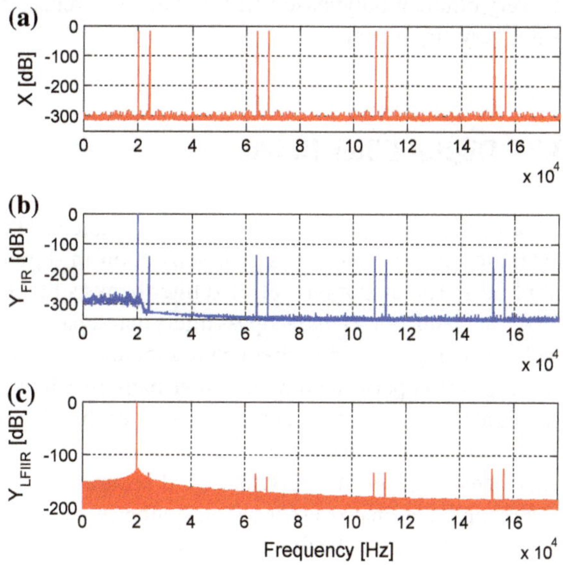

case of applying an FIR filter it is very efficient to implement the system using a polyphase circuit. The block diagram of such a solution is shown in Fig. 3.52. This is a FIR based signal interpolator for $R = 8$ with periodically switched coefficients and filter order $L = 455$. In this case the interpolator requires $L + 1$ multiplication and addition operations per one input signal sample. This kind of filter structure is easily and efficiently realized by the DSP.

Table 3.13 Comparison of parameters of interpolators

Filter type	Filter order, L	MAC operations per sample, N_{MAC}
Parks-McClellan FIR	455	456
Elliptic LF IIR, $N = 2048$, $Nov = 1024$	$2 * 8$	384

This FIR filter based interpolator is compared with an interpolator based on the LF IIR filter as shown in Fig. 3.50. Spectra of interpolators signals for $f_{syg} = 19.98$ kHz sinusoidal signal are shown in Fig. 3.53. Figure 3.53a shows the spectrum of the input signal after upsampler, Fig. 3.53b shows the spectrum of the FIR interpolator output signal and Fig. 3.53c shows the spectrum of the LF IIR interpolator output signal. The spectra of the output signals of both interpolators are very similar. Table 3.13 shows comparison of the parameters of both interpolators. The interpolator with the LF IIR filter requires a little less MAC operations than with FIR filter. Also, an additional advantage of this solution is the achieving a zero phase shift in the output signal.

The LF IIR filters are suitable for the interpolator, however in comparison with the very efficient polyphase FIR filter, the difference in quantity of MAC operations is not very significant.

3.7 Digital Filter Banks

Filter banks can be used for separation of signals. They can be used for harmonics separation in APF and for signal separation in digital crossover in class-D audio amplifiers. The author has selected filter banks suitable for these applications.

A filter bank is a set of bandpass filters that separates the input signal into subbands, each one carrying a single frequency subband of the original signal. The process of decomposition performed by the filter bank is called analysis (meaning analysis of the signal in terms of its components in each subband); the output of analysis is referred to as a subband signal with as many subbands as there are filters in the filter bank. The set of filters for signal separation is called an analysis filter bank. The reconstruction process is called synthesis, meaning reconstitution of a complete signal resulting from the filtering process. The set of filters for signal reconstruction is called a synthesis filter bank. Using analysis filter banks it is possible to decompose signal spectra into a number of directly adjacent frequency bands and recombine the signal spectra by means of a synthesis filter bank. In most cases signals are separated into more than two subband signals. The background of filter banks is described in digital signal processing literature, in particular the following books can be recommended [17, 36, 82]. The general form of the M-channel filter bank is

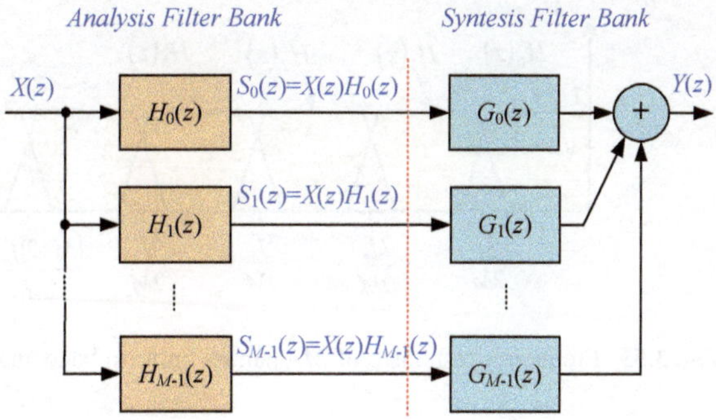

Fig. 3.54 An M-channel band analysis and synthesis filter bank

shown in Fig. 3.54, where: M is the number of subbands. The output signal of filter bank $Y(z)$ can be calculated by the equation

$$Y(z) = X(z) \sum_{k=0}^{M-1} (H_k(z)G_k(z)) \quad . \tag{3.35}$$

It is possible to simplify this equation to

$$Y(z) = X(z)F(z) \quad , \tag{3.36}$$

where: $F(z)$ denotes the quality of signal reconstruction.

If $|F(e^{j\Omega}| = 1$ for all frequencies, the filter bank is without amplitude distortion. When $F(e^{j\Omega})$ has linear phase (constant group delay), the filter bank is without phase distortion. If $F(z)$ is pure delay, the filter bank is called perfect reconstruction. A filter bank with amplitude and/or phase distortion which can be kept arbitrarily small is called a filter bank with almost perfect reconstruction. Another function important for the discussed filter banks is power complementary [82]. This ensures representation of the whole input signal spectrum in subbands. For an M-channel power complementary filter bank the square sum of transfer functions $|H_k(e^{j\omega})|$ module is equal to $K = const$, typically $K = 1$

$$\sum_{k=0}^{M-1} |H_k(e^{j\omega})|^2 = K \quad . \tag{3.37}$$

The typical frequency responses of M-channel overlapping uniform band analysis and synthesis filter banks are shown in Fig. 3.55. The theory of filter banks is well described by Vaidyanathan [82], Fliege [36] and many others.

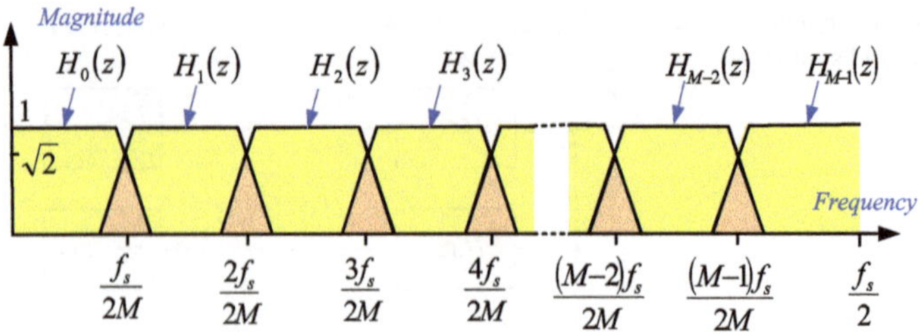

Fig. 3.55 Frequency responses of M-channel uniform band analysis and synthesis filter banks

3.7.1 Strictly Complementary Filter Bank

For the strictly complementary (SC) FIR filter bank the sum of the set of transfer functions $[H_0(z), H_1(z), \ldots, H_{M-1}(z)]$ is equal to pure delay

$$\sum_{k=0}^{M-1} H_k(z) = cz^{n_0} \quad , c \neq 0 \ , \tag{3.38}$$

where: c—is constant. In this case the synthesis filter bank is simplified to a simple sum. It is a good solution for the system under consideration, where synthesis is performed by adding two acoustic waves, one from the tweeter and the second from the woofer/midrange. For a two channel ($M = 2$) strictly complementary linear phase FIR filter bank the design procedure is very simple. If the low-pass filter $H_0(z)$ is a transfer function of linear phase filter type 1 (order of filter N is even) then

$$H_0(z) + H_1(z) = z^{N/2} \quad . \tag{3.39}$$

Transfer functions of both filters $H_0(z)$ and $H_1(z)$ can be described by equations

$$\begin{cases} H_0(z) = b_{00} + \cdots + b_{0N/2}z^{-N/2} + \cdots + b_{0N}z^{-N} \\ H_1(z) = b_{10} + \cdots + b_{1N/2}z^{-N/2} + \cdots + b_{1N}z^{-N} \end{cases} \tag{3.40}$$

then by substitution in Eq. 3.40 this becomes

$$\begin{aligned} b_{00} + \cdots + b_{0N/2}z^{-N/2} + \cdots + b_{0N}z^{-N} + \cdots \\ +b_{10} + \cdots + b_{1N/2}z^{-N/2} + \cdots + b_{1N}z^{-N} = z^{-N/2} \quad . \end{aligned} \tag{3.41}$$

High-pass filter $H_1(z)$ coefficient can be calculated by the formula (3.41)

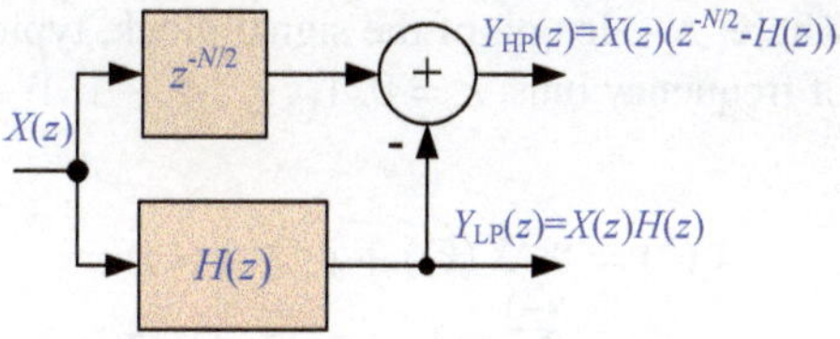

Fig. 3.56 Block diagram of 2-channel strictly complementary analysis filter bank

$$\begin{cases} b_{10} = -b_{00} \\ \dots \\ b_{1N/2} = 1 - b_{0N/2} \\ \dots \\ b_{1N} = -b_{0N} \quad , \end{cases} \tag{3.42}$$

Although a strictly complementary linear phase FIR two channel filter bank can be realized by two separate FIR filters, another solution is possible by directly using the equation

$$H_0(z) = H(z) \quad \text{and} \quad H_1(z) = z^{N/2} - H(z) \quad . \tag{3.43}$$

A block diagram of such a solution is depicted in Fig. 3.56. The MATLAB program for the coefficients calculating of strictly complementary FIR filter is shown in Listing 3.5.

Listing 3.5 The calculation of strictly complementary FIR filter coefficients

```
function [b,a]=stricomp(b,a)
dlug=length(b);
N=dlug-1;
if rem(N,2)~=0
   error('Filter order should be even.')
end
b_srodek=b(N/2+1);
b=-b;
b(N/2+1)=1-b_srodek;
```

3.7.2 *DFT Filter Bank*

Spectrum analysis of signal in power electronics is an important measuring technique. The usual method for spectrum analysis is the DFT and its efficient implementation, the fast Fourier transform (FFT). For real discrete signal $x(n)$ the DFT kth spectrum bin is described by the equation

$$X(k) = \sum_{n=0}^{N-1} (x(n) W_N^{kn}) \quad , \tag{3.44}$$

where: N—length of the signal block, typically equal to signal period, k—number of frequency bins, $k = 0, 1, \ldots, N-1$, $W_N = e^{-j2\pi/N}$.

$$X(k) = \Re(X(k)) + j \cdot \Im(X(k)) = \sum_{n=0}^{N-1} (x(n)\cos(2\pi kn/T)) - j\sum_{n=0}^{N-1} (x(n)\sin(2\pi kn/T)) \quad , \tag{3.45}$$

and

$$|X(k)| = \sqrt{\Re(X(k))^2 + \Im(X(k))^2} \quad , \qquad \varphi(k) = \arctan\left(\frac{\Im(X(k))}{\Re(X(k))}\right) \quad . \tag{3.46}$$

The DFT can be used as an analysis filter bank. A block diagram of this bank is depicted in Fig. 3.57. Typically, for every sample of input signal N—point, DFT is calculated as shown in Fig. 3.58. Simplified magnitude characteristics of this filter bank are shown in Fig. 3.59.

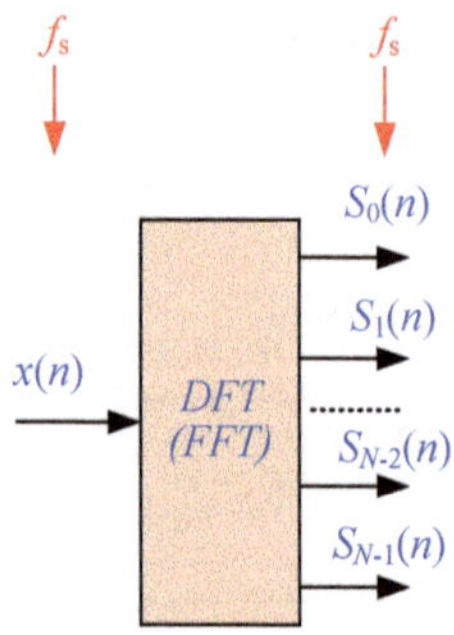

Fig. 3.57 DFT filter bank

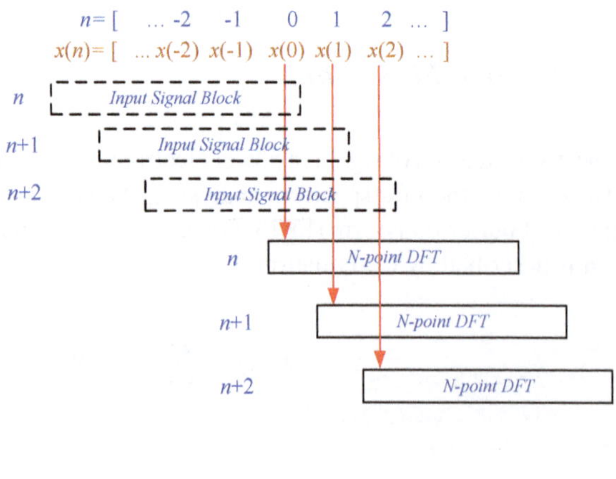

Fig. 3.58 Block diagram of signal flow for typical DFT

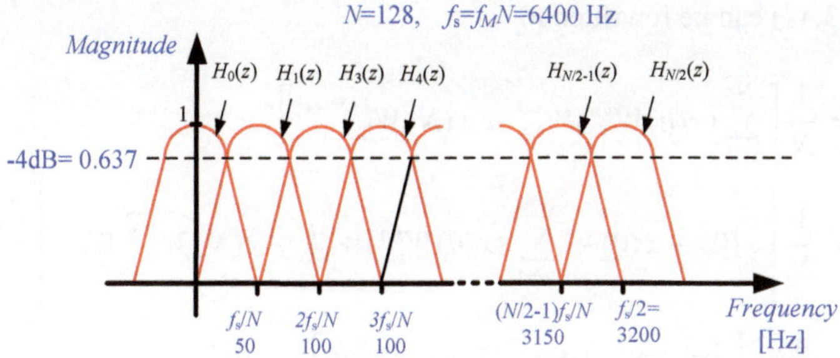

Fig. 3.59 Simplified frequency responses of N-channel band analysis DFT filter bank

The common drawbacks of the DFT filter bank include:

- N^2 complex multiplication—can be decreased by using FFT to $N \log N$,
- poor frequency response,
- large computation power required for the DSP,
- need for coherent sampling.

In many applications it is not necessary to evaluate all the bands of the spectrum $X(k)$ of the analyzed signal $x(n)$. This occurs for example in harmonic compensation systems in which it is necessary to evaluate one or more harmonics. Thus setting all the bands of the spectrum is superfluous when one only to designate or more. In this case, it is useful to employ the so-called Goertzel algorithm [38]. Discussion of this algorithm can be found in many publications, among others by Oppenheim et al. [52] and by Zieliński [88].

3.7.3 Sliding DFT Algorithm

In control systems there is often a need to determine the spectrum for the upcoming sequence of the samples. Therefore, in this case, the application of the iterative determination of the spectrum seems to be a good solution. The current value of a single kth bin of the spectrum is determined by the formula

$$X_k(0) = \sum_{n=0}^{N-1} (x(n) W_N^{kn}) \quad , \tag{3.47}$$

where: $k = 0, 1, \ldots, N-1$, and value of the next sample is determined by

$$X_k(1) = \sum_{n=1}^{N} (x(n) W_N^{k(n-1)}) \quad . \tag{3.48}$$

Equation (3.47) can be rewritten to

$$
\begin{aligned}
X_k(1) &= \frac{1}{N}\left[\sum_{n=1}^{N}(x(n)W_N^{kn})W_N^{-k} + x(N)W_N^{k(N-1)}\right] = \\
&= \frac{1}{N}\left[x(0) - x(0) + \sum_{n=1}^{N-1}(x(n)W_N^{kn})W_N^{-k} + x(N)\overbrace{W_N^{kN}}^{=1}W_N^{-k}\right] = \\
&= \frac{W_N^{-k}}{N}\left[\sum_{n=0}^{N-1}(x(n)W_N^{kn}) + x(N) - x(0)\right] = \\
&= W_N^{-k}\left[\frac{1}{N}\sum_{n=0}^{N-1}(x(n)W_N^{kn}) + \frac{1}{N}(x(N) - x(0))\right] = \\
&= W_N^{-k}\left[X_k(0) + \frac{1}{N}(x(N) - x(0))\right] \quad .
\end{aligned}
\tag{3.49}
$$

Hence the Eq. (3.49) allows the iterative calculation of the spectrum for each sample of the input signal $x(n)$. Leaving aside the scaling factor $1/N$ this equation can be written as

$$X_k(n) = W_N^{-k}\left[X_k(n-1) + x(n) - x(n-N)\right] \quad . \tag{3.50}$$

This algorithm is called recursive DFT or sliding DFT, and is well described by Jacobsen and Lyons [41, 42] and many others [13, 28, 49, 50, 61, 88]. For calculating a few bins the sliding DFT algorithm is more effective than ordinary DFT. It is very simple and efficient, especially in a coherent sampling case. The z-domain transfer function for the kth bin of the sliding DFT filter is described by the equation

$$H_{SDFT}(z) = \frac{W_N^{-k} - W_N^{-k}z^{-N}}{1 - W_N^{-k}z^{-1}} = \frac{e^{j2\pi k/N} - e^{j2\pi k/N}z^{-N}}{1 - e^{j2\pi k/N}z^{-1}} \quad . \tag{3.51}$$

The block diagram of a single-bin sliding DFT filter is depicted in Fig. 3.60. The spectrum of one harmonic component of input signal is thus calculated

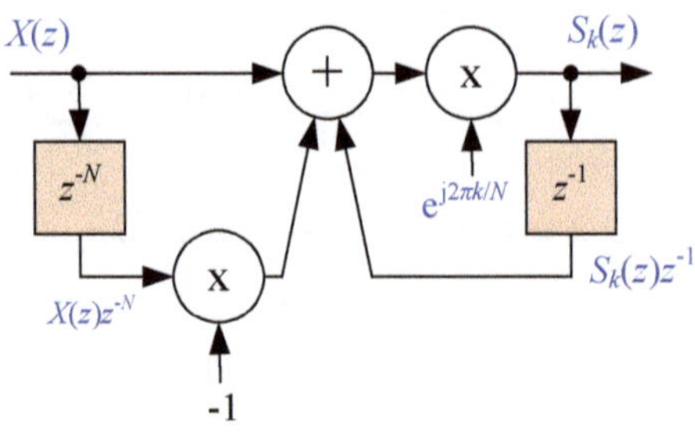

Fig. 3.60 Block diagram of single-bin sliding DFT filter

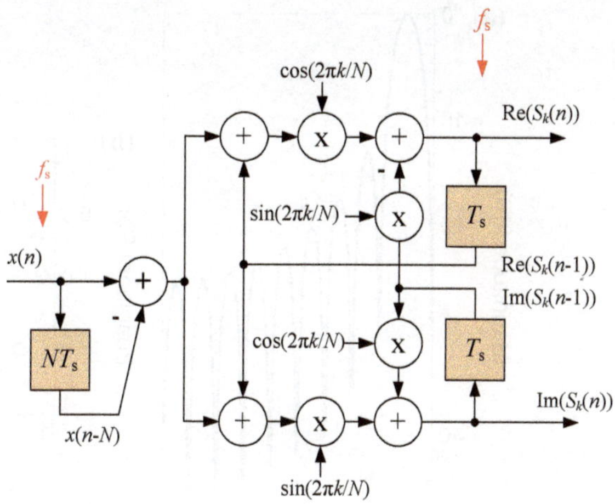

Fig. 3.61 Block diagram of realization of a k_{th} single-bin sliding DFT filter

$$\begin{aligned}S_k(n) &= e^{j2\pi k/N}(x(n) - x(n-N) + S_k(n-1)) = \\ &= (\cos(2\pi k/N) + j\sin(2\pi k/N))(x(n) - x(n-N) + \mathrm{Re}(S_k(n-1)) + j\mathrm{Im}(S_k(n-1))) = \\ &= \cos(2\pi k/N)(x(n) - x(n-N) + \mathrm{Re}(S_k(n-1))) - \mathrm{Im}(S_k(n-1))\sin(2\pi k/N)) + \\ &\quad j(\sin(2\pi k/N_M))(x(n) - x(n-N) + \mathrm{Re}(S_k(n-1)) + \mathrm{Im}(S_k(n-1))\cos(2\pi k/N)) \ .\end{aligned} \tag{3.52}$$

Unfortunately, the typical microprocessor does not have a built-in complex numbers arithmetic, hence Eq. (3.52) is transformed into a system of equations

$$\begin{cases} S_{kr}(n) = \cos(2\pi k/N)(x(n) - x(n-N) + S_{kr}(n-1)) - S_{ki}(n-1)\sin(2\pi k/N) \\ S_{ki}(n) = \sin(2\pi k/N)(x(n) - x(n-N) + S_{kr}(n-1)) + S_{ki}(n-1)\cos(2\pi k/N) \end{cases} , \tag{3.53}$$

where: $S_{kr}(n) = \mathrm{Re}(S_k(n))$ and $S_{ki}(n) = \mathrm{Im}(S_k(n))$.

The realization diagram of a kth single-bin sliding DFT filter is presented in Fig. 3.61.

The magnitude frequency characteristic of a single-bin sliding filter for $N = 20$ and $k = 1$ is shown in Fig. 3.62a. The passband and stopband are very poor but they are adequate for coherent sampled signals. The sliding DFT filter is only marginally stable because its pole resides on the z-domain's unit circle as shown in Fig. 3.62b. By using damping factor r it is possible to force the pole to be at a radius of r inside the unit circle. The transfer function of this solution is described by equation

$$H_{SDFT}(z) = \frac{1 - r^N z^{-N}}{1 - re^{j2\pi k/N}z^{-1}} , \tag{3.54}$$

Guaranteed-stable sliding DFT filter structure is depicted in Fig. 3.63. It is useful for low resolution fixed-point calculation, for example in fixed-point digital signal processors (DSP) and field programmable gate array (FPGA) circuits. For floating-point DSP, such as SHARC, it is achievable using the circuit in Fig. 3.60.

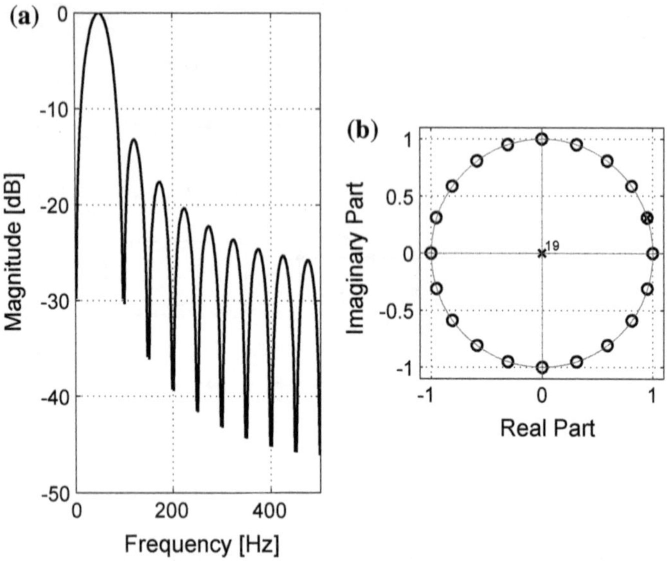

Fig. 3.62 Sliding DFT characteristics for $N = 10$, $k = 1$ and $f_s = 1000\,\text{Hz}$: **a** magnitude, **b** z-domain pole/zero location

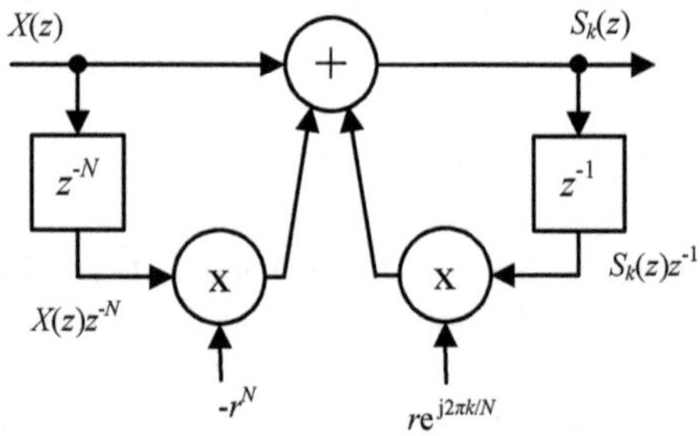

Fig. 3.63 Block diagram of guaranteed-stable sliding DFT filter

Using the single-bin SDFT filter it is possible to build an analysis filter bank for selected frequency bins. A block diagram of the N—channel analysis filter bank is shown in Fig. 3.64. Of course, implementation of such a bank for all DFT frequencies has no sense. However, for selected frequencies (such as power line harmonics) application of the filter bank can be very effective. The simplified frequency characteristic is shown in Fig. 3.59.

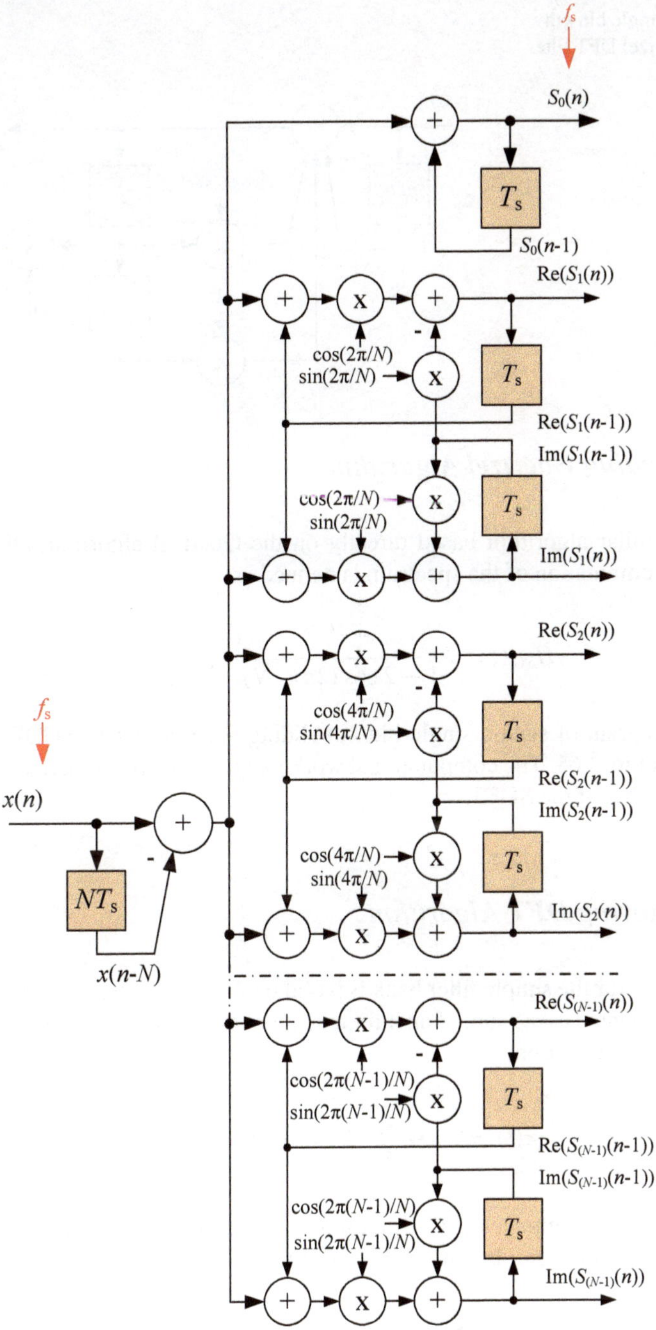

Fig. 3.64 Block diagram of N—channel SDFT analysis filter bank

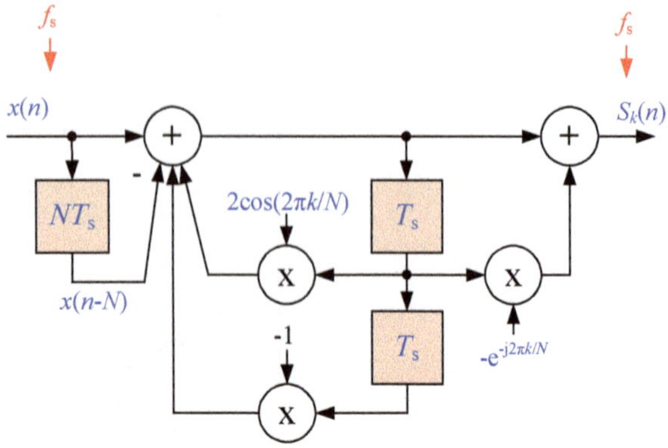

Fig. 3.65 Single bin kth sliding Goertzel DFT filter

3.7.4 *Sliding Goertzel Algorithm*

Another similar algorithm based directly on the Goertzel algorithm allowing for recursive computation of the spectrum is defined as

$$H_{SG}(z) = \frac{(1 - e^{j2\pi k/N} z^{-1})(1 - z^{-N})}{1 - 2\cos(2\pi k/N)z^{-1} + z^{-2}} \quad . \tag{3.55}$$

A block diagram of such a single bin kth sliding Goertzel DFT (SGDFT) filter is depicted in Fig. 3.65. The computational workload of the sliding Goertzel DFT filter is less than that of the SDFT.

3.7.5 *Moving DFT Algorithm*

Another idea for the simple filter bank is based on the Fourier series. Periodic signal can be represented as the sum of an infinite number of sinusoidal components, which is described by equation

$$x(t) = X_0 + \sum_{k=1}^{\infty} X_k \sin(2\pi kt + \varphi_k) \quad , \tag{3.56}$$

where: X_0—DC component, X_k—amplitude of kth component, φ_k—phase (argument) of kth component. Equation (3.56) can be rewritten

$$x(t) = X_0 + \sum_{k=1}^{\infty} (A_k \cos(2\pi kt) + B_k \sin(2\pi kt)) \quad , \tag{3.57}$$

and

$$\begin{aligned} X_0 &= \frac{1}{T}\int_0^T x(t)\mathrm{d}t \\ X_k &= \sqrt{A_k^2 + B_k^2} \\ \varphi_k &= \arctan\frac{A_k}{B_k} \quad . \end{aligned} \tag{3.58}$$

Coefficients A_k and B_k are determined by equations

$$\begin{cases} A_k = \dfrac{2}{T}\int_0^T x(t)\cos(2\pi kt)\mathrm{d}t \\ B_k = \dfrac{2}{T}\int_0^T x(t)\sin(2\pi kt)\mathrm{d}t \quad , \end{cases} \tag{3.59}$$

Moving Fourier transform

$$x(t) = X_0 + \sum_{k=1}^{\infty} X_k(t)\sin(2\pi kt + \varphi_k(t)) \quad , \tag{3.60}$$

where: X_0—DC component, $X_k(t)$—amplitude of kth component, $\varphi_k(t)$—phase (argument) of kth component.

$$x(t) = X_0 + \sum_{k=1}^{\infty}(A_k(t)\cos(2\pi kt) + B_k(t)\sin(2\pi kt)) \quad , \tag{3.61}$$

where:

$$\begin{cases} A_k(t) = \dfrac{2}{T}\int_{t-T}^{t} x(t)\cos(2\pi k\tau)\mathrm{d}\tau \\ B_k(t) = \dfrac{2}{T}\int_{t-T}^{t} x(t)\sin(2\pi k\tau)\mathrm{d}\tau \quad , \end{cases} \tag{3.62}$$

Discrete version of moving DFT transform (MDFT)

$$x(n) = X_0 + \sum_{k=1}^{N} X_k(n)\sin(2\pi kn/N + \varphi_k(n)) \quad , \tag{3.63}$$

and

$$\begin{cases} A_k(n) = \dfrac{2}{N} \sum\limits_{n=k-N+1}^{k} \cos(2\pi kn/N) \\ B_k(n) = \dfrac{2}{N} \sum\limits_{n=k-N+1}^{k} \sin(2\pi kn/N) \quad , \end{cases} \tag{3.64}$$

The value of coefficients $A_k(n)$ and $B_k(n)$ can be calculated by recursive equations

$$\begin{cases} A_k(n) = A_k(n-1) + \dfrac{2}{N}(x(n) - x(n-N)) \cos(2\pi kn/N) \\ B_k(n) = B_k(n-1) + \dfrac{2}{N}(x(n) - x(n-N)) \sin(2\pi kn/N) \quad . \end{cases} \tag{3.65}$$

Finally the kth component is determined by

$$y_k(n) = A_k(n)\cos(2\pi kn/N) + B_k(n)\sin(2\pi kn/N) \quad . \tag{3.66}$$

A block diagram of a moving DFT analysis filter bank for one component is depicted in Fig. 3.66. Frequency response for such a filter for $k = 1$, $N = 32$ and $f_s = 1600\,\text{Hz}$ is shown in Fig. 3.67. In this case the passband center frequency is 50 Hz, the gain is equal to 1, the phase shift of 50 Hz is equal to 0. Just like the previous filter banks based on the DFT algorithm, the circuit does not have very good filtration properties and is better suited for harmonic filtering in systems with coherent sampling.

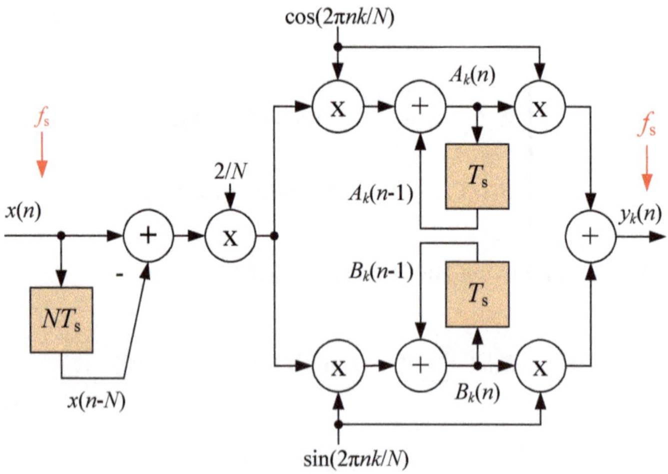

Fig. 3.66 Block diagram of moving DFT filter

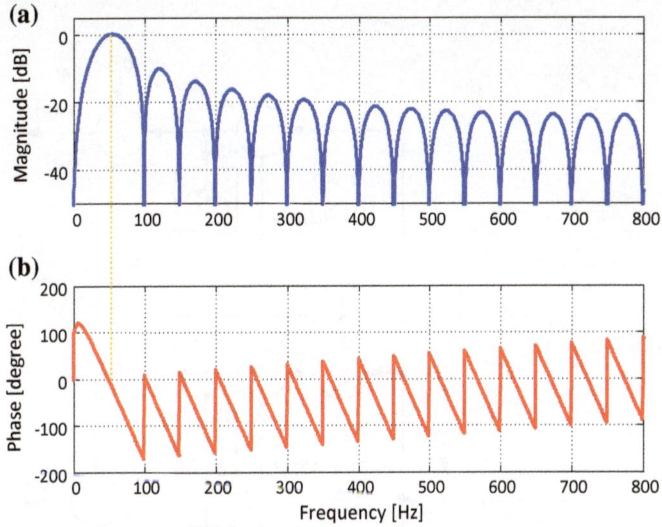

Fig. 3.67 Frequency response of MDFT filter for: $k = 1$, $N = 32$, $f_s = 1600\,\text{Hz}$: **a** magnitude, **b** phase

Using the filter in Fig. 3.66 it is possible to build an analysis filter bank for selected components. A block diagram of an N-channel moving DFT analysis filter bank is depicted in Fig. 3.68.

3.7.6 Wave Digital Lattice Filter Bank

Lattice wave digital filters are very well suited for building a filter bank. Figure 3.69 depicts analysis and synthesis filter banks. A potential connection between the filter banks is indicated by a dotted line. Especially attractive are filter banks using bireciprocal lattice wave digital filters. A two-channel analysis filter bank is used to separate signal into two subband signals for $f/f_s = 0.25$. A subband coding filter bank consists of an analysis filter bank followed by a synthesis filter bank. The analysis and synthesis filter banks are maximally decimated filter banks. Figure 3.70a, b shows the two channel version, also called the quadrature mirror filter (QMF) bank. If the filter banks are connected there is a following relation between sampling rates $f_{s1}/2 = f_{s2}$. The corresponding synthesis filter bank recombines the subband signals to obtain the original signal again. The lattice wave digital filter bank with recovery of effective pseudopower [22] has additional advantages: greater dynamic range, low level of rounding noise, and broader singing margin under looped conditions. The pairs of two complementary filters are shown in Fig. 3.70.

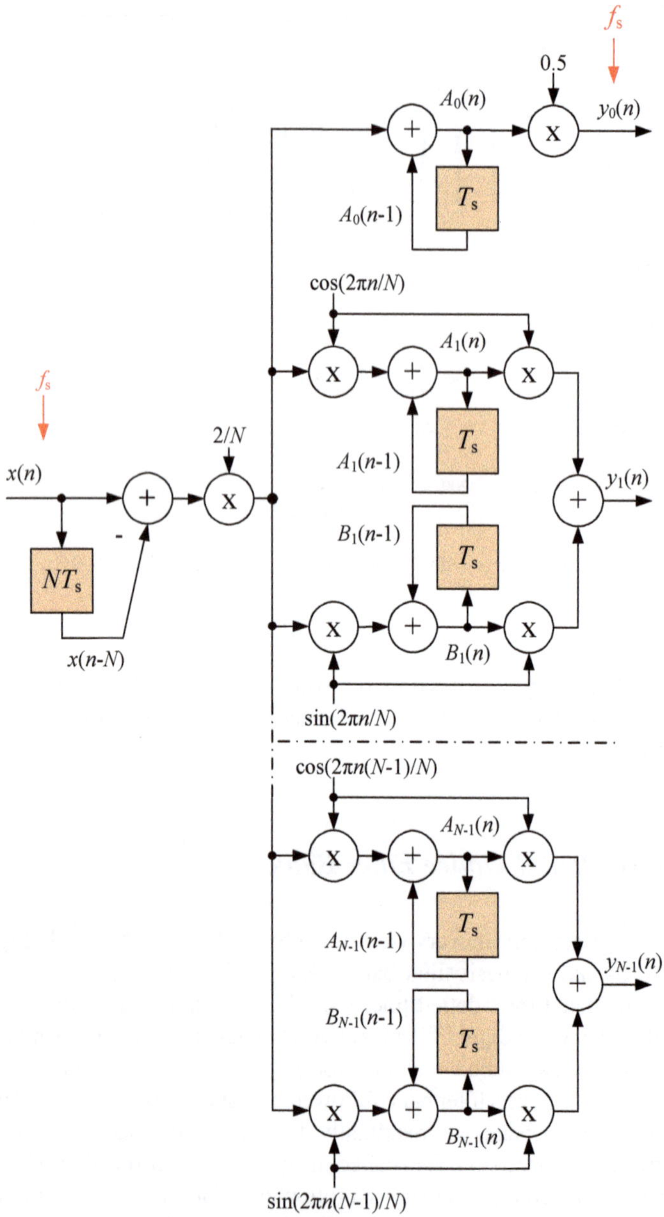

Fig. 3.68 Block diagram of N—channel moving DFT analysis filter bank

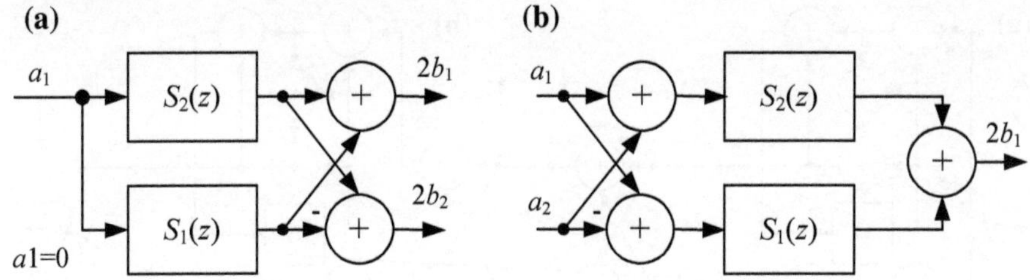

Fig. 3.69 Lattice wave digital filter banks: **a** analysis filter bank, **b** synthesis filter bank

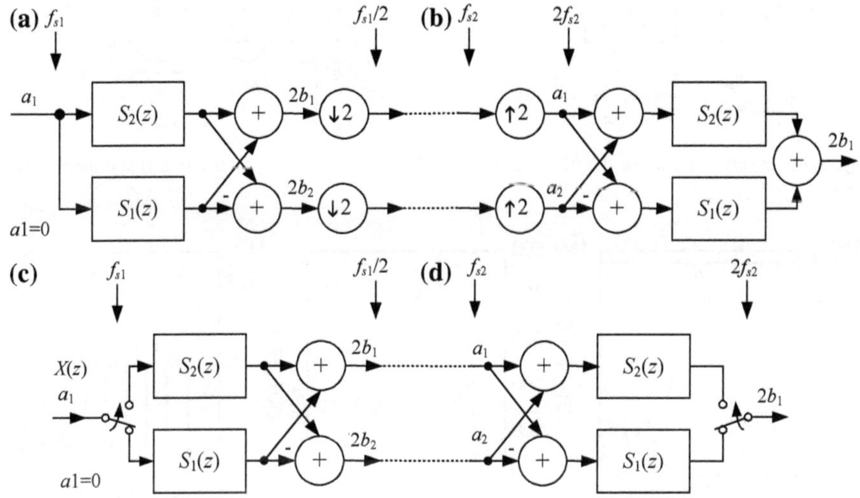

Fig. 3.70 Filter banks with bireciprocal lattice wave digital filters: **a** analysis filter bank with decimation, **b** synthesis filters with interpolation, **c** polyphase analysis filter bank, **d** polyphase synthesis filter bank

Figure 3.70c, d shows polyphase implementation of the filter banks. In these filters downsampler and upsampler are realized by simple switches.

Using the methods presented by Gazsi [37] it is possible to design bireciprocal lattice wave digital filter with coefficients useful for implementation for low resolution fixed point arithmetic. One example is 7-order elliptic filter with binary values of coefficients $\gamma_1 = 0.01011001$b, $\gamma_3 = 0.010001$b, $\gamma_5 = 0.00011$b, and $\gamma_1 = 0.34765625$, $\gamma_2 = 0.09375$, $\gamma_3 = 0.265625$ in decimal code respectively. This filter is used for building analysis and synthesis filter banks. A diagram with full details of the filter banks with recovery of effective pseudopower are shown in Fig. 3.71. The transfer functions of filter branches are:

$$S_2(z) = \frac{\gamma_1 + z^{-1}}{1 + \gamma_1 z^{-1}} \quad , \quad S_1(z) = \frac{\gamma_3 + z^{-1}}{1 + \gamma_3 z^{-1}} \frac{1 - \gamma_5 + z^{-1}}{1 + (1 - \gamma_5) z^{-1}} \quad . \tag{3.67}$$

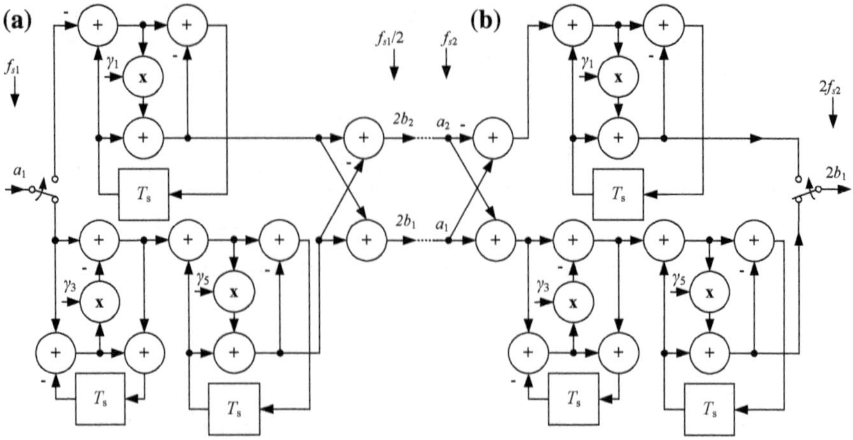

Fig. 3.71 Block diagram of the filter banks using BWDF: **a** analysis filter bank, **b** synthesis filter bank

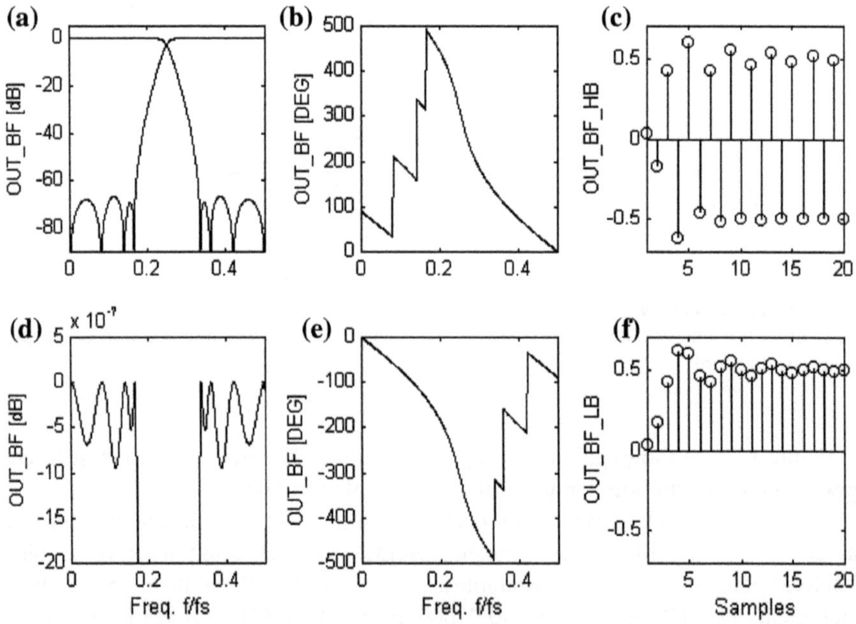

Fig. 3.72 Frequency responses of the synthesis lattice filter: **a** amplitude characteristics of low-pass and high-pass inputs, **d** passband low-pass and high-pass inputs, **b** phase characteristics of high-pass input, **e** phase characteristics low-pass input, amplitude responses on unit step input signal: **c** high-pass input, **f** low-pass input

The main advantages of the filter are: simplicity, the filter speed is decreased by the decimation fold $M = 2$. For realization of the complete lattice filter bank only seven multipliers and twenty two adders are needed. Two main versions of the filter realization are possible: fixed-point and floating-point. Using filter coefficients in the binary form in the fixed-point version, it is possible to replace the multiplier by a shifter, making it very simple and quick. For these filter banks the author made a simulation using the MATLAB program. A block diagram of the synthesis digital lattice filter is shown in Fig. 3.71b. For measuring filter frequency response, the unit impulse method was applied. The impulse responses for both of the filter inputs were calculated. In the low-pass input checking procedure was applied on the filter input a_2, a block with zero samples and on the second input a_1, a block with unit impulse signal (a single nonzero sample). During the high-pass input checking procedure the input was swapped. For every response $N = 2048$ samples were stored and after that, the FFT of both results was calculated. The results of these are in Fig. 3.72. The amplitude characteristics of both inputs are shown in Fig. 3.72a, and passband characteristics are shown in Fig. 3.72d. The amplitude frequency responses are mirror images of each other about $f/f_s = 0.25$. The passband losses are very small, close to 0.003 dB. The phase characteristics of the high-pass input and of the low-pass input are shown in Fig. 3.72b, e respectively.

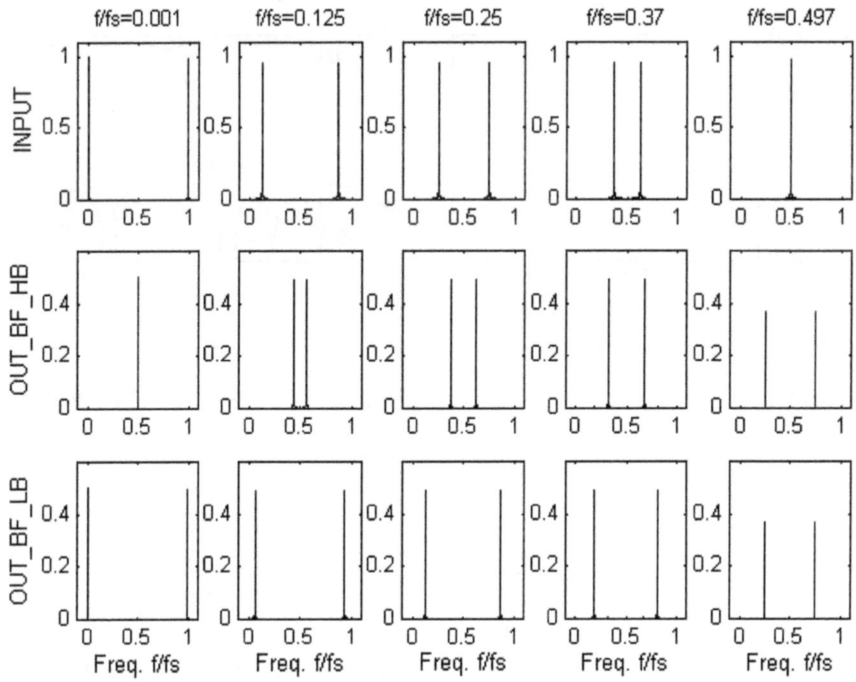

Fig. 3.73 Frequency amplitude responses of the synthesis filter bank on high-pass and on low-pass sinusoidal input signals with frequencies f/f_S: 0.001, 0.125, 0.25, 0.37, 0.497

To prepare the amplitude synthesis filter response for a unit step, one of the synthesis inputs was excited with a unit step pulse signal and another with zero samples. For preparing a second response the inputs were swapped. The results of this are shown in Fig. 3.72. The result of low-pass input (Fig. 3.72f) is typical but the response of the high-pass input (Fig. 3.72c) is different than that expected, after transient responses to the filter output, signal with frequency $f = 0.5 f_s$ occurs. This effect is explained in Fig. 3.73 in which the frequency amplitude responses of the synthesis filter bank on high-pass and on low-pass sinusoidal input signals are shown. In the first row are the FFT of input signals, in the second row are FFT of the responses of high-pass input and in the third row are FFTs of the responses of low-pass input. The block diagram of the analysis digital filter bank is shown in Fig. 3.71a. This filter was implemented in the program MATLAB too. For the analysis filter bank, the quickest measuring method using the unit input impulse can not be applied because the downsampler is placed in this case in front of the other filter elements and the impulse would affect only one branch of the filter. For checking filter characteristics, a method with sinusoidal signal on the input and maximum amplitude detector on the output was used. This method was applied to the frequency range of $f/f_s = 0{-}0.5$ with step 2/N using $N = 1024$ sample block. The first 50 samples of every response were zeroed for damping transient distortion. The

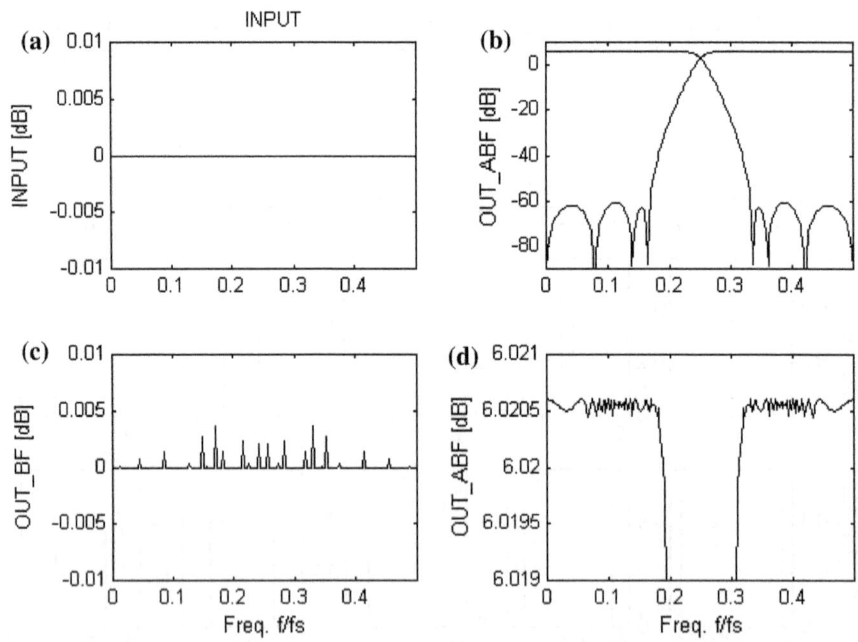

Fig. 3.74 Frequency characteristics of the digital lattice filter bank obtained in sinusoidal input signal with amplitude detector on the outputs: **a** input signal, **b** amplitude of analysis filters, **c** amplitude of the filter bank output, **d** passband of analysis filters

obtained amplitude characteristics are shown in Fig. 3.74. The characteristics of the analysis filter outputs are shown in Fig. 3.74b. The amplitude frequency responses are similar to synthesis filter characteristics, and are mirror images of each other about the $f/f_s = 0.5$. The passband characteristics are shown in Fig. 3.74d, the passband losses are near to 0.0001dB. In Fig. 3.74c the output signal of the synthesis filter is a reconstructed version of the input signal. The reconstruction error in this method is less than 0.005dB. In the second method the same input signal was applied, but the amplitude detector was replaced by a FFT. For every sinusoidal signal the maximum amplitude of a FFT response was found. This method was applied for the frequency range of $f/f_s = 0{-}0.5$ with step 4/N using $N = 2048$ sample block. The characteristics obtained (Fig. 3.75) are similar to those in the above methods. Analysis filter bank can be tested by unit impulse if the test circuit will use a version of the bank without the input switch, as shown in Figs. 3.47 and 3.70a.

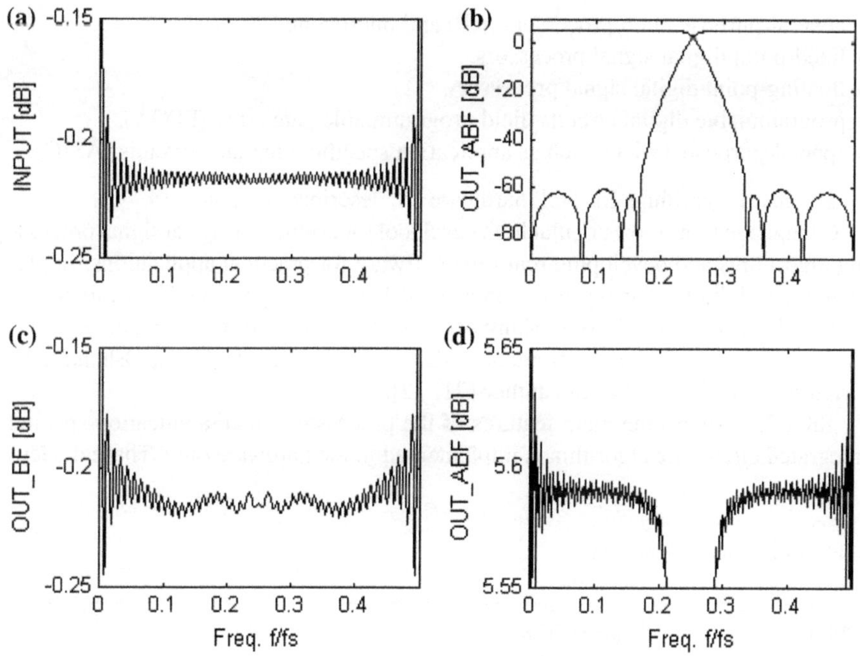

Fig. 3.75 Frequency characteristics of the digital lattice filter bank obtained in sinusoidal input signal with FFT on the outputs: **a** input signal, **b** amplitude of analysis filters, **c** amplitude of the filter bank output, **d** passband of analysis filters

3.8 Implementation of Digital Signal Processing Algorithms

Many power electronics control circuits have constraints on latency; that is, for the system to work, the control circuit operation must be completed within some fixed time, and deferred processing is not viable. Therefore for such application the requirements for the control system are the highest. A specifically optimized architecture for digital signal processing calculation is s feature of the digital signal processor (DSP). Most general-purpose microprocessors can execute digital signal processing algorithms successfully, but they are not designed for the intensive calculations and use of them requires much greater hardware and software resources than using a DSP. However, manufacturers of microprocessors are continually modifying their products to approaching the capabilities of digital signal processors. Some important features of a DSP are described below. Currently for digital signal processing computation it is possible to consider five main types of digital devices:

- general-purpose microprocessors (μP) and microcontrollers (μC),
- fixed-point digital signal processors,
- floating-point digital signal processors,
- programmable digital circuits, field programmable gate array (FPGA),
- special-purpose devices such as application specific integrated circuits (ASIC).

The main algorithms for DSP hardware are described in Table 3.14

C language is the most popular high-level tool for evaluating digital signal processing algorithms and developing real-time software for practical applications. Implementation of digital signal processing using C language is presented by Embree and Kimble [27], Press et al. [60]. Many aspects of digital algorithm implementations using digital signal processors are considered by: Wanhammar [86], Oshana [55], Orfanidis [54], Bagci [11] and author [71, 72].

Table 3.15 shows the main features of the processors. In an application-specific integrated circuit the algorithm is implemented in the hardware only. These devices

Table 3.14 Main DSP algorithms

Algorithm	Formula
FIR filter	$y(n) = \sum_{k=0}^{N} b_k x(n-k)$
IIR filter	$y(n) = \sum_{k=0}^{N} b_k x(n-k) + \sum_{k=1}^{M} a_k y(n-k)$
Discrete convolution	$y(n) = \sum_{k=0}^{N-1} x(k)h(n-k)$
Correlation	$y(n) = \sum_{k=0}^{N-1} w(k)x(n+k)$
DFT	$Y(n) = \sum_{k=0}^{N-1} x(k)(\cos(2\pi nm/N) - \mathrm{j}\sin(2\pi nm/N))$

Table 3.15 Summary of DSP hardware implementation

	μP and μC	Fixed-point DSP	Floating-point DSP	FPGA	ASIC
Flexibility	Programming	Programming	Programming	Programming	None
Processing speed	Low-medium	High	High	High	High
Support for multiplication and accumulation	None-rare	Yes	Yes	Possible	Possible
Reliability	Medium-high	High	High	Medium	High
Resolution	Low	Low-medium	Medium-high	Low-medium	Medium
Additional peripheries: counters, PWMs, A/Ds	Yes	Possible	Possible	Yes	Yes
Design time	Medium-long	Short	Short	Medium	Long
Power consumption	Low	Low	Medium-high	Low-medium	Low
Design cost	Low-medium	Low	Low	Medium	High
Unit cost	Low-medium	Low-medium	Medium-high	Low-medium	Low

are designed to perform a fixed-function or set of functions. These devices run exceedingly fast in comparison to a programmable solution, but they are not as flexible. If the algorithm is a stable and well defined function that needs to run really fast with low power consumption, an ASIC may be a good solution.

Field-programmable gate arrays are programmable digital devices and it is possible to reprogram them in the field. These devices are not as flexible as microprocessors. FPGA producers have prepared special libraries for the implementation of digital signal processing algorithms.

General purpose microprocessors and microcontrollers are the most versatile solution. Such a solution is now available for a lot of μP and μC families, as are many integrated software tools for their programming. The disadvantage of universal processors is their poor computing performance for signal processing applications.

Among the many publications on DSPs and implementation of digital signal processing algorithms, special mention can be given to books written by: Chassaing [14, 15], Kuo and Lee [43], Dahnoun [24], Wanhammar [86], Orfandis [53, 54], Dabrowski [20]. Below the features of digital signal processors will be further presented.

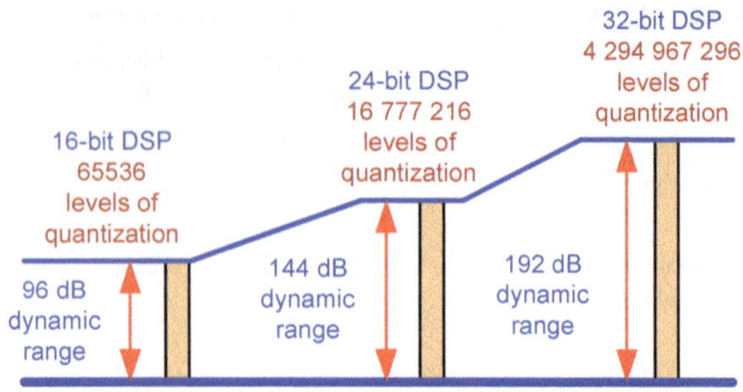

Fig. 3.76 Dynamic ranges of fixed-point digital signal processors

3.8.1 *Basic Features of the DSP*

In this section are discussed the major hardware components which allow a very efficient implementation of digital signal processing algorithms. These elements are not usually found in universal microprocessors and must be replaced by additional software. For a fixed-point DSP the number of bits determines the dynamic range of signal processing. Figure 3.76 shows the dynamic range of a typical fixed-point DSP.

3.8.1.1 Multiplication and Accumulation

As shown in Table 3.14 in digital signal processors the multiply-accumulate (MAC) operation is a basic operation that computes the product of two numbers and adds that product to an accumulator. The MAC operation is described by equation

$$y(n+1) = a(n)x(n) + y(n) \quad . \tag{3.68}$$

The block diagram of a typical DSP multiplier with accumulator (MAC) is shown in Fig. 3.77, with input operands $x(n)$ and $a(n)$ having b-bit resolution, and the output having $2b$-bit resolution. Accumulation is made using $2b$-bit resolution and finally this makes it possible to calculate the basic Eq. (3.68) with better accuracy than b-bit

$$\overbrace{y(n+1)}^{2b-bit} = \overbrace{\overbrace{a(n)}^{b-bit}\,\overbrace{x(n)}^{b-bit} + \overbrace{y(n)}^{2b-bit}}^{2b-bit} \quad . \tag{3.69}$$

This is a method which extends the dynamic range while keeping the cost of the system and its power consumption within reasonable limits. This kind of multiplication with accumulation is a typical solution in DSP. Furthermore, although it used to be rarely embedded in cores of microprocessors and microcontrollers, nowadays it is more and more commonly used in this way. It is worth noting that it is also possible to implement MAC operations with FPGA and ASIC circuits.

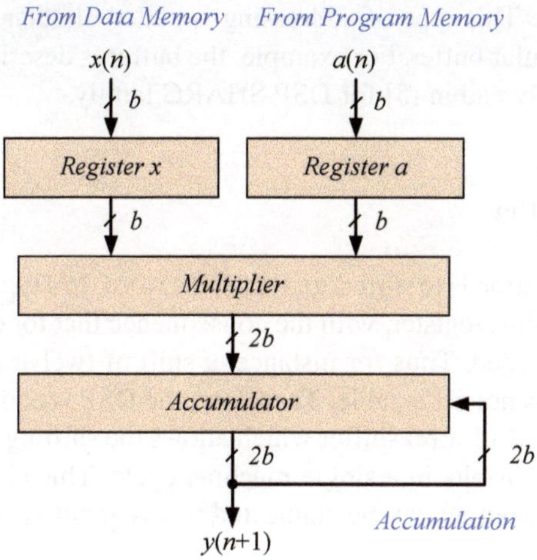

Fig. 3.77 Block diagram of multiplier with accumulator

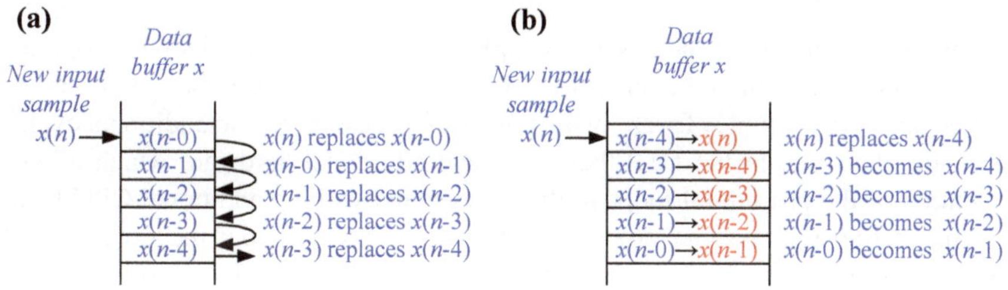

Fig. 3.78 Implementation of delay line: **a** with shifting of samples, **b** with pointer manipulation using circular addressing

3.8.1.2 Circular Addressing

In most digital signal processing algorithms convolution is one the major algorithms, and therefore moving samples in the buffer is one of the basic operations. Figure 3.78 shows the two basic variants of such an operation. In the first one (Fig. 3.78a), all samples are shifted in the data buffer. This is a very uneconomical solution, because in moving the samples the operation consumes processor time unproductively. In a better solution (Fig. 3.78b) only the pointer of the beginning of the buffer is modified, without having to move the samples. Circular addressing uses pointer manipulation to add the new samples to the buffer by overwriting the oldest available samples hence reusing the memory buffer. When the pointer reaches the last location of the delay line it needs to wrap back to the beginning of the line. This solution is widely used in digital signal processors and it is supported by appropriate hardware that allows the performing of operations on the addresses in the background, parallel to the main program. This type of addressing is called circular addressing and the buffer is called a circular buffer. For example, the buffer is described in [62] for DSP TMS320C6000 family and in [5] for DSP SHARC family.

3.8.1.3 Barrel Shifter

The next type of operation is to shift digital bits in a word. In a typical microprocessor it is performed by a shift register, with the consequence that for each shift of one bit one clock cycle is needed. Thus for instance, a shift of twelve bits requires twelve clock cycles, which is not acceptable. Therefore, the DSP is equipped with a matrix shift system also called a Barrel shifter which allows the shifting or rotating of a data word by any number of bits in a single machine cycle. This is implemented like a multiplexer, and each output can be connected to any input, depending on the shift distance.

3.8.1.4 Hardware-Controlled Loop

For subsequent operations necessary to implement convolution the loop is necessary for the implementation of the repetitions. In a typical microprocessor operations are carried out by repetition software, which makes it necessary to add additional cycles to handle the loop. In the DSP there is added additional hardware which allows execution of the procedural loops in the background with no additional CPU load. As a result, only digital signal processing operations are executed in the loop, without additional losses associated with handling machine cycles of the loop. This feature is also called zero overhead looping—using dedicated hardware to take care of counters and lattices in loops.

3.8.1.5 Saturation Arithmetic

Another important feature of the calculation unit used in control systems is saturation arithmetic. This is important for fixed-point arithmetic. The system should behave like an analog one, when the signal reaches the lower or upper limit. In analog circuits the lower and the upper limit is equal with respect to negative supply voltage and positive supply voltage. In typical ALU, a change of the signal sign can occur when the signal overflows, which may result in serious consequences for the control circuit. Therefore programmers must ensure that the signal limit is not exceeded, which often requires extra programming effort. A simple program for adding two variables with checking overflows and limiting the signal output is presented in Listing 3.6.

Listing 3.6 Software saturation

```
# define max 0x7fff
# define min 0x8000
int x,u,y;
long y_temp;
...
y_temp = x + u;
if y_temp > max
   y = max;
else if y_temp < min
   y = min;
else
   y=(int)y_temp;
```

The DSP has additional hardware for arithmetic saturation, typically controlled by a bit in a special control register. Therefore, when arithmetic saturation is switched on, the programmer does not have to worry about checking the signal overflows and the processor does not need any additional program for checking overflows. The program for adding two variables is simple as shown in Listing 3.7.

Listing 3.7 Hardware saturation

```
# define max 0x7fff
# define min 0x8000
int x,u,y;
...
y = x + u;
```

Figure 3.79 shows a signal for incrementing $y = y + 1$, for a 16-bit two's complement binary code (U2), when the signal value does not exceed the value 7FFF. It is similar with a decrementing signal $y = y - 1$, when the value of the signal does not exceed the value 8000.

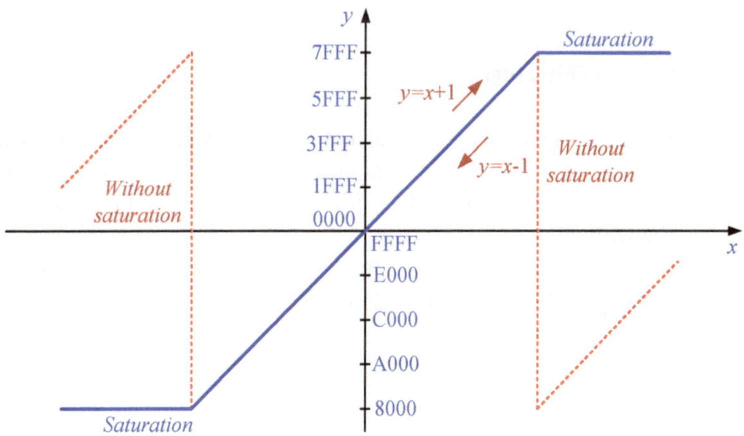

Fig. 3.79 Arithmetic saturation

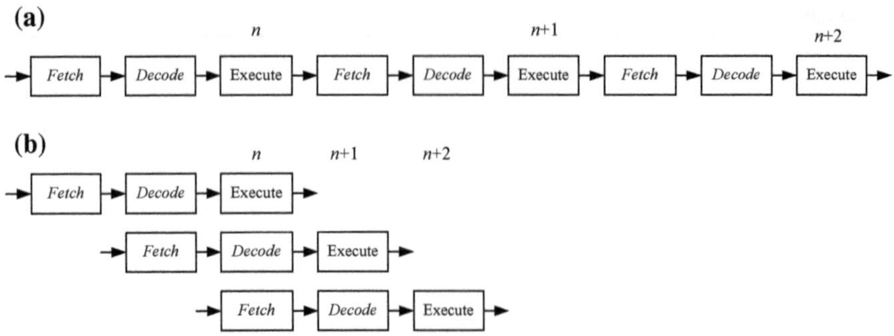

Fig. 3.80 Instruction execution: **a** without pipeline, **b** with pipeline

3.8.1.6 Pipelined Architecture

In typical processors, instruction execution consists of three phases: fetch, decode and execute. This process is shown in Fig. 3.80a, where execution of the instruction needs three processor cycles. In the DSP there is a well developed operational parallelism, to accelerate the processor operations, with simultaneously performed fetch, decode and execute operations as shown in Fig. 3.80b. With this modification processors can work three times faster, for the linear flow of the program. Of course, in the event of a program branch, the whole effect is lost. Therefore, in the digital signal processor, delayed branches have been introduced, which allow the use of fetched and decoded instructions. The DSP pipelines are even more sophisticated and powerful, in that they allow also reduction of efficiency losses in the processor due to branches, hardware-controlled loop, interrupts etc.

3.8.1.7 Parallel Architecture

Nowadays DSPs are designed with parallel architecture, which allows for executing more than one operation at a time. For example, while one instruction is performing a multiplication and accumulation another instruction can be moving data and other resources on the DSP chip.

Another way of achieving parallel architecture is to use a processor with multiple cores. This technique is most commonly used in ARM microprocessors. A limitation of this approach is that it requires a careful division of digital signal processing into separate threads, which is not always easy and may sometimes even not be achievable.

3.9 Selected Microcontrollers Suitable for Power Electronics Control Circuits

On the market there are now very many different processors and it is very hard to choose the right one. In the author's opinion the most important are the following features:

- separate program and data memories (Harvard architecture), allows the DSP to fetch code without affecting the performance of the calculations,
- pipelined architecture, with DSPs executing instructions in stages so more than one instruction can be executed at a time. For example, while one instruction is doing a multiply another instruction can be fetching data with other resources on the DSP chip,
- single cycle operation,
- multiplier with extended resolution of accumulation, multiplier-accumulator (MAC unit),
- barrel shifter—single cycle matrix shifter,
- hardware-controlled looping, to reduce or eliminate the overhead required for looping operations,
- memory-address calculation unit, hardware-controlled circular addressing,
- saturation arithmetic, in which operations that produce overflows will accumulate at the maximum (or minimum) values,
- parallel architecture, parallel instruction set,
- support for fractional arithmetic,
- efficient and fast interrupt system.

In contrast to what general purpose processors provide, the above features enable rapid execution of calculations with adequate precision. Table 3.16 shows the selected DSP useful for the implementation of a control circuit for power electronics circuits.

Figure 3.81 depicts a digital control circuit for the three-phase circuit. Typically analog signals representing three-phase voltages and currents are sampled. As has already been proven in Chap. 2, the best solution is to use simultaneous sampling of signals [25]. The application of A/D converters with successive approximation eliminates the time delays of the signals. While the 16-bit resolution allows for a large dynamic range signals. While the 16-bit resolution of A/D converters allows for a large dynamic range signals. Then, the data from the A/D converters go to the digital signal processor (DSP), wherein the control algorithm is implemented. Basic features of the DSP are described above. The calculation results are then sent to the PWM modulators, which, through systems that generate dead time control the transistors of the inverter. As before, the 16-bit resolution provides a large dynamic range of the signals. In addition, the system should be also equipped with a digital phase locked loop (PLL) circuit (to synchronize with the mains supply), binary input/output and communication interfaces. To supervise the operation of the entire system is essential watchdog too. It would be good that could be able to fulfill such a control system using a single microcontroller, however, by the author's knowledge such system is

Table 3.16 Selected DSP suitable for power electronics control circuits

Name	TMS320F283x	TMS320C67x	SHARC	MC56F84xxx
Architecture	Harvard	Floating-point VLIW	Enhanced Harvard	Dual Harvard
Fixed-point	32-bit	32-bit	32/64-bit	32-bit
Floating-point	IEEE single-precision	IEEE double-precision	32/40-bit IEEE	No
PWM	18 PWM, 150 ps	Yes	16 PWM	24 PWM, 312 ps
A/D	2 × 12-bit, 80 ns, 2 × SH	None	None	2 × 12-bit high speed
MAC	32 × 32 bit or dual 16 × 16 bit MAC	Two ALUs fixed-point, 32-bit fixed-point multipliers with 64-bit product	80-bit accumulation	32 × 32-bit with 32-bit or 64-bit result
Shifter	Barrel	Barrel	Barrel	32-bit
Signal processing performance	300 MHz 600 MFLOPS	350 MHz 2100 MFLOPS	450 MHz 2700 MFLOPS	100 MHz 100 MIPS
ROM	256 k × 16 flash memory	No	Up to 4 Mb	256 kB
RAM	34 k × 16	256 k	1–4 Mb	Up to 32 kB
Input/output	88 GPIO pins	Yes	Up to 16-bit	General purpose I/O
Special addressing modes	Circular addressing	Yes	32 hardware circular buffer	Parallel instruction set with unique DSP addressing modes
Hardware loop	Yes	Yes	Six nested levels of zero-overhead looping in hardware	Hardware DO and REP loops
Producer	Texas Instruments	Texas Instruments	Analog Devices	Freescale

not currently available. The following microcontrollers more or less close to the presented system. The closest to the ideal is microcontroller TMS320F28377D.

It should also be noted that there is also a solution that uses a microprocessor and field-programmable gate array (FPGA). This can be achieved by using two separate systems or FPGA with built-in microprocessor. However, discussion of such systems is beyond the scope of this book.

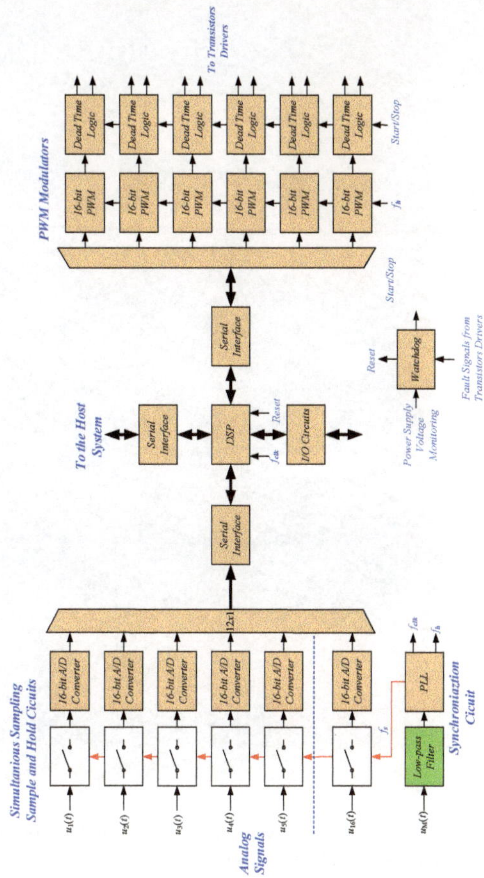

Fig. 3.81 Block diagram of control circuit for three-phase power circuit

3.9.1 TMS320F28335

One of the most interesting DSP families is the TMS320F28x family, also called digital signal controller (DSC), from Texas Instruments [75, 76]. A typical representative of this family is the TMS320F28335 system. It is a complete system with many useful features in a single silicon chip. Therefore it is especially good for power electronics applications. The core of the processor consists of an IEEE-754 single-precision floating-point unit. Especially useful features for power electronics applications are: 16-channel 12-bit A/D converter with 80-ns conversion rate and two sample-and-hold circuits, 18 PWM outputs, clock and system control with dynamic PLL ratio changes, 256K × 16 flash memory and 34K × 16 SARAM memory. The instruction cycle of the processor is equal to 6.67 ns for a processor clock rate equal to 150 MHz. For simplifying the design process a ControlCARD module with TMS320F28335 is used. The ControlCARD is a small 100 pin DIMM (dual in line "memory module") style vertical plug-in board based on the F28335. These ControlCARDs have all the

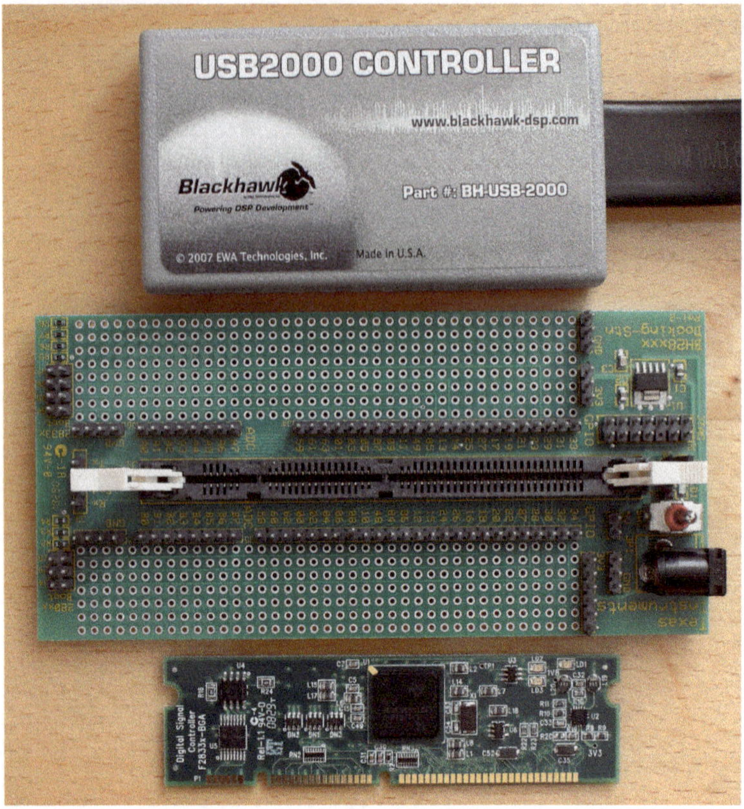

Fig. 3.82 System for developing software and hardware: F28335 control card, CC28xxx docking station and USB2000 Controller

necessary circuits: clock, supply LDO, decoupling, pull-ups, etc., to provide reliable operation for the DSC device (Fig. 3.82). This reference design is very robust and is meant for operation in noisy electrical environments (especially important in power electronics). It includes the following features:

- all general purpose input/output (GPIO), A/D converter and other key signal routed to gold edge connector fingers,
- clamping diode protection at A/D converter input pins,
- anti-aliasing filter (noise filter) at A/D converter pins,
- galvanic isolated UART communications.

The docking station is a very basic small mother board which accepts any member of the plug-in control card family. It provides the required 5 V power supply and gives the user access to all the GPIO and ADC signals. The system for developing software is shown in Fig. 3.82. It consists of the F28335 control card, CC28xxx docking station and USB2000 Controller or XDS100 USB—JTAG emulators. Thanks to using the emulator it is possible to develop the software comfortably. The emulator has access to all the processor registers and memories and its use makes it possible to program internal flash memory. For software development Code Composer Studio™ v7 is used. Additionally Texas Instruments has prepared a lot of supporting tools, such as controlSUITE™, Baseline Software Setup, DSP2833x Header Files, etc. For this processor family there have been created very useful teaching tools [76, 79].

It should also be noted that the TMS320F2000 family has the support of the PSIM and MATLAB. The MATLAB and PSIM can generate code that is ready to run on the hardware boards based on Texas Instruments TMS320F2000 family.

3.9.2 TMS320F2837xD

Texas Instruments continues to develop the TMS320F2000 processor family. Currently, the best equipped of its family member is TMS320F2837xD [80, 81]. In respect of the TMS320F28335, the TMS320F28377D has the following important improvements:

- two CPU and two control law accelerator (CLA) real-time control co-processors,
- 800 MIPS,
- 1 MB Flash, 204 KB SRAM,
- 24 PWM channels,
- 16 high-resolution pulse width modulator (HRPWM) channels,
- 4x A/D converters, single sample-and-hold (SH) on each A/D converter,
- 16-bit/12-bit modes A/D converters,
- 1.1 MSPS per ADC for 16-bit mode, 3.5 MSPS per ADC for 12-bit mode,
- USB 2.0 (MAC + PHY),
- 169 individually programmable, multiplexed general-purpose input/output.

Such microcontrollers are getting closer to the ideal, but it should be noted that with the increase of their functions so does the number of pages of description increase. Typical full description of these microcontrollers has a volume of several thousand pages. Texas Instruments is trying to remedy this problem by creating additional applications, such as controlSUITE™ and powerSUITE.

3.9.3 Digital Signal Processor—TMS320C6xxx Family

The Texas Instruments TMS320C6000 family is a high performance fixed and floating point DSP range. Those processors are characterised by high speed and multiple arithmetic units, which can operate simultaneously. As a result they achieve high performance. For example, a low cost member of this family with fixed/floating point achieves 3648 MIPS (million instructions per second) at 456 MHz and 2746 MFLOPS (million floating-point instructions per second) at 456 MHz [77].

Fig. 3.83 DSP TMS320C6713 evaluation module with 16-bit A/D converter ADS8364 evaluation module

This high speed of calculation is possible to obtain through the use of parallel arithmetic units and the use of very long instruction words (VLIW), 256 bits. The DSP core consists of two data paths with four functional units. So there are eight parallel functional units. To take advantage of all the arithmetic units great care is needed when writing programs. Creating programs in assembly language is quite complex and in principle for this family programs are written in C. For this purpose Code Composer Studio has been developed, a highly efficient C compiler and an assembly optimizer, as well as an environment for creating optimal programs for parallel arithmetic units. However, programming these processors is more complicated in comparison to the SHARC processors. The TMS320C6000 family is described in Texas Instrument publications [77, 78], a very useful teaching manual [79], and in independent books [12, 14, 24]. Typically the TMS320C6000 family does not have particularly useful peripherals power electronic applications, so their use is possible only together with external peripherals. In the author's opinion there is no big problem with input signal acquisition. It is possible to find prepared modules from Texas Instruments. For example, Fig. 3.83 shows the DSP TMS320C6713 family evaluation module with a 6-channel, 16-bit simultaneous sampling A/D converter—ADS8364 evaluation module. This solution is supported by Texas Instruments software.

Now the family of processors TMS320C6000 migrates to the direction of multi-core solution. For example the TMS320C665x has up to two cores and TMS320C667x has up to eight cores.

3.9.4 Digital Signal Processors—SHARC Family

In the author's opinion, a classic and probably the most programmer-friendly is the SHARC DSP family from Analog Devices [1–4, 47]. This is because these processors have a very logical and clear structure. The assembler is a very simple and effective so-called algebraic assembler. A block diagram of the SHARC core is shown in Fig. 3.84 [5, 6]. The processor structure uses enhanced Harvard architecture, and consists of two sets of buses, one for data memory (DM) and a second for program memory (PM). PM and DM buses are capable of supporting 2 × 64-bit data transfers between memory and the core at every processor cycle.

Address buses are controlled by two address calculators DAG 1 and DAG 2. The DAGs are used for indirect addressing and implementing circular data buffers in hardware. Circular buffers allow efficient programming of delay lines and other

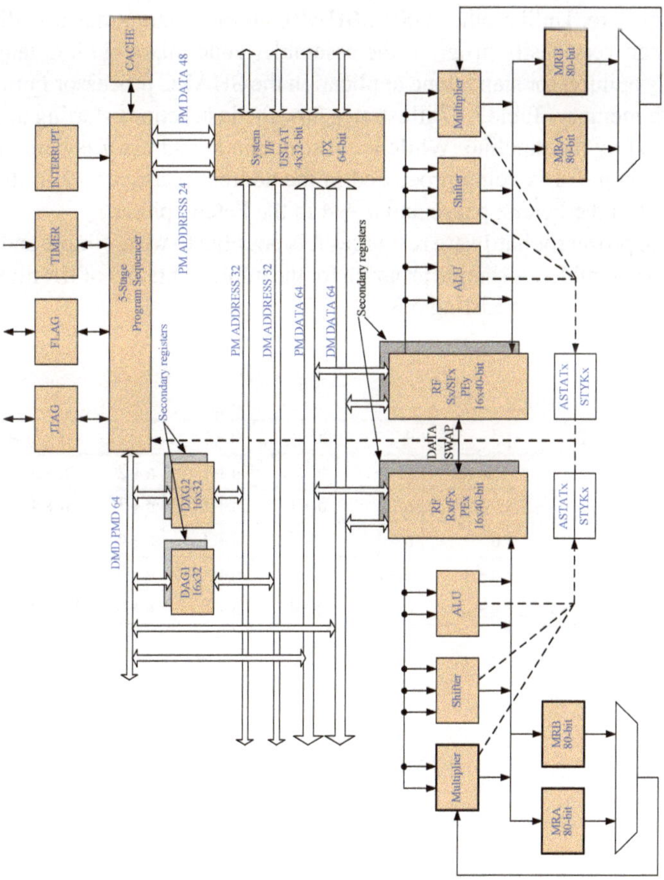

Fig. 3.84 Simplified block diagram of SHARC DSP core ADSP-21367/8/9

data structures required in digital signal processing, and are commonly used in digital filters (Table 3.14) and Fourier transform (Table 3.14). The two DAGs contain sufficient registers to allow the formation of up to 32 circular buffers (16 primary register sets, 16 secondary). The DAGs automatically handle address pointer wrap-around, reduce overheads, increase performance, and simplify implementation. The SHARC has two computation units (PEx, PEy), each of which comprises: an ALU, multiplier with Barrel shifter, and 16×40-bit data register file. These computation units support IEEE 32-bit single precision floating-point, 40-bit extended precision floating-point, and 32-bit fixed-point data formats. These units perform all operations in a single cycle. The three units within each processing element are arranged in parallel, maximizing computational throughput. Single multifunction instructions execute parallel ALU and multiplier operations. The processor includes an instruction cache that enables a three-bus operation for fetching an instruction and four data values. The cache is selective and only the instructions whose fetches conflict with PM bus data accesses are cached. The cache memory allows full-speed operation of processor core. Unlike other DSPs, SHARC has a programmer-friendly assembler, so it can very easily program the assembler code mixed with C language. In the author's opinion for standalone application the SHARC processor family should have a flash memory. Table 3.17 illustrates how the instructions starting at address n are processed by the pipeline. While the instruction at address n is being executed, the instruction $n+1$ is being processed in the address phase, $n+2$ in the Decode phase, $n+3$ in the Fetch2 phase and $n+4$ in the Fetch1 phase.

Using the processor hardware resources it is possible to write simple and effective programs in assembler. A sample program for the implementation of IIR filter second-order section is shown in Listing 3.8.

Table 3.17 SHARC pipeline

Cycles	1	2	3	4	5	6	7	8	9
Execute					n	n+1	n+2	n+3	n+4
Address				n	n+1	n+2	n+3	n+4	n+5
Decode			n	n+1	n+2	n+3	n+4	n+5	n+6
Fetch2		n	n+1	n+2	n+3	n+4	n+5	n+6	n+7
Fetch1	n	n+1	n+2	n+3	n+4	n+5	n+6	n+7	n+8

Listing 3.8 IIR filter second-order section

```
/* IIR Biquad Stage */
/* DM(I0,M1), DM(I1,M1) - data buffers in RAM */
/* PM(I8,M8) - buffer for coefficients in
program memory */
B1=B0;
/* first data */
F12=F12-F12, F2 = DM(I0,M1), F4 = PM(I8,M8);
Lcntr=N, do (pc,4) until lce; /* loop body */
/* parallel instructions */
F12=F2*F4, F8=F8+F12, F3 = DM(I0,M1), F4 = PM(I8,M8);
/* parallel instructions */
F12=F3*F4, F8=F8+F12, DM(I1,M1)=F3, F4 = PM(I8,M8);
/* parallel instructions */
F12=F2*F4, F8=F8+F12, F2 = DM(I0,M1), F4 = PM(I8,M8);
/* parallel instructions */
F12=F3*F4, F8=F8+F12, DM(I1,M1)=F8, F4 = PM(I8,M8);
/* last MAC, delayed return */
RTS(db), F8=F8+F12,
Nop;
Nop;
```

SHARC processors are equipped with comprehensive PWM modulators, so they can be used for controlling power electronics output circuits. However, acquisition of analog input signals should be realized by external modules.

3.10 Conclusions

This chapter has considered selected digital signal processing algorithms useful for a power electronics control circuit. Special attention has been paid to implementation aspects using digital signal processors. The author has presented an overview of the characteristics of microprocessors useful for implementing digital signal processing. The Chapter has included descriptions of wave digital filters and a modified wave digital filter with the author's modifications. There has also been shown an effective application of wave digital filters in multirate circuits. In spite of their good characteristics, the filters described in this Chapter are not commonly used. The presented methods and circuits are used in applications described in Chaps. 5 and 6.

References

1. Analog Devices (1994) ADSP-21000 family application handbook, vol 1. Analog Devices, Inc.
2. Analog Devices (1999) Interfacing the ADSP-21065L SHARC DSP to the AD1819A AC-97 soundport codec. Analog Devices, Inc.
3. Analog Devices (2003) ADSP-21065L EZ-KIT lite evaluation system manual. Analog Devices, Inc..
4. Analog Devices (2004) ADSP-2106x SHARC Processor user's manual. Analog Devices, Inc.
5. Analog Devices (2005) ADSP-2136x SHARC processor hardware reference. Rev 1.0, Analog Devices, Inc.

6. Analog Devices (2007) ADSP-21364 Processor EZ-KIT lite evaluation system manual. Rev 3.2, Analog Devices, Inc.
7. Arriens HL (2006) (L)WDF Toolbox for MATLAB reference guide. Technical report, Delft University of Technology, WDF Toolbox RG v1 0.pdf
8. Arriens HL (2006) (L)WDF Toolbox for MATLAB, user's guide. Technical report, Delft University of Technology, WDF Toolbox UG v1 0.pdf
9. Aziz SA (2004) Efficient arbitrary sample rate conversion using zero phase IIR. In: Proceedings of AES 116th convention, Berlin, Germany. Audio Engineering Society
10. Aziz SA (2007) Sample rate converter having a zero phase filter. United State Patent, Patent No: US 7,167,113 B2
11. Bagci B (2003) Programming and use of TMS320F2812 DSP to control and regulate power electronic converters. Master's thesis, University of Applied Science Cologne
12. Bateman A, Paterson-Stephens I (2002) The DSP handbook: algorithms, applications and design techniques. Prentice Hall, London
13. Bruun G (1978) Z-transform DFT filters and FFT's. IEEE Trans Acoust Speech Sig Process 26(1):56–63
14. Chassaing R (2005) Digital signal processing and applications with the C6713 and C6416 DSK. Wiley, New York
15. Chassaing R, Reay D (2008) Digital signal processing and applications with the C6713 and C6416 DSK. Wiley, New York
16. Chen WK (ed) (1995) The circuits and filters handbook. IEEE Press, Boca Raton
17. Crochiere RE, Rabiner LR (1983) Multirate digital signal processing. Prentice Hall, Inc., Upper Saddle River
18. Czarnach R (1982) Recursive processing by noncausal digital filters. IEEE Trans Acoust Speech Sig Process 30(3):363–370
19. Dabrowski A (1988) Pseudopower recovery in multirate signal processing (Odzysk pseudomocy użytecznej w wieloszybkościowym przetwarzaniu sygnaów), vol 198. Wydawnictwo Politechniki Poznanskiej, Poznan (in Polish)
20. Dabrowski A (1997) Multirate and multiphase switched-capacitor circuits. Chapman & Hall, London
21. Dabrowski A (ed) (1997) Digital signal processing using digital signal processors. Wydawnictwo Politechniki Poznańskiej, Poznań (in Polish)
22. Dabrowski A, Fettweis A (1987) Generalized approach to sampling rate alteration in wave digital filters. IEEE Trans Circ Syst Theor 34(6):678–686
23. Dabrowski A, Sozanski K (1998) Implementation of multirate modified wave digital filters using digital signal processors. XXI Krajowa Konferencja Teoria Obwodów i Układy Elektroniczne, KKTUIE98, Poznan
24. Dahnoun N (2000) Digital signal processing implementation using the TMS320C6000 DSP platform. Pearson Education Limited
25. Data Translation (2009) Benefits of simultaneous data acquisition modules. Technical report, Data translation
26. Delft University of Technology (2012) (L)WDF Toolbox for Matlab. Technical report, Delft University of Technology
27. Embree PM, Kimble B (1991) C language algorithms for digital signal processing. Prentice Hall Inc., Upper Saddle River
28. Farhang-Boroujeny B, Lee Y, Ko C (1996) Sliding transforms for efficient implementation of transform domain adaptive filters. Sig Process 52(1):83–96. Elsevier
29. Fettweis A (1971) Digital filter structures related to classical filter networks. AEU, Band 25, Heft 2:79–89
30. Fettweis A (1972) Pseudo-passivity, sensitivity, and stability of wave digital filters. IEEE Trans Circ Theor 19(6):668–673
31. Fettweis A (1982) Transmultiplexers with either analog conversion circuits, wave digital filters, or SC filters—a review. IEEE Trans Commun 30(7):1575–1586
32. Fettweis A (1986) Wave digital filters: theory and practice. Proc IEEE 74(2):270–327

33. Fettweis A (1989) Modified wave digital filters for improved implementation by commercial digital signal processors. Sig Process 16(3):193–207
34. Fettweis A, Levin H, Sedlmeyer A (1974) Wave digital lattice filters. Int J Circ Theor Appl 2(2):203–211
35. Fettweis A, Nossek J, Meerkotter K (1985) Reconstruction of signals after filtering and sampling rate reduction. IEEE Trans Acoust Speech Sig Process 33(4):893–902
36. Flige N (1994) Multirate digital signal processing. Wiley, New York
37. Gazsi L (1985) Explicit formulas for lattice wave digital filters. IEEE Trans Circ Syst 32(1):68–88
38. Goertzel G (1958) An algorithm for the evaluation of finite trigonometric series. Am Math Mon 65:34–35
39. Hamming R (1989) Digital filters. Dover Publications Inc., New York
40. Izydorczyk J, Konopacki J (2003) Analog and digital filters. Wydawnictwo Pracowni Komputerowej, Gliwice (in Polish)
41. Jacobsen E, Lyons R (2003) The sliding DFT. IEEE Sig Process Mag 20(2):74–80
42. Jacobsen E, Lyons R (2004) An update to the sliding DFT. IEEE Sig Process Mag 21:110–111
43. Kuo SM, Lee BH (2001) Real-time digital signal processing, implementation, applications, and experiments with the TMS320C55X. Wiley, New York
44. Kurosu A, Miyase S, Tomiyama S, Takebe T (2003) A technique to truncate IIR filter impulse response and its application to real-time implementation of linear-phase IIR filters. IEEE Trans Sig Process 51(5):1284–1292
45. Lawson S (1995) Wave digital filters. In: Chen W-K (ed) The circuits and filters handbook. IEEE Press, Boca Raton, pp 2634–2657
46. Lawson S, Mirzai A (1990) Wave digital filters. Ellis-Horwood, New York
47. Ledger D, Tomarakos J (1998) Using The low cost, high performance ADSP-21065L digital signal processor for digital audio applications. Revision 1.0, Analog Devices, Norwood, USA
48. Lyons R (2004) Understanding digital signal processing, 2nd edn. Prentice Hall, Upper Saddle River
49. Lyons R, Bell A (2004) The swiss army knife of digital networks. IEEE Sig Process Mag 21(3):90–100
50. Mitra S (2006) Digital signal processing: a computer-based approach. McGraw-Hill, New York
51. Mouffak A, Belbachir M (2012) Noncausal forward/backward two-pass IIR digital filters in real time. Turk J Electr Eng Comput Sci 20(5):769–789
52. Oppenheim AV, Schafer RW (1999) Discrete-time signal processing. Prentice Hall, Englewood Cliffs
53. Orfanidis SJ (1996) ADSP-2181 experiments. http://www.ece.rutgers.edu/~orfanidi/ezkitl/man.pdf. Accessed Dec 2012
54. Orfanidis SJ (2010) Introduction to signal processing. Prentice Hall, Inc., Upper Saddle River
55. Oshana R (2005) DSP software development techniques for embedded and real-time systems. Newnes
56. Owen M (2007) Practical signal processing. Cambridge University Press, Cambridge
57. Pasko M, Walczak J (1999) Signal theory. Wydawnictwo Politechniki Slaskiej, Gliwice (in Polish)
58. Powell SR, Chau PM (1991) A technique for realizing linear phase IIR filters. IEEE Trans Sig Process 39(11):2425–2435
59. Proakis JG, Manolakis DM (1996) Digital signal processing, principles, algorithms, and application. Prentice Hall, Inc., Englewood Cliffs
60. Press WH, Teukolsky SA, Vetterling WT, Flannery BP (2007) Numerical recipes: the art of scientific computing, 3rd edn. Cambridge University Press, Cambridge
61. Rabiner LR, Gold B (1975) Theory and application of digital signal processing. Prentice Hall, Inc., Englewood Cliffs
62. Rao D (2001) Circular buffering on TMS320C6000. Application report, SPAR645A, Texas Instruments

63. Sozanski K (1999) Design and research of digital filters banks using digital signal processors. PhD thesis, Technical University of Poznan (in Polish)
64. Sozanski K (2002) Implementation of modified wave digital filters using digital signal processors, In: Conference proceedings of the 9th international conference on electronics, circuits and systems, ICECS 2002, pp 1015–1018
65. Sozański (2003) Active power filter control algorithm using the sliding DFT. In: Workshop proceedings of the signal processing 2003, Poznan, Poland, pp 69–73
66. Sozański K (2004) Harmonic compensation using the sliding DFT algorithm. In: Conference proceedings of the 35rd annual IEEE power electronics specialists conference, PESC 2004, Aachen, Germany
67. Sozański K (2008) Improved shunt active power filters. Przeglad Elektrotechniczny (Electr Rev) 45(11):290–294
68. Sozański K (2010) Digital realization of a click modulator for an audio power amplifier. Przeglad Elektrotechniczny (Electr Rev) 2010(2):353–357
69. Sozański K (2012) Realization of a digital control algorithm. In: Benysek G, Pasko M (eds) Power theories for improved power quality. Springer, London, pp 117–168
70. Sozański K (2013) A linear-phase IIR filter for audio signal interpolator. In: Conference proceedings of the signal processing, algorithms, architectures, arrangements and applications, SPA 2013, Poznan, Poland, pp 65–69
71. Sozański K (2015) Selected problems of digital signal processing in power electronic circuits, In: Conference proceedings of SENE 2015, Lodz, Poland
72. Sozański K (2016) Signal-to-noise ratio in power electronic digital control circuits. In: Conference proceedings of the signal processing, algorithms, architectures, arrangements and applications, SPA 2016. Poznan University of Technology, pp 162–171
73. Sozański K, Strzelecki R, Fedyczak Z, (2001) Digital control circuit for class-D audio power amplifier. In: Conference proceedings of the 2001 IEEE 32nd annual power electronics specialists conference, PESC 2001, pp 1245–1250
74. Tantaratana S (1995) Design of IIR filters. In: Chen WK (ed) The circuits and filters handbook. IEEE Press, Boca Raton
75. Texas Instruments (2008) TMS320F28335/28334/28332, TMS320F28235/28234/28232 digital signal controllers (DSCs). Data manual, Texas Instruments, Inc.
76. Texas Instruments (2010) C2000 Teaching materials, tutorials and applications. SSQC019, Texas Instruments, Inc.
77. Texas Instruments (2011) TMS320C6745/C6747 DSP technical reference manual. SPRUH91A, Texas Instruments, Inc.
78. Texas Instruments (2011) TMS320C6746 fixed/floating-point DSP. Data sheet, SPRS591, Texas Instruments, Inc.
79. Texas Instruments (2012) C6000 Teaching materials. SSQC012, Texas Instruments, Inc.
80. Texas Instruments (2016) TMS320F2837xD Dual-Core Delfino Microcontrollers. Data sheet, Texas Instruments, Inc.
81. Texas Instruments (2016) The TMS320F2837xD architecture: achieving a new level of high performance. Technical brief, Texas Instruments, Inc.
82. Vaidyanathan PP (1992) Multirate systems and filter banks. Prentice-Hall Inc, Englewood Cliffs
83. Venezuela RA, Constantindes AG (1982) Digital signal processing schemes for efficient interpolation and decimation. IEE Proc Part G 130(6):225–235
84. Vesterbacka M (1997) On implementation of maximally fast wave digital filters. Disertations no. 487, Linköping University
85. Vesterbacka M, Palmkvist K, Wanhammar L (1996) Maximaly fast, bit-serial lattice wave digital filters. In: Proceedigs of DSP workshop 1996, Loen, Norway. IEEE, pp 207–210
86. Wanhammar L (1999) DSP integrated circuit. Academic Press, London
87. Willson AN, Orchard HJ (1994) An Improvement to Powell and Chau linear phase IIR filter. IEEE Trans Sig Process 42(10):2842–2848

88. Zieliński TP (2005) Digital signal processing: from theory to application. Wydawnictwo Komunikacji i Lacznosci, Warsaw (in Polish)
89. Zieliński TP, Korohoda P, Rumian R (eds) (2014) Digital signal processing in telecommunication, basics, multimedia, transmission. Wydawnictwo Naukowe PWN, Warsaw (in Polish)
90. Zolzer U (2008) Digital audio signal processing. Wiley, New York
91. Zolzer U (ed) (2002) DAFX–digital audio effects. Wiley, New York

Chapter 4
Selected Simulation Methods and Programs for Power Electronics Circuits

4.1 Introduction

Analysis, synthesis, and design of electrical and electronic circuits often constitute a complex mathematical and computational problem. Using computers for this task has facilitated substantial progress in the field. Thanks to the development of computers and software, these problems can be solved using numeric and symbolic methods. The latter case has been described among others by Small and Hosack [37] and others [1, 4, 5, 38]. Special programs for use in electrical engineering (including electronics) have been created for solving these problems, such as **Spice, PSIM, NAP2** etc. These programs enable the solving of circuit problems; they require one to input circuit topography and circuit parameters, but a detailed knowledge of the mathematical methods used for its analysis is not necessary. In a sense, the user does not have to fully understand the resulting phenomena.

Spice (**S**imulation **P**rogram with **I**ntegrated **C**ircuit **E**mphasis) is a general-purpose circuit simulation program for nonlinear DC, nonlinear transient, and linear AC analyses. Circuits may contain resistors, capacitors, inductors, mutual inductors, independent voltage and current sources, four types of dependent sources, lossless and lossy transmission lines (two separate implementations), switches, and the five most common semiconductor devices: diodes, BJTs, JFETs, MESFETs, and MOSFETs. Spice was created at the beginning of the seventies at the University of Berkeley. Fortunately, Spice source code was from the beginning public domain software and was distributed by the University of Berkeley [43] for a nominal charge (to cover the cost of magnetic tape). Pspice [34] is a popular version of Spice designed by MicroSim Corporation, later available from Orcad (now Cadence), and runs under the PC Windows and Macintosh platforms. The program was originally designed for analog circuit simulation, however, it can be used for power electronics and digital circuits too [17]. Nevertheless, for impulse application, convergence errors can be met. The creation of the Spice program was considered a milestone by IEEE [10]. Today a few large scale IC manufacturers continue to develop the process of Spice-based circuit simulation programs. Among these are ADIsimPE at Analog Devices,

K. Sozański, *Digital Signal Processing in Power Electronics Control Circuits*,
Power Systems, DOI 10.1007/978-1-4471-7332-8_4

LTspice at Linear Technology [15], and TINA at Texas Instruments (now Design-Soft) [46]. TINA has a rare function—symbolic analysis of analog circuits. The symbolic analysis produces the closed form expression of the transfer function, equivalent resistance, impedance, and response of analog linear networks. In DC and AC analysis mode, TINA derives formulas in full-symbolic or semi-symbolic form. It is very useful for linear analog circuits. There are also special programs for simulation of power electronics circuit, for example GeckoCIRCUITS is the circuit simulator for modeling power electronics systems [7].

The second group of software tools supporting the process of circuits analysis, synthesis and design are versatile general purpose software tools for numeric and symbolic calculations: commercial **MATLAB®**, **Mathematica, Maple, Mathcad, Magma** and non-commercial **GNU Octave, Scilab, SageMath, SymPy**. Application of these software tools, unlike programs such as SPICE, does not exempt the user from knowing Kirchhoff's laws, and he must formulate a mathematical problem which it is necessary to solve.

MATLAB® (Matrix Laboratory) from MathWorks company is numeric computation software for engineering and scientific calculations [22]. Matlab has a rich set of integrated graphic capabilities. At the beginning, MATLAB was designed to give easy access to LINPACK and EISPACK without having to learn Fortran. It was developed in the late 1970s by Cleve Moler, the chairman of the computer science department at the University of New Mexico. After the initial success it was commercialized by the MathWorks company. Now MATLAB is a very popular software tool for engineering and scientific calculations and is a reference tool for this area of application.

GNU Octave is a high-level language designed for numerical computations [9]. It is typically used for such problems as solving linear and nonlinear equations, numerical linear algebra, statistical analysis, and for performing other numerical experiments. Octave helps in solving linear and nonlinear problems numerically, and for performing other numerical calculations. It uses a language mostly compatible with MATLAB. It may also be used as a batch-oriented language for automated data processing. GNU Octave is freely redistributable software under the terms of the GNU General Public License.

SageMath is a free open-source mathematics software system licensed under the GPL (General Public License) [35]. Sage is mathematical software with features covering many aspects of mathematics, including algebra, combinatorics, numerical mathematics, number theory, and calculus. It builds on top of many existing open-source packages: NumPy, SciPy, matplotlib, Sympy, Maxima, GAP, FLINT, R and many more. Access to their combined power is through a common, Python-based language or directly via interfaces or wrappers. The development of Sage is supported by both volunteer work and grants, and creates a viable, free, open source alternative to Magma, Maple, Mathematica, and MATLAB. Another software tool, under a BSD (Berkeley Software Distribution) license **SymPy** [45] is the Python library for symbolic mathematics. It aims to become a full-featured computer algebra system (CAS) while keeping the code as simple as possible in order to be comprehensible and easily extensible. SymPy is written entirely in Python.

Scilab [36] is another open source option for numerical computing and a high-level, numerically oriented programming language. It can be used for signal processing, statistical analysis, image enhancement, fluid dynamics simulations, numerical optimization, and modeling, simulation of explicit and implicit dynamic systems and symbolic manipulations. In some ways, Scilab may be the best of the MATLAB alternatives. Since the syntax of Scilab is similar to MATLAB, Scilab includes a source code translator for assisting the conversion of code from MATLAB to Scilab. Scilab also includes a free package called Xcos hybrid dynamic systems modeler and simulator, which can be compared to Simulink® from the MathWorks [22].
Mathematica, Maple, Mathcad, Magma are general purpose commercial mathematical symbolic and numeric computation programs [16, 18–20]. They are used in much scientific, engineering, mathematical, and computing fields. They also cover other aspects of technical computing, including visualization, data analysis, matrix computation, and connectivity.

4.2 Simulation Using MATLAB®

The use of MATLAB® for solving the problems of electrical circuits analysis is described in a lot of publications, among others [2, 3, 6, 11–14, 25, 39, 41, 42]. More than 1500 books related to MATLAB can be found on MathWorks web page [24].

Figure 4.1 shows methods of electrical analog circuit time domain analysis. Especially noteworthy is the application of the Laplace transform, which is very well realized in MATLAB [13, 44].

4.2.1 DC and AC Analysis of Analog Circuits

MATLAB has support for complex numbers with several built-in functions available (Table 4.1). The imaginary unit is denoted by i or the j symbol preferred in electrical engineering. To create the complex variables $z_1 = 3 + 2\mathrm{j}$ and $z_2 = \frac{A}{\sqrt{2}}\,\mathrm{e}^{\mathrm{j}1.5\pi}$, in the MATLAB environment simply enter z1 = 7 + 2 * j and z2 = A/sqrt(2) * exp(j * 1.5 * pi) respectively.

The basic variable type of MATLAB is matrices. Matrix operations are very efficient and are the most valuable advantage of MATLAB [8, 22]. Scalars and vectors are special cases of matrices having sizes 1 × 1, 1 × N or N × 1. In MATLAB there are the following rules for matrix data input:

- elements of a row are separated by blanks or commas,
- each row ends with a semicolon,
- a list of elements must be enclosed in square brackets [ ].

Table 4.1 Basic MATLAB functions for manipulating complex numbers

Mathematics	MATLAB
Re(z)	real(z)
Im(z)	imag(z)
$\|z\|$	abs(z)
$\angle z$	angle(z)
z^*	con(z)

An example of creating variables in MATLAB is shown in Listing 4.1.

Listing 4.1 Creating variables

```
x = 1.3e5; % scalar
y = [2 0.3 1e-3 2 5]; % row vector
t = 1:1:10; % row vector t=[1 2 3 4 5 6 7 8 9 10]
z = [2; 4; 6; 8; 10] % column vector
A = [4 3 2 1 0; 1 3 5 7 9] % 2 x 5 matrix
C = [1:5;7:-1:1] % 5 x 7 matrix
B = zeros(4,5) % 4 x 5 matrix with contents 0
C = ones(3,15) % 3 x 15 matrix with contents 1
```

Operations on the matrix and complex numbers can be used to solve linear circuits. A MATLAB application for solving an exemplary electrical circuit is shown in Fig. 4.2. For the considered RC circuit it is possible to write equations:

$$U_1(\mathrm{j}\omega) = E_1(\mathrm{j}\omega) = \frac{A}{\sqrt{2}}\mathrm{e}^{(\mathrm{j}\phi)} \quad , \tag{4.1}$$

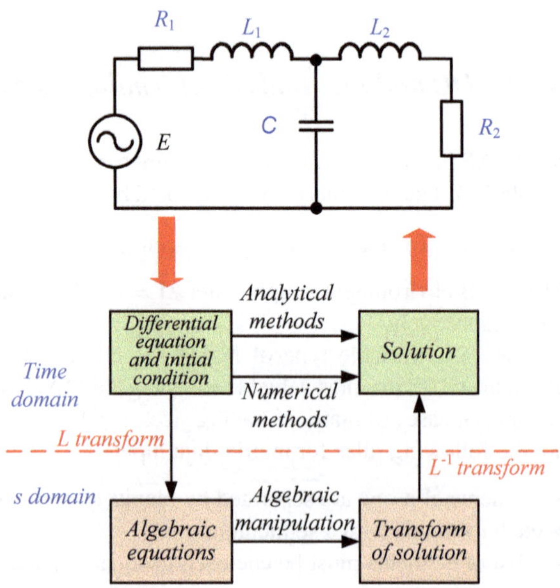

Fig. 4.1 Circuit analysis methods

Fig. 4.2 An exemplary RC circuit

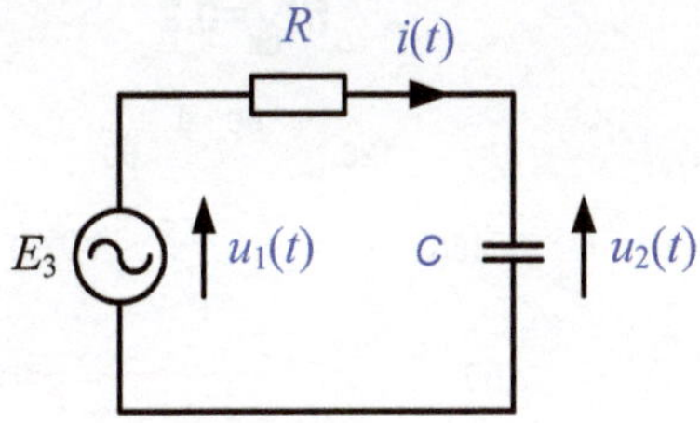

and

$$U_2(\mathrm{j}\omega) = \frac{A}{\sqrt{2}} e^{(\mathrm{j}\phi)} \frac{1 - \mathrm{j}\omega RC}{1 + \omega^2 R^2 C^2} \quad . \tag{4.2}$$

Based on the above Eqs. (4.1) and (4.2), the MATLAB program created for calculating frequency response of the RC circuit for the frequency range $0.002 f_{cr}$ do $100 f_{cr}$ and with the frequency step $df = 0.002 f_{cr}$ is shown in Listing 4.2.

Listing 4.2 Frequency response calculation of the RC circuit

```
% Frequency response of RC circuit
clear all; close all;
l_width = 2; f_size = 12;
C=1; R=1; fcr=1/(2*pi*R*C);
df=0.002; f = (0:df:100-df)*fcr;
n=1:1:50000; frel=0.002*n;
omega=2*pi*f; fi=0;
A=1*sqrt(2); U1=A/(2^0.5)*exp(j*fi);
U2=(U1*1)./(1+j*omega*R*C);
figure('Name','Compass','NumberTitle','off');
 wekt1=[U1(1) U2(20)]; wekt2=[U1(1) U2(500)]; wekt3=[U1(1) U2(1000)];
 wekt4=[U1(1) U2(2500)];
 subplot(221), compass(wekt1,'r'), set(gca,'FontSize',f_size),
   title('f/f_{cr}=0.2'),
 subplot(222), compass(wekt2,'r'), set(gca,'FontSize',f_size),
   title('f/f_{cr}=1'),
 subplot(223), compass(wekt3,'r'), set(gca,'FontSize',f_size),
   title('f/f_{cr}=2'),
 subplot(224), compass(wekt4,'r'), set(gca,'FontSize',f_size),
   title('f/f_{cr}=5');
   print('compass.pdf','-dpdf');
f_size = 16;
figure('Name','Nyquist','NumberTitle','off');
 plot(U2/U1,'r','linewidth',l_width), set(gca,'FontSize',f_size),
   xlabel('Re(H(j\omega))'); ylabel('Im(H(j\omega))');
   grid on;
   print('nyquist.pdf','-dpdf');
figure('Name','|H|, phase','NumberTitle','off');
  H_magdB=20*log10(abs(U2/U1));
  phase=angle(U2)*180/pi;
  subplot(211), semilogx(frel, H_magdB,'r'),set(gca,'FontSize',f_size),
    title('(a)'),
    ylabel('A(\omega) [dB]'); grid on;
  subplot(212), semilogx(frel, phase,'r'),set(gca,'FontSize',f_size),
    title('(b)'),
    ylabel('\phi(\omega) [deg]'), xlabel('\it f/f_{cr}'); grid on;
    print('Mag_and_phase_log.pdf','-dpdf');
```

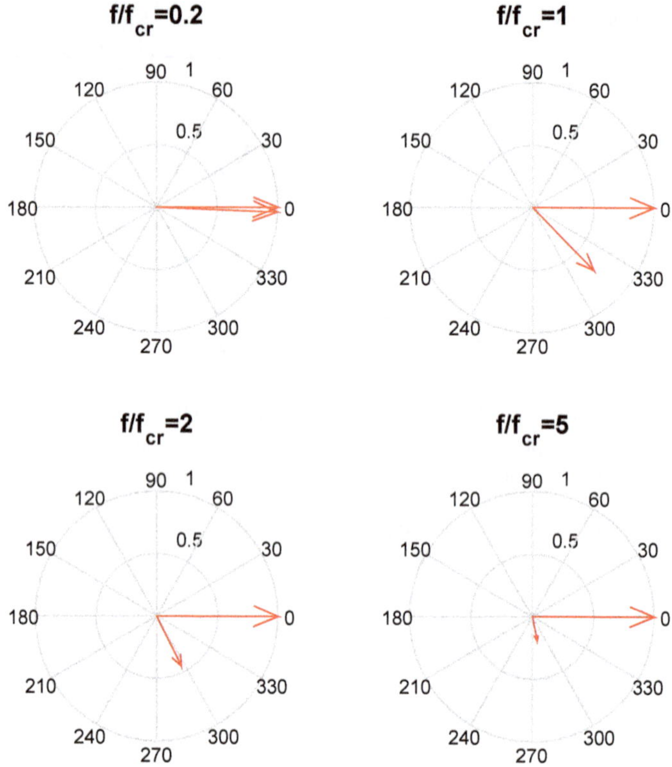

Fig. 4.3 Vector graphs of voltages $U_1(\mathrm{j}\omega)$ and $U_2(\mathrm{j}\omega)$ for selected values of f/f_{cr}

MATLAB has a useful function for this case, function **compass(U, V)**, which draws a graph that displays the vectors with components U, V. Figure 4.3 shows vector diagrams of the considered circuit for selected frequencies.

For the considered circuit the frequency transfer function can be obtained from equation

$$H(\mathrm{j}\omega) = \frac{U_2(\mathrm{j}\omega)}{U_1(\mathrm{j}\omega)} = \frac{1}{1+\mathrm{j}\omega RC} = \frac{1-\mathrm{j}\omega RC}{1+\omega^2R^2C^2} \quad . \tag{4.3}$$

The frequency transfer function can be expressed as a sum of the real and imaginary parts

$$\begin{aligned} H(\mathrm{j}\omega) = |H(\mathrm{j}\omega)|\mathrm{e}^{\mathrm{j}\phi(\omega)} &= |H(\mathrm{j}\omega)|\cos(\phi(\omega)) + \mathrm{j}|H(\mathrm{j}\omega)|\sin(\phi(\omega)) = \\ &= \mathrm{Re}(H(\mathrm{j}\omega)) + \mathrm{j}\mathrm{Im}(H(\mathrm{j}\omega)) \quad . \end{aligned} \tag{4.4}$$

where: $\mathrm{Re}(H(\mathrm{j}\omega))$, $\mathrm{Im}(H(\mathrm{j}\omega))$—real part and imaginary part of transfer functions.

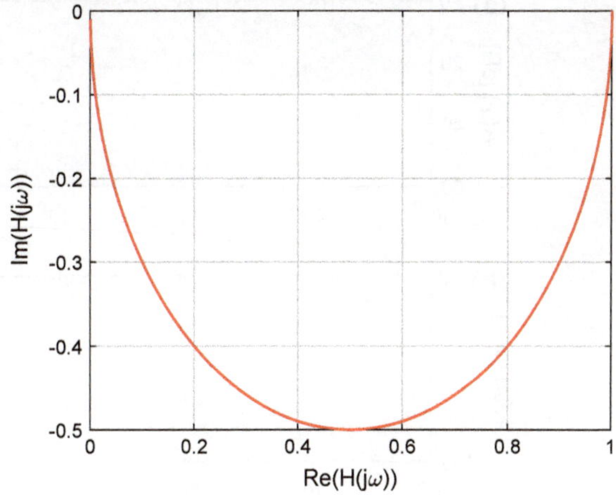

Fig. 4.4 Nyquist plot of RC circuit

A Nyquist plot is a parametric plot of a frequency response of the RC circuit shown in Fig. 4.4. The real part $\mathrm{Re}(H(\mathrm{j}\omega))$ of the transfer function is plotted on the X axis. The imaginary part $\mathrm{Im}(H(\mathrm{j}\omega))$ of the transfer function is plotted on the Y axis.

In MATLAB, magnitude and the phase angle can be calculated by function **abs(x)** and function **angle(x)** respectively. Frequency responses of the considered RC circuit are shown in Fig. 4.5. MATLAB has support for the analysis of analog circuits using the Laplace transform, e.g., functions **laplace, freqs, lsim**. For example using freqs function it is possible to determine the frequency response of a circuit defined by transfer function

$$H(s) = \frac{U_2(s)}{U_1(s)} = \frac{b_1 s^N + b_2 s^{N-1} + \ldots + b_{N+1}}{b_1 s^M + b_2 s^{M-1} + \ldots + b_{M+1}} . \tag{4.5}$$

For the considered RC circuit it is possible to find Laplace transfer functions by substituting $s = \mathrm{j}\omega$ to frequency transfer functions (4.3)

$$H(\mathrm{j}\omega) = \frac{U_2(\mathrm{j}\omega)}{U_1(\mathrm{j}\omega)} = \frac{1}{1 + \mathrm{j}\omega RC} \overset{s=\mathrm{j}\omega}{\Longleftrightarrow} H(s) = \frac{U_2(s)}{U_1(s)} = \frac{1}{1 + sRC} . \tag{4.6}$$

The program calculates and graphs the frequency response of the RC circuit which is placed in Listing 4.3. The determined frequency characteristic of the RC circuit shown in Fig. 4.6.

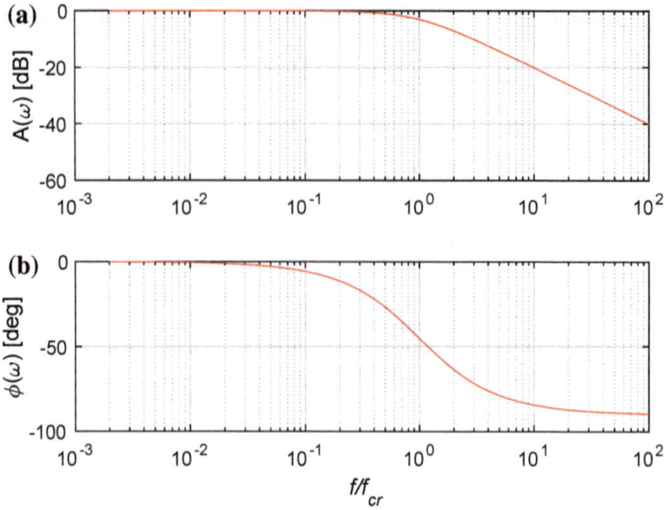

Fig. 4.5 Frequency responses of the considered RC: **a** magnitude, **b** phase

Listing 4.3 Frequency response calculation of the RC circuit using Laplace transformation

```
% RC Laplace
clear all; close all;
l_width = 2; f_size = 16;
C=1; R=1;
b=1; % numerator
a=[R*C 1]; % denumerator
w = logspace(-2,1,200); % generate logarithmically spaced vectors
h = freqs(b,a,w);
mag = abs(h); phase = angle(h); phasedeg = phase*180/pi;
figure('Name','RC␣Frequency␣response','NumberTitle','off');
  subplot(2,1,1), loglog(w,mag,'r','linewidth',l_width),
        set(gca,'FontSize',f_size), grid on
        title('(a)'), ylabel('Magnitude');
  subplot(2,1,2), semilogx(w,phasedeg,'b','linewidth',l_width),
        set(gca,'FontSize',f_size),grid on, title('(a)'),
        xlabel 'Frequency␣(rad/s)', ylabel('Phase␣[deg]');
  print('Mag_and_phase_RC_freqs.pdf','-dpdf');
```

4.2.2 DC and AC Nodal and Loop Analysis of Circuits

Nodal analysis is based on Kirchhoff's current law (KCL), the algebraic sum of all the current at any node in the circuit equals zero

$$I = YV \quad . \tag{4.7}$$

where: Y—matrix of nodal admittance, V—matrix of nodal voltages, I—vector of node currents.

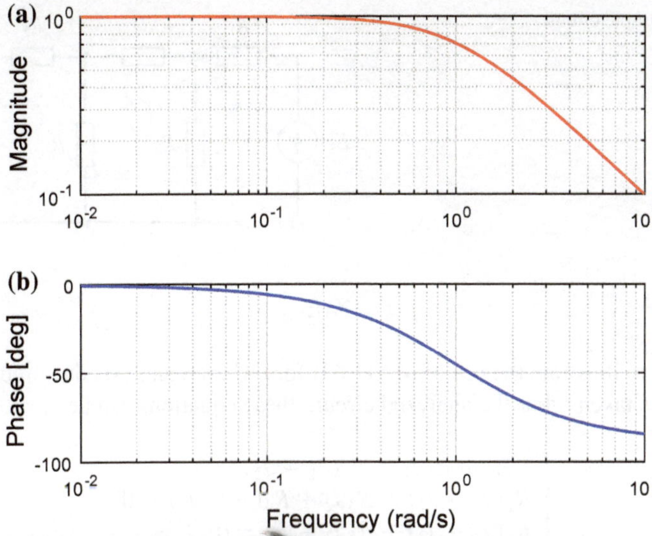

Fig. 4.6 Frequency response of RC circuit

For a given matrix Y and vector I, vector V can be calculated by the matrix equation

$$V = Y^{-1}I \quad , \tag{4.8}$$

where Y^{-1}—denotes an inverse of matrix Y.

In MATLAB, it is possible to compute V using the command

$$\mathrm{V} = \mathrm{Y}\backslash\mathrm{I} \quad . \tag{4.9}$$

The second method uses Kirchhoff's voltage law (KVL), the sum of all the voltages around any closed path in a circuit equals zero. Loop analysis is a method for obtaining loop currents. In the method, the equations describing the circuit have the form

$$E = JZ \quad , \tag{4.10}$$

where: Z—matrix of mesh impedance, J—vector of loop currents, E—voltages vector. The vector of loop currents can be obtained by solving the matrix equation

$$J = Z^{-1}E \quad . \tag{4.11}$$

Similar to the case in the previous method, in MATLAB it is possible to compute J using the command

Fig. 4.7 Example circuit

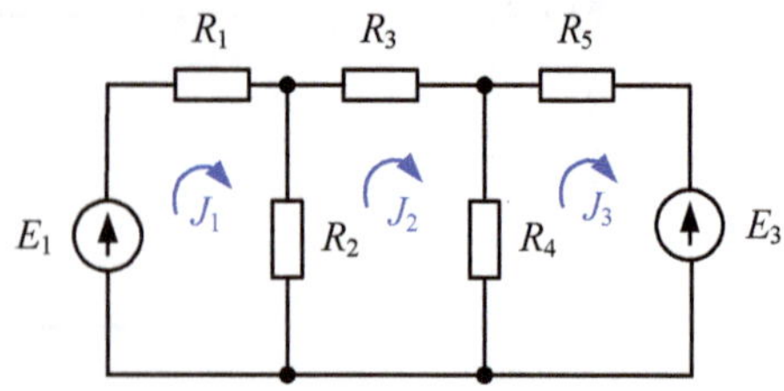

$$\mathrm{J} = \mathrm{Z} \backslash \mathrm{E} \quad . \tag{4.12}$$

In order to illustrate the use of MATLAB for DC loop analysis the circuit shown in Fig. 4.7 is taken. For the analyzed circuit, these equations can be created

$$\begin{cases} -E_1 + R_1 J_1 + R_2(J_1 - J_2) = 0 \\ R_2(J_2 - J_1) + R_3 J_2 + R_4(j_2 - J_3) = 0 \\ R_4(J_3 - J_2) + R_5 J_3 + E_3 = 0 \quad , \end{cases} \tag{4.13}$$

after transforming them for loop analysis the following equations are obtained

$$\begin{cases} (R_1 + R_2)J_1 - R_2 J_2 + 0J_3 = E_1 \\ -R_2 J_1 + (R_2 + R_3 + R_4)J_2 - R_4 J_3 = 0 \\ 0J_1 - R_4 J_2 + (R_4 + R_5)J_3 = -E_3 \quad . \end{cases} \tag{4.14}$$

The program for calculating loop currents of the example circuit is shown in Listing 4.4.

Listing 4.4 Loop analysis

```
R=[ 2.5 1.8 3 1 7] % value of resistors R1 to R5
E=[12 0 -5] % value of voltage sources E1, E2, E3
Z=[R(1)+R(2),-R(2),0; -R(2), R(2)+R(3)+R(4),-R(4); 0,-R(4), R(4)+R(5)];
J=Z\E % calculation of loop currents
```

The result of the program execution is as follows:

```
R =  2.5  1.8  3  1  7
E =  12  0  5
J =
      3.164796219728294e+00
      8.936798582398112e-01
     -5.132900177200237e-01
```

A similar solution is possible for AC circuit analysis [2].

4.2.3 Transient Analysis of Analog Circuits

For time domain analysis, the linear analog circuit can be described by the ordinary differential equation. To solve the equations, numerical or analytic methods should be used. These problems are described in many books about electrical circuit analysis [40, 41]. In MATLAB, there are a few functions for numerical solving of differential equations **ode23, ode45, ode113, ode15s, ode23s, ode23t, ode23tb, ode15i**. The **ode23** solve non-stiff differential equations using the low order method, while the **ode45** solve non-stiff differential equations using the medium order method. In general, **ode45** is the best function to apply as a first tray function for solving differential equations [22].

RL circuits are often used in power electronics circuits. In order to illustrate the use of MATLAB for transient analysis, the RL circuit shown in Fig. 4.8 is used. In the circuit, the switch is in position 1 and at $t = 0$, the switch moves from position 1 to 2. Using Kirchhoff's voltage law (KVL), it is possible to create the equations

$$\begin{aligned} &E_2 - u_L(t) - u_R(t) = 0 \quad , \\ &u_L(t) = L\frac{\mathrm{d}i(t)}{\mathrm{d}t} \quad , \\ &E_2 - L\frac{\mathrm{d}i(t)}{\mathrm{d}t} - i_L(t)R = 0 \quad , \end{aligned} \tag{4.15}$$

and finally it is possible to obtain the differential equation

$$\frac{\mathrm{d}i(t)}{\mathrm{d}t} = \frac{E_2}{L} - \frac{R}{L}i_L(t) \quad , \tag{4.16}$$

The solution of the differential equation is

$$i_L(t) = \frac{E_2}{R} + \left(i_L(0) - \frac{E_2}{R}\right)\mathrm{e}^{-\frac{t}{\tau}} \quad , \tag{4.17}$$

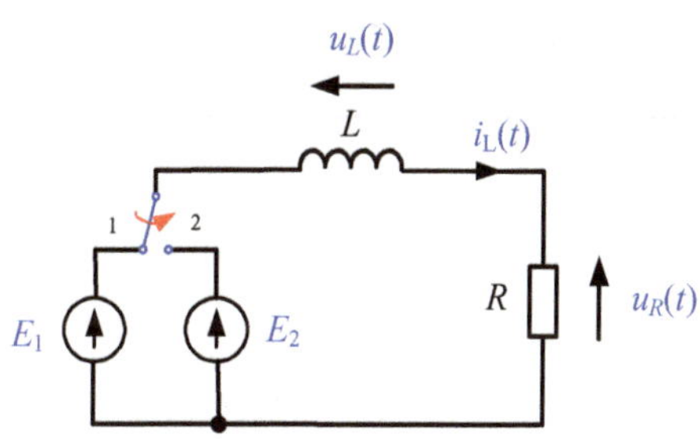

Fig. 4.8 Example RL circuit

where $\tau = \frac{L}{R}$—the circuit time constant, $i_L(0)$—inductor initial current.

Assuming that the current at $t = 0$ is $i_L(0) = \frac{E_1}{R}$ it is possible to calculate the current

$$i_L(t) = \frac{E_2}{R} + \left(\frac{E_1}{R} - \frac{E_2}{R}\right) \mathrm{e}^{-\frac{t}{\tau}} . \tag{4.18}$$

A program for calculating step response outputs of the RL circuit is shown in Listing 4.5. The responses are calculated by two methods, analytical, using Eq. (4.18), and numerical, using differential equation (4.16). The differential equation is defined in the separate m-file, as shown in Listing 4.6. The name of the m-file, is diff_RL.m. It is also possible to define the differential equation by the **inline** function. The step responses of the RL circuit are depicted in Fig. 4.9.

Listing 4.5 Program for calculating the RL circuit transient response using analytical and numerical solutions

```
clear all; close all; clc;
global E2 R L tau
l_width = 2; f_size = 16;
E2=1.0; R=1.0; L=100e-3; tau=L/R; E1=-1;
i_L0 = E1/R; % initial (conditions) inductor current
N=2^12; t0 = 0; tf = 500e-3;
[t, i_Ldiff] = ode45('diff_RL',[t0 tf],i_L0); % solution with flexible step
t=t'; i_Ldiff=i_Ldiff';
i_Lana = E2/R+(i_L0-E2/R)*exp(-t/tau); % the analytical solution
% plot two solutions
figure('Name','RL Transient response','NumberTitle','off');
subplot(211), plot(t,i_Ldiff,'b',t,i_Lana,'r','linewidth',l_width)
set(gca,'FontSize',f_size),title('(a)'),
legend('i_{Lana}(t)','i_{Ldiff}(t)');grid on;
ylabel('Inductor current, i_{L}(t)');
subplot(212), plot(t,i_Ldiff-i_Lana,'linewidth',l_width);
set(gca,'FontSize',f_size), grid on, title('(b)'),
xlabel('Time [s]'), ylabel('i_{Lana}(t)-i_{Ldiff}(t)');
```

Listing 4.6 The RL circuit differential equation function

```
function dy = diff_RL(t,y)
global E2 R L tau
dy = E2/L - y/tau;
end
```

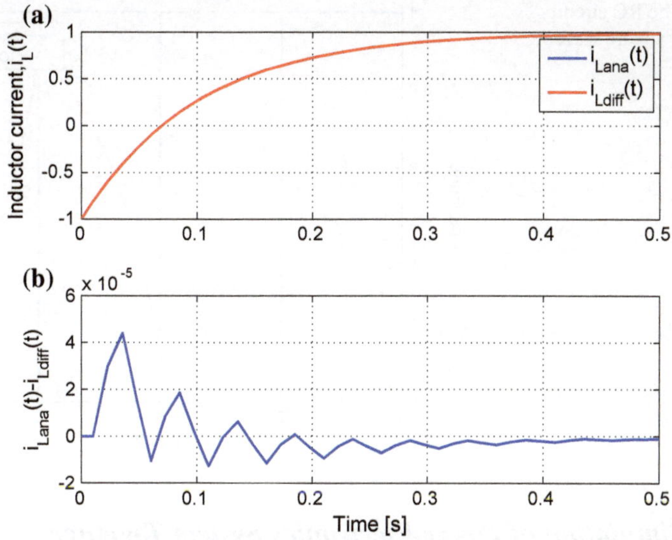

Fig. 4.9 Step response of the RL circuit: **a** currents, **b** difference of currents

Using the Laplace transform, comparable simulation tasks can be realized much easier. MATLAB function **lsim** can simulate the time response of the dynamic system to arbitrary inputs. Listing 4.7 shows an example program for calculating the time response of the RC circuit determined by the transfer function (4.6) to a periodic square wave input. The result of such a simulation is depicted in Fig. 4.10.

Listing 4.7 The RC circuit square wave response calculating program using Laplace transformation

```
clear all; close all;
l_width = 2; f_size = 16;
C=1; R=1; N=8000; dt=0.001;
b=1; % numerator
a=[R*C 1]; % denumerator
H=tf(b,a)  % RC transfer function
t=(0:N-1)*dt;
x=[ones(1,N/4) zeros(1,N/4) ones(1,N/4) zeros(1,N/4)];
y=lsim(H,x,t);
figure('Name','RC␣impuse␣response','NumberTitle','off');
plot(t,x,'b',t,y,'r','linewidth',l_width),
set(gca,'FontSize',f_size), grid on
ylabel('Amplitude'), xlabel 'Time␣[s]'; legend('Square␣wave','RC␣response');
print('RC_imp_resp.pdf','-dpdf');
```

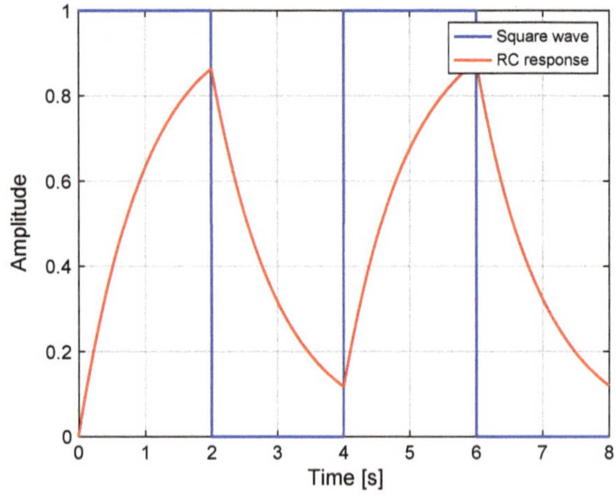

Fig. 4.10 The RC circuit response for square wave

4.2.4 *Simulation of Power Electronics System Together with Digital Control Circuit*

In this section the example of a hysteresis current controller is used. It is simulated using MATLAB. The circuit diagram is depicted in Fig. 4.11a, consists of a digital control part, a power part realized by two ideal switches Q_1, Q_4 and the RL circuit. The response of the RL circuit is calculated using Eq. (4.16). In order to increase the accuracy of the simulation RR times oversampling is applied. The simulation block diagram is depicted in Fig. 4.11b. The MATLAB program for calculating square signal response is shown in Listing 4.8. During the simulation, the following parameters are used $U_{C1} = 400$ V, $U_{C2} = -400$ V, $R = 0.2\ \Omega$, $L = 5.5$ mH, $h = 0.1$, $f_s = 50$ kHz, $f_c = 50$ kHz, with an oversampling ratio $RR = 100$. The MATLAB implementation of the current controller is shown in Listing 4.9. The result of the simulation is shown in Fig. 4.12.

Listing 4.8 The MATLAB implementation of current controller

```
clear all; close all; clc;
l_width = 2; f_size = 16;
Uc1=400; Uc2=-400; R=0.3; L=5.5e-3; tau=L/R;
N=2000; fs=50000; Ts=1/fs; f=50;
RR=100; kr=0.025; h=0.10;
t=(0:N-1)*Ts;
i_ref = square(2*pi*f*t,50);
i_out=f_inv_RL_sol_hist(i_ref,h,kr,R,L,Uc1,Uc2,RR,Ts);
trr=(0:length(i_out)-1)*Ts/RR;
figure('Name','Currents','NumberTitle','off');
plot(trr,i_out,t,i_ref/kr,'r','linewidth',l_width),
set(gca,'FontSize',f_size), grid on;
ylabel('i_{ref}(t)/k_r,i_{out}(t)'), xlabel('Time␣[s]');
legend('i_{ref}(t)/k_r','i_{out}(t)');
```

Listing 4.9 Hysteresis current controller using solution of differential equation

```
function i_out=f_inv_RL_sol_hist(i_ref,h,kr,R,L,Ep,En,RR,Ts)
tau=L/R; t=(0:RR-1)*Ts/RR; N=length(i_ref);
% initialization of variables
i_outRR=zeros(1,RR); i_out=zeros(1,RR*N); u_inv=ones(1,N)*Ep;
for n=1:N
  e=i_ref(n)-i_outRR(RR)*kr;
  if n==1
      ; % start
  elseif u_inv(n-1)==Ep&&e<-h
    u_inv(n)=En; % switch on En
  elseif u_inv(n-1)==En&&e>h
    u_inv(n)=Ep; % switch on Ep
  else
     u_inv(n)=u_inv(n-1);
  end
  i_outRR = (exp(-t./tau)*(i_outRR(RR)-u_inv(n)/R))+u_inv(n)/R;
  i_out(1,(((n-1)*RR+1):(n*RR)))=i_outRR; %fill the output vector
end
```

Another method for simulating the exemplary current controller is to use a digital model of the output circuit. For this purpose, the voltage-to-current transfer function of the RL circuit is created

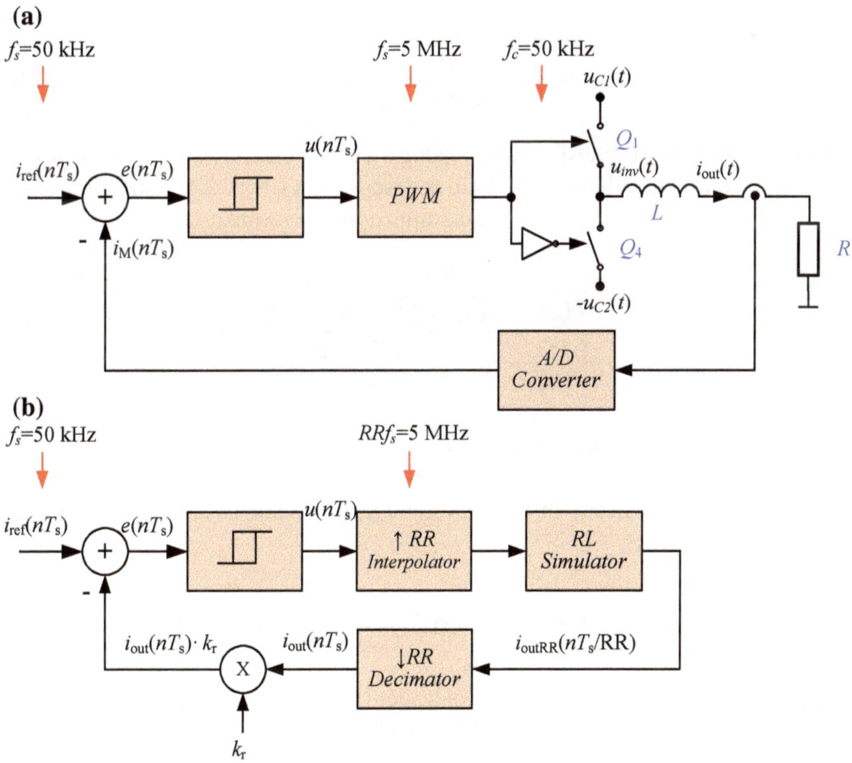

Fig. 4.11 The digital hysteresis current controller: **a** circuit diagram, **b** simulation diagram

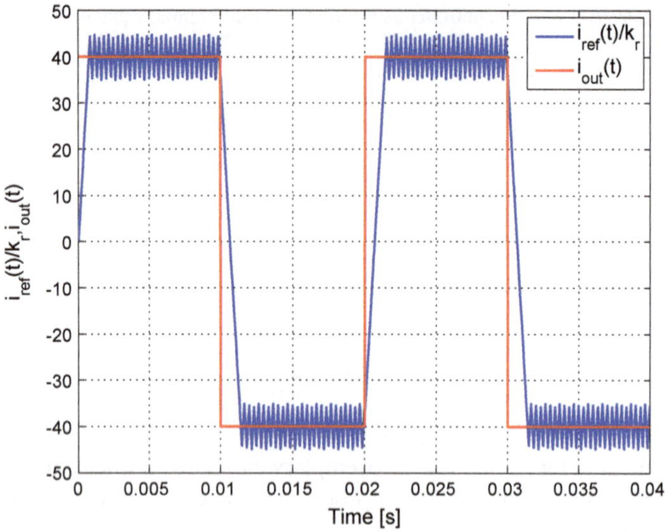

Fig. 4.12 The output current of the digital hysteresis current controller

$$H(s) = \frac{I(s)}{U(s)} = \frac{1}{R + sL} \quad . \tag{4.19}$$

Then, using the bilinear transform, the analog transfer function (4.19) is transformed into a digital one. The bilinear transform is described in Chap. 3. Finally, the digital transfer function of the RL circuit is determined by the formula

$$H(z) = \frac{I(z)}{U(z)} = \frac{\dfrac{T_s}{RT_s + 2L} + \dfrac{T_s}{RT_s + 2L} z^{-1}}{1 - \dfrac{RT_s - 2L}{RT_s + 2L} z^{-1}} \quad , \tag{4.20}$$

where T_s—the transfer function $H(z)$ sampling period.

Assuming the parameters of the RL circuit are the same as above this gives the digital transfer function

$$H(z) = 1.8181719 \cdot 10^{-5} \frac{1 - z^{-1}}{1 - 0.9999891 z^{-1}} \quad . \tag{4.21}$$

MATLAB implementation of current controller (Fig. 4.11b) using a digital model of the output circuit is shown in Listings 4.10 and 4.11.

Listing 4.10 The MATLAB implementation of current controller using digital model

```
clear all; close all; clc;
l_width = 2; f_size = 16;
Uc1=400; Uc2=-400; R=0.3; L=5.5e-3;
N=2000; fs=50000; Ts=1/fs; f=50;
RR=50; kr=0.025; h=0.10; fs_rr=fs*RR; Ts_rr=1/fs_rr;
t=(0:N-1)*Ts;
i_ref = square(2*pi*f*t,50);
%% conversion of the RL circuit to digital form
num_s=[1]; den_s=[L R]; % voltage to current transfer function
[num_d den_d]=bilinear(num_s,den_s,fs_rr)
i_out=f_inv_RL_filter_hist(i_ref,h,kr,num_d,den_d,Uc1,Uc2,RR,Ts);
trr=(0:length(i_out)-1)*Ts/RR;
figure('Name','Currents','NumberTitle','off');
plot(trr,i_out,t,i_ref/kr,'r','linewidth',l_width),
set(gca,'FontSize',f_size), grid on;
ylabel('i_{ref}(t)/k_r,i_{out}(t)'), xlabel('Time [s]');
legend('i_{ref}(t)/k_r','i_{out}(t)');
```

Listing 4.11 Hysteresis current controller using digital model of the RL circuit

```
function i_out=f_inv_RL_filter_hist(i_ref,h,kr,num_d,den_d,Ep,En,RR,Ts)
N=length(i_ref);
% initialization of variables
Zf=[]; i_outRR=zeros(1,RR); i_out=zeros(1,RR*N); u_inv=ones(1,N)*Ep;
for n=1:N
  e=i_ref(n)-i_outRR(RR)*kr;
  if n==1
      ; % start
  elseif u_inv(n-1)==Ep&&e<-h
    u_inv(n)=En; % switch on En
  elseif u_inv(n-1)==En&&e>h
    u_inv(n)=Ep; % switch on Ep
  else
     u_inv(n)=u_inv(n-1);
  end
  for k=1:RR
    [i_outRR(k) Zf]=filter(num_d,den_d,u_inv(n),Zf);
  end
  i_out(1,(((n-1)*RR+1):(n*RR)))=i_outRR; %fill the output vector
end
```

4.2.5 Simulation of the Power Electronics System Using Simulink®

MATLAB is a powerful multi-paradigm numerical computing environment. It has several dozen toolboxes for solving a huge number of problems in the fields of engineering and science. One such tool is Simulink®, which is a block diagram environment for multidomain simulation and model-based design [21]. The application of Simulink shortens by a multiple factor the preparation time for simulation of systems. Simulink consists of dozens of toolboxs too. One such is Powersys—a toolbox

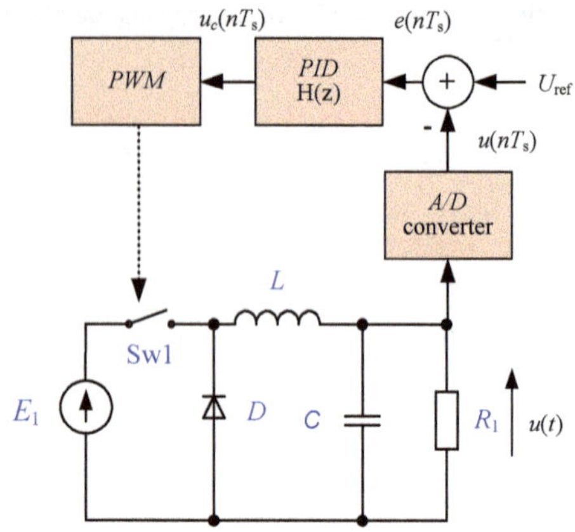

Fig. 4.13 The DC/DC converter

designed for power circuit simulation. Simulink program is accompanied by a number of sample applications. On the Mathworks web page [23], it is also possible to find many examples developed by users.

A circuit diagram of an exemplary DC/DC converter with digital control is depicted in Fig. 4.13. A Simulink model has been made for this circuit, and is shown in Fig. 4.14.

4.3 Simulation Using PSIM

PSIM is a specialized electrical engineering simulation and design tool for research and product development in power electronics systems with digital control systems from Powersim Inc. [28, 33]. In the author's opinion, one of the most comfortable programs for simulating power electronics circuits is the PSIM program. The PSIM greatly reduces development time.

4.3.1 Simulation Using the Typical PSIM Blocks

The PSIM simulation environment consists of the schematic program PSIM, two simulator engines, a PSIM engines, and the new SPICE engine [29], and the waveform processing program SIMVIEW1 [27, 33]. A circuit is represented in PSIM in four blocks: power circuit, control circuit, sensors, and switch controllers. The power circuit consists of switching devices, RLC branches, transformers, and coupled inductors. The control circuit is represented as a block diagram. Analog and digital components, logic components, and nonlinear components are used in the control circuit. To measure currents and voltages in the power circuit sensors are used.

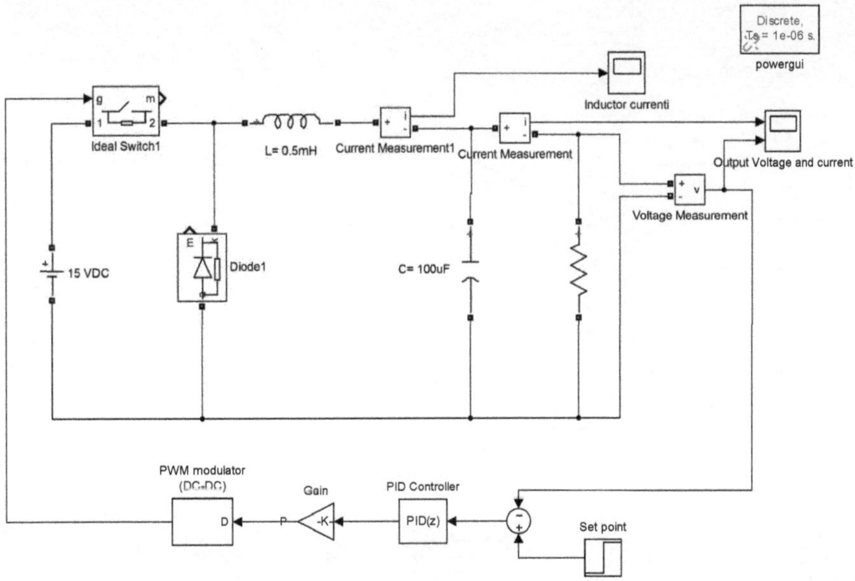

Fig. 4.14 The simulation diagram of the DC/DC converter

Figure 4.15 shows a diagram of the exemplary power electronics circuit simulated by PSIM. This is a three-phase shunt active power filter (APF) with a three-level inverter. The APF consists of a digital control circuit based on a sliding DFT algorithm. The result of the simulation research for the step change of load current is shown in Fig. 4.16. There are shown waveforms of three current lines $i_M(t)$, the load $i_L(t)$ and compensation $i_C(t)$.

4.3.2 Simulation Using C Code

As shown in Fig. 4.15, entering the digital control circuit using PSIM blocks may be cumbersome. Much greater freedom in the creation of the control system can be achieved by using C code. Therefore, it is possible to avoid many of the limitations associated with the use of defined blocks. Using C code for realizing the digital control algorithm the user can implement more sophisticated control algorithms. In PSIM it is only possible using **Simplified C Block**, **C Block**, **Simple DLL Block** and **General DLL Block** [26–28]. Those blocks can be used in either the power circuit or the control circuit.

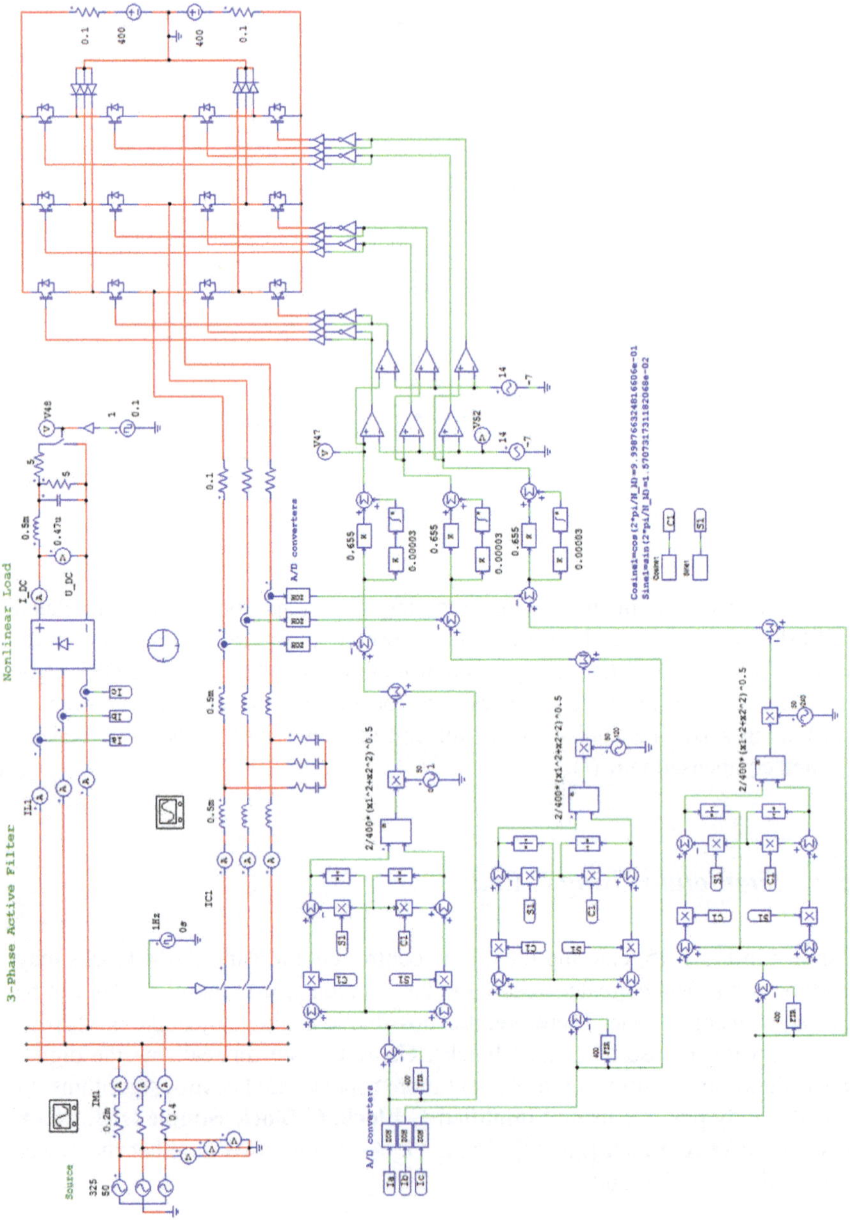

Fig. 4.15 The simulation diagram of the three-level shunt APF

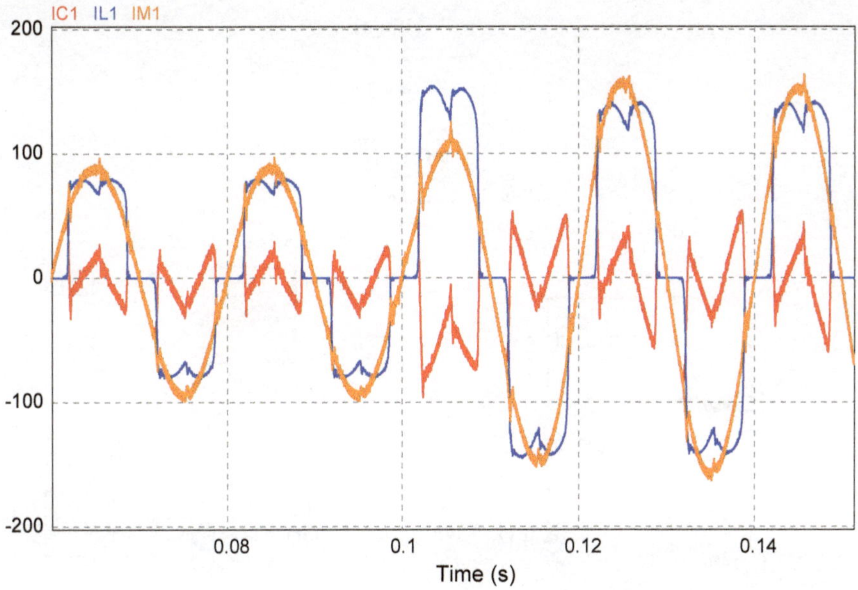

Fig. 4.16 The result of the APF simulation using PSIM, waveforms of currents $i_C(t)$, $i_L(t)$, $i_M(t)$

In **Simplified C Block** C code can be entered directly without any compiling. A C interpreter engine will interpret and execute the C code at runtime. This block makes it very easy to support custom C codes, and to define and modify the functionality of a block. The block is executed at each simulation time step *delt*. Unlike the DLL blocks, the Simplified C Block allows users to enter the C code directly without compiling the code. In comparison to the C Block, the Simplified C Block is easier to use.

What is also important to notice is that the value of the time step should be able to obtain coherent sampling of signals during simulation. For example, for processing signals with a frequency equal to 50 Hz, the value of the time step, should be equal to 20 ms divided by the integer number.

In order to illustrate the use of Simplified C Block, the hysteresis digital controller has been applied. The circuit shown in Fig. 4.17 is defined by a Simplified C Block with two inputs and two outputs. In the block, the following variables are valid: t, $delt$—current time, passed from PSIM and time step, passed from PSIM; input $x1$, $x2$—input variables passed from PSIM; output $y1$, $y2$—output variables passed back to PSIM. The C code is presented in Listing 4.12. The output current is assigned to the variable $x1$, the reference current is assigned to the variable $x2$ and the transistor control pulses are assigned to the variables $y1$ and $y2$. The results of the simulation are shown in Fig. 4.18.

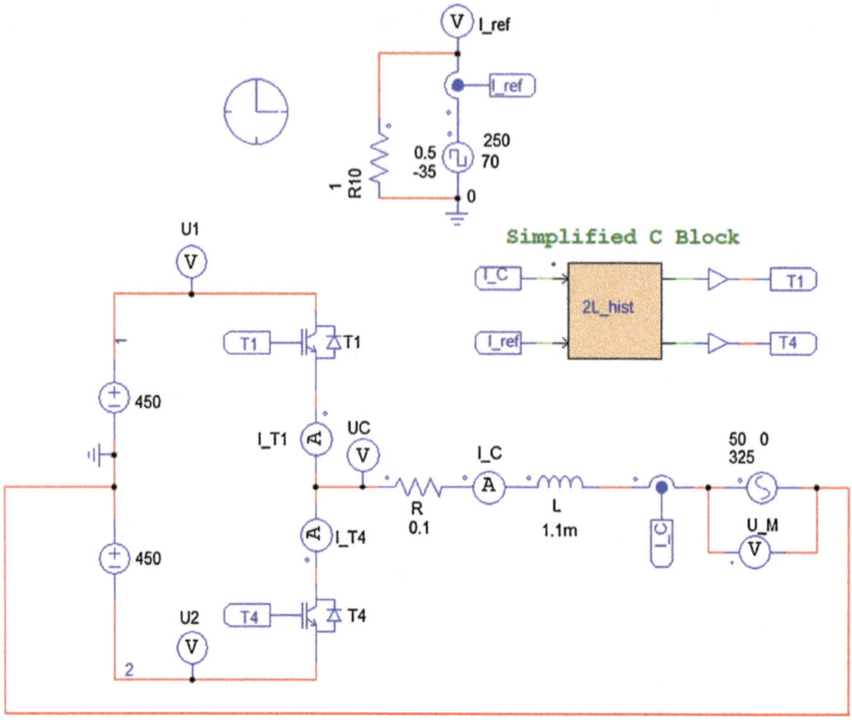

Fig. 4.17 The diagram of hysteresis current controller

Listing 4.12 Program for hysteresis current controller

```
//2L_hist3.C
//Simplified C Block
double e=0; //error
double iref=0; //setpoint current
double iout=x1; //output current
double h=3.5; //half of hysteresis band
static int S=0; //state of inverter
// if(t==delt){y1=1; y2=0; S=1;} //reset
iref=x2;
e=iref-iout; // the error calculation
if (-e>h) {
    y1=0; y2=1; S=0; // change the state
}
else if (e>h) {
    y1=1; y2=0; S=1; // change the state
}
else if(S==1) {
    y1=1; y2=0;
}
else {
    y1=0; y2=1;
}
```

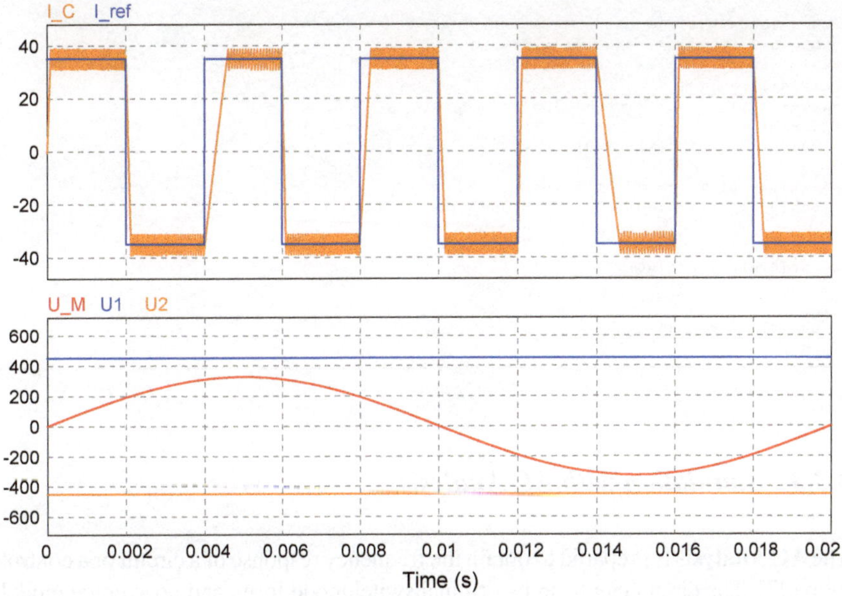

Fig. 4.18 The results of the simulation of the current controller, response for square wave

Another solution for simulation using C code is an external DLL (dynamic link library). There are two types of such blocks **General DLL Block** allows users to define an arbitrary number of inputs/outputs and **Simple DLL Block** with 1 input/out up to 25 inputs/outputs [26–28]. They allow users to write code in C/C++, compile it into DLL using external compilers, e.g., Microsoft Visual C/C++ Express, Bloodshed Dev-C++, etc., and link it with PSIM. As in the previous case, a DLL block receives values from PSIM as inputs, performs the calculation, and sends the results back to PSIM. PSIM calls the DLL routine at each simulation time step. There is, however, an exception, when the inputs of the DLL block are connected to the discrete elements the DLL block is called only at the discrete sampling times.

The example of the hysteresis digital current controller shown in Fig. 4.17 is used, where the Simplified C Block should be replaced by the General DLL Block. The controller C code for making DLL using Microsoft Visual C++ 2010 Express compiler is presented in Listing 4.13.

Listing 4.13 C code for realization of the hysteresis current controller using DLL

```
//for Microsoft  Visual C++ 2010 Express
#include "stdafx.h"
#include <math.h>
using namespace std;
__declspec(dllexport) void simuser (double t, double delt, double *in, double *out)
{
double e=0; //error
double iref=0; //setpoint current
```

```
double iout=in[0]; //output current
double h=3.5; //half of hysteresis band
static int S=0; //state of inverter
iref=in[1];
e=iref-iout; // the error calculation
if (-e>h) {
   out[0]=0; out[1]=1; S=0; // change the state
}
else if (e>h) {
   out[0]=1; out[1]=0; S=1; // change the state
}
else if(S==1) {
   out[0]=1; out[1]=0;
}
else {
   out[0]=0; out[1]=1;
}
```

4.3.3 Simulation with AC Analysis

The **AC Analysis** is prepared to obtain the frequency response of a circuit or a control loop [27]. The circuit can be in its original switchmode form, and no average model is required. However, the average model takes a much shorter time to perform the AC analysis.

The principle of the AC analysis is that a small AC excitation signal is injected into the system input as the perturbation, and the signal at the same frequency is extracted at the output. To obtain accurate AC analysis results, the excitation source amplitude must be properly set. The amplitude of AC sweep should be small enough for the perturbation to stay in the linear region and the amplitude must be large enough not to affect the output signal by numerical errors. The **AC Analysis** for open-loop circuit consists of the following steps:

- place AC sinusoidal voltage source (**Sine** source) at input, as the excitation source for the AC sweep,
- place **AC Sweep Probe** at the desired output,
- place the **AC Sweep** block on the schematic,
- run the simulation.

Figure 4.19 shows an exemplary circuit for AC analysis. The items added for the AC analysis are highlighted in red. The AC analysis is quite time consuming and took dozens of minutes. The results of such analysis is depicted in Fig. 4.20

The AC analysis can also be used to find the frequency response of a closed-loop system. In such a case, it is necessary to apply **AC Sweep Probe (loop)**.

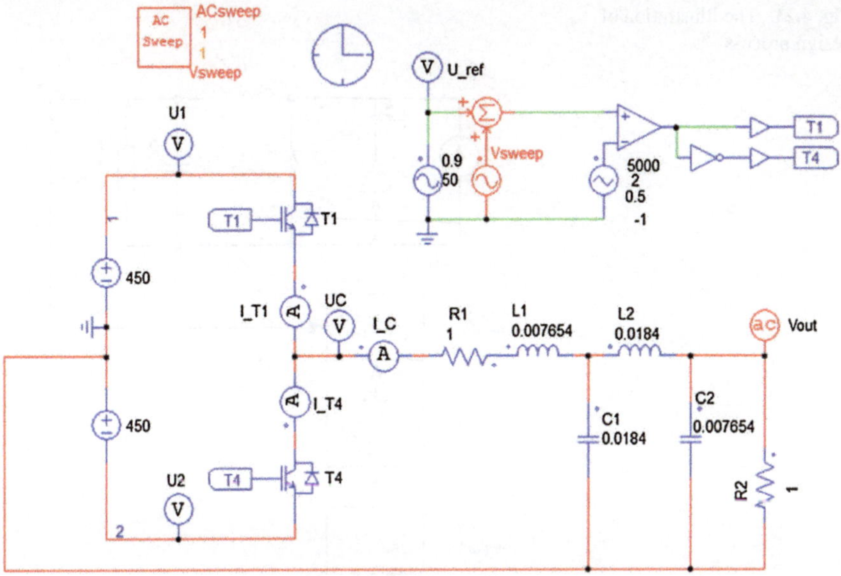

Fig. 4.19 The inverter with PWM, AC analysis circuit diagram

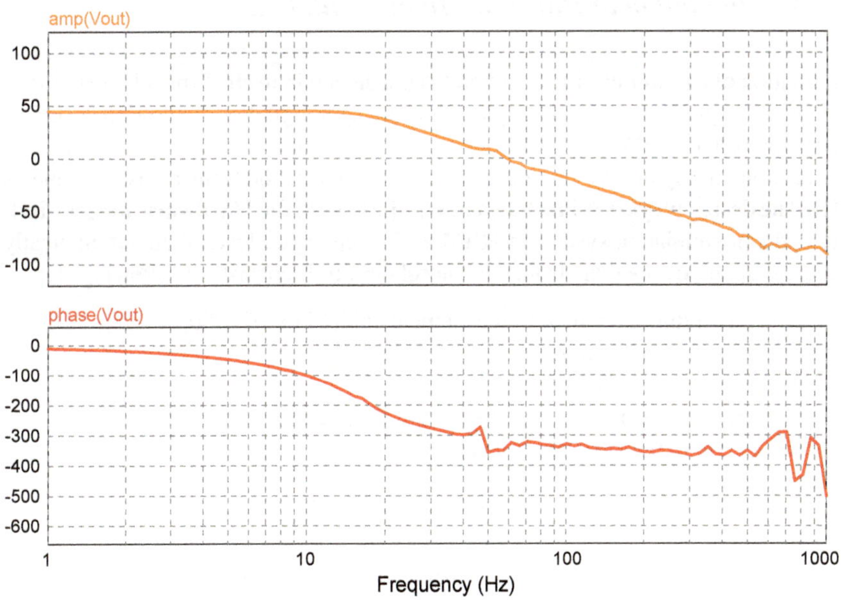

Fig. 4.20 The open-loop frequency response of the inverter with PWM

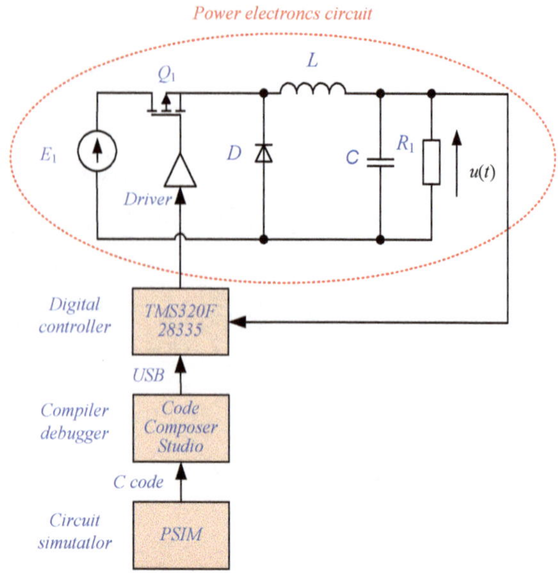

Fig. 4.21 The illustration of design process

4.3.4 Simulation to Hardware Implementation

Simulation of a power electronics circuit is not an end in itself. Typically then, a digital control algorithm design is implemented using a digital signal processor (DSP), microcontroller or FPGA. The Powersys company enables direct cooperation [32] with a large family of DSP microcontrollers TMS320F2000 from Texas Instruments (TI). The design process is illustrated in Fig. 4.21. With the **Hardware Target Module** one can simulate a system in PSIM at the schematic level, then automatically generate the hardware code from the control circuit. It consists of following steps:

- creating a circuit diagram with the elements of TMS320F2000,
- simulation and verification,
- generating code for the DSP target,
- compiling code using Code Composer Studio (CCS),
- loading code to the DSP target using CCS,
- running code in the DSP target using CCS,
- monitoring waveforms with PSIM's DSP oscilloscope,
- verification,
- uploading code to DSP target (for flash release DSP version) using CCS.

SimCoder™ is an add-on option of the PSIM software [30]. It generates C code from PSIM schematics. With specific hardware target libraries, the C code generated by SimCoder can run directly on target hardware platforms. One of them is the universal **DSP Development Board** (DSP board) [31], which is designed for product development of power electronics that uses Texas Instruments C28xxx family DSPs.

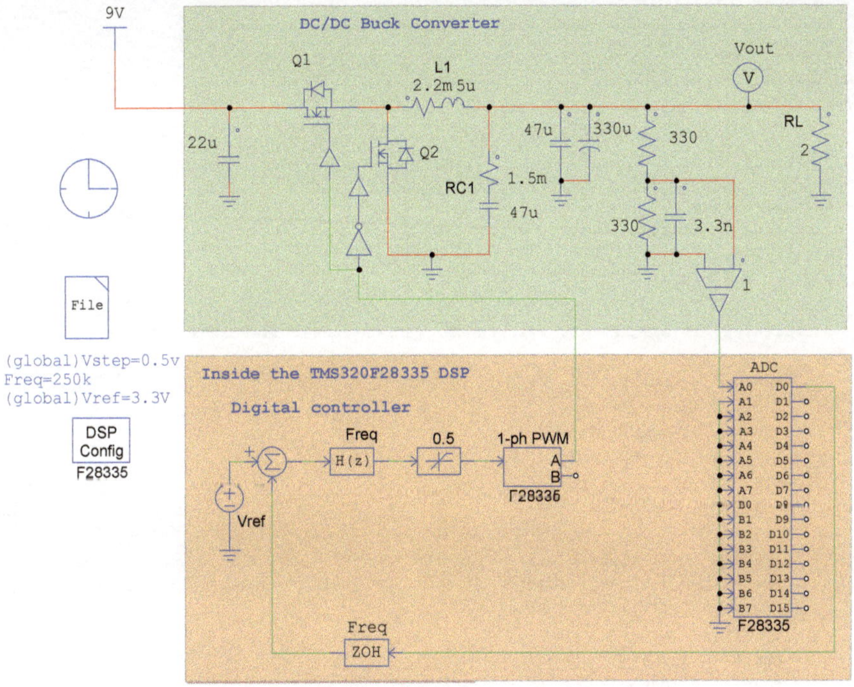

Fig. 4.22 The diagram of DC/DC buck converter

Applied on the DSP board are TI C2000 DSP family control cards—the control card F28335 is shown in Fig. 3.82. The DSP board contains all necessary signal conditioning circuits for a lot of power electronics applications.

Figure 4.22 depicts a diagram of an exemplary simple simulation circuit with DC/DC buck converter controlled by the TMS320F28335 DSP. The diagram consists of two elements of the DSP hardware **1-phase PWM** modulator and **ADC** converter. The digital controller with transfer function $H(z)$ is also realized by the DSP. The results of the simulation of an entire DC/DC buck converter are shown in Fig. 4.23. After the successful results of the simulation, C code for TMS320F28335 DSP is generated, it is described in the Listing 4.14. In the next step the C code is used by Code Composer Studio for programming TMS320F28335 DSP.

Listing 4.14 C code generated by PSIM for realization of the digital controller using TMS320F28335

```
/*********************************************************************
// This code is created by SimCoder Version 10.0 for F2833x Hardware Target
//
// SimCoder is copyright by Powersim Inc., 2009--2016
//
// Date: February 16, 2017 07:04:00
```

```
*************************************************************************************/
#include     <math.h>
#include     "PS_bios.h"
typedef float DefaultType;
#define GetCurTime() PS_GetSysTimer()
interrupt void Task();
// Parameters in parameter file _ParamFile1
DefaultType Vstep = 0.5;
#define Freq    250000.0
DefaultType Vref = 3.3;
interrupt void Task()
{
    DefaultType fLIMIT_UPPER2, fFILTER_D2, fSUM4, fZOH3, fPSM_ADC1, fVDC3;

    PS_MaskIntr(M__INT3);

    fPSM_ADC1 = PS_GetDcAdc(0);
    fVDC3 = Vref;
    fZOH3 = fPSM_ADC1;
    fSUM4 = fVDC3 - fZOH3;
    {
        static  DefaultType fIn[2] = {0, 0}, fOut[2] = {0, 0};
        fFILTER_D2 = 0.007 * fSUM4 + (-(0.01)) * fIn[0] - (-(1.0)) * fOut[0]
        + 0.005 * fIn[1] - 0 * fOut[1];
        fIn[1] = fIn[0];
        fIn[0] = fSUM4;
        fOut[1] = fOut[0];
        fOut[0] = fFILTER_D2;
    }
    fLIMIT_UPPER2 = (fFILTER_D2 > 0.5) ? 0.5 : fFILTER_D2;
    // Start of changing PWM2(1ph) registers
    // Set Duty Cycle
#ifdef PWM_IN_CHECK
    if (fLIMIT_UPPER2 <= 0) {
        PWM_CMPA(2) = 0;
    } else if (fLIMIT_UPPER2 >= (1 + 0)) {
        PWM_CMPA(2) = PWM_TBPRD(2);
    } else {
#else    // PWM_IN_CHECK
    {
#endif
        DefaultType _val = ((fLIMIT_UPPER2 - 0) * (1.0/1));
        PWM_CMPA(2) = (int)(PWM_TBPRD(2) * _val);
    }
    // End of changing PWM2(1ph) registers
    PS_ExitPwmIntr(2, M__INT3);
}
void Initialize(void)
{
    PS_SysInit(30, 10);
    PS_StartStopPwmClock(0);     // Stop Pwm Clock
    PS_InitTimer(0, 0);
    PS_InitPwm(2, 1, (double)Freq*1, (0E-6)*1e6, PWM_POSI_ONLY);
    // pwnNo, waveType, frequency, deadtime, outtype
    PS_SetPwmPeakOffset(2, 1, 0, 1.0/1);
    PS_SetPwmIntrType(2, ePwmNoAdc, 1, 0);
```

```
    PS_SetPwmVector(2, ePwmNoAdc, Task);
    PS_SetPwmTzAct(2, eTZHighImpedance);
    PS_SetPwm2RateSH(0);
    PS_StartPwm(2);

    PS_ResetAdcConvSeq();
    PS_SetAdcConvSeq(eAdcCascade, 0, 2.0);
    PS_AdcInit(0, !0);

    PS_StartStopPwmClock(1);      // Start Pwm Clock
}
void main()
{
    Initialize();
    PS_EnableIntr();    // Enable Global interrupt INTM
    PS_EnableDbgm();
    for (;;) { // empty loop
    }
}
```

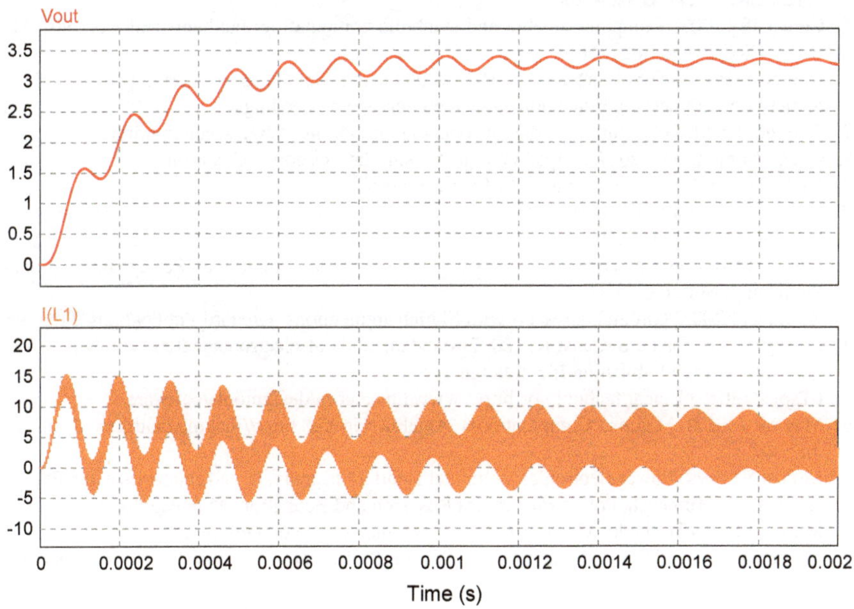

Fig. 4.23 The result of simulation of DC/DC buck converter, output voltage $v_{out}(t)$, inductor current $i_{L1}(t)$

4.4 Conclusions

This chapter has presented selected methods and programs suitable for simulating digital control circuits in power electronics systems. Special attention has been paid to the simulation of power electronics circuits and to power electronics circuits together with digital control circuits. The selected problems and examples have been drawn from the author's years of experience in this field.

References

1. Ari N (2009) Symbolic computation techniques for electromagnetics: with MAXIMA and MAPLE. LAP Lambert Academic Publishing, Saarbrücken
2. Attia JO (1999) Electronics and circuit analysis using Matlab. CRC Press, Boca Raton
3. Blajberg F, Ionel DM (2017) Renewable energy devices and systems with simulations in Matlab and ANSYS. CRC Press, Boca Raton
4. Cohen JS (2002) Computer algebra and symbolic computation: elementary algorithms. AK Peters/CRC Press, Boca Raton
5. Cohen JS (2003) Computer algebra and symbolic computation: mathematical methods. AK Peters/CRC Press, Boca Raton
6. Conrad WR (1995) Solving RL and RC circuits using Matlab. In: Proceedings of American Society for Engineering Education, annual conference proceedings
7. Getreuer P (2009) Writing fast Matlab code. matopt.pdf. http://www.getreuer.info
8. Gecko-Simulations (2017). http://www.gecko-simulations.com/index.html
9. GNU Octave. https://www.gnu.org/software/octave/
10. IEEE (2017) List of IEEE milestones. http://ethw.org/Milestones:List_of_IEEE_Milestones
11. Irwin JD, Nelms RM (2015) Basic engineering circuit analysis, 11th edn. Wiley, Hoboken
12. Jain A (2010) Power electronics: devices, circuits and Matlab simulations. Penram International Publishing, Mumbai
13. Karris ST (2003) Circuit analysis II with Matlab applications. Orchard Publications, Fremont
14. Lonngren KE, Savov SV, Jost RJ (2007) Fundamentals of electromagnetics with MATLAB, 2nd edn. SciTech Publishing Inc, Raleigh
15. LTspice (2017) Linear technology. http://www.linear.com/designtools/software/
16. Magma (2017) Magma. Computational Algebra Group. http://magma.maths.usyd.edu.au/magma/
17. Mohan N (1999) Power electronics: computer simulation, analysis, and education using PSpice schematics. Minnesota Power Electronics Research and Education, Minneapolis
18. Maplesoft (2017) Maple. Maplesoft. http://www.maplesoft.com/solutions/education/
19. Mathematica (2017) Mathematica. Wolfram Research. https://www.wolfram.com/mathematica/
20. Mathcad (2017). http://www.ptc.com/engineering-math-software/mathcad
21. Matlab (2016) Simulink user's guide. Matlab&Simulink 2016b. MathWorks
22. Matlab (2017) Matlab documentation. MathWorks. http://www.mathworks.com
23. Matlab (2017) Matlab file exchange. http://www.mathworks.com/matlabcentral/fileexchange/
24. Matlab (2017) MATLAB and Simulink based books. https://www.mathworks.com/support/books.html
25. Patil M, Rodey P (2015) Control systems for power electronics: a practical guide. Springer, New Delhi
26. Powersys (2004) Tutorial. PSIM simulation software, Powersys Inc., How to use the DLL block

27. Powersys (2016) PSIM user's guide. Version 11.0. Powersys Inc.
28. Powersys (2016) Tutorial. Powersys Inc., How to use general DLL block
29. Powersys (2016) SPICE module user's guide. Version 11.0. Powersys Inc.
30. Powersys (2016) SimCoder user's guide. Version 11.0. Powersys Inc.
31. Powersys (2016) DSP development board user's manual. Version 11.0. Powersys Inc.
32. Powersys (2016) Tutorial. Powersys Inc., Auto code generation for F2833X target
33. Powersys (2017). https://powersimtech.com/products/psim/
34. Pspice (2017). http://www.pspice.com/
35. SageMath (2017). http://www.sagemath.org/index.html
36. Scilab (2017). http://www.scilab.org/scilab/about
37. Small DB, Hosack JM (1991) Exploration in calculus with a computer algebra system. McGraw-Hill College, New York
38. Szczesny R (1999) Computer simulation of power electronic systems. Wydawnictwa Politechniki Gdaskiej (in Polish)
39. Shaffer R (2006) Fundamentals of power electronics with Matlab. Charles River Media, Independence
40. Shenkman AL (2005) Transient analysis of electric power circuits handbook, Springer, Boston
41. Sobierajski M, Labuzek M (2005) Programming in Matlab for electrical engineering. Oficyna Wydawnicza Politechniki Wroclawskiej, Wroclaw (in Polish)
42. Sozaski K (1998) Zastosowanie programu Matlab w elektrotechnice. In: Stryjski R (ed) Magnucki K. Wybrane metody komputerowe stosowane w technice. WSP, Zielona Gra (in Polish)
43. SPICE (2017). http://bwrcs.eecs.berkeley.edu/Classes/IcBook/SPICE/
44. Sturm RD, Kirk DE (2000) Contemporary linear systems using Matlab. Brooks/Cole Publishing Company, Pacific Grove
45. SymPy (2017). http://www.sympy.org/en/index.html
46. TINA (2017) DesignSoft. https://www.tina.com/

Chapter 5
Selected Active Power Filter Control Algorithms

5.1 Introduction

In the early days of active power filters (APF), at the end of the 1980s and early 1990s, hybrid control circuits were used, which consisted of both analog and digital components. For example, an integrated circuit AC vector processor AD2S100 from Analog Devices [1] was used by the author in such control circuits [55]. In subsequent years a slow transition to fully digital control systems occurred. These systems are currently most commonly used. The use of digital control systems made it possible to use more complex digital signal processing algorithms. Therefore this chapter is devoted to the selected digital signal processing algorithms designed to control APFs.

This chapter describes author's modifications of selected APF control algorithms [48]. To begin with, harmonic detectors are considered: IIR filter, LWDF, sliding DFT [39, 41] and sliding Goertzel, moving DFT. Then the author's implementation of a classical control circuit based on modified $p - q$ algorithm is discussed [49].

Dynamic distortion in APF makes it impossible to fully eliminate line harmonics. In some cases, the line current *THD* ratio for systems with APF compensation can reach a value of a dozen or so percent. So the problems of active power filter dynamics should be investigated. The power loads can be divided into two main categories: predictable loads and noise-like loads. Most loads belong to the first category. For this reason, it is possible to predict current values in subsequent periods, after a few periods of observation. The author has proposed APF models suitable for analysis and simulation of this phenomena. The author has found a solution to these problems [41, 44, 47, 48]. For predictable line current changes, it is possible to develop a predictable control algorithm that allows for significant reduction in APF dynamics compensation errors. The following sections describe the author's modification using a predictive circuit to reduce dynamic compensation errors [41, 44, 48].

Subsequent sections include control circuits with filter banks which allow the selection of compensated harmonics. The considered filter banks are based on moving DFT and $p - q$ algorithms [50, 51].

K. Sozański, *Digital Signal Processing in Power Electronics Control Circuits*,
Power Systems, DOI 10.1007/978-1-4471-7332-8_5

For unpredictable line current changes, the author has developed a multirate APF [44, 48]. The presented multirate APF has a fast response to sudden changes in the load current. Therefore using multirate APF it is possible to decrease the *THD* ratio of line current even for unpredictable loads.

5.2 Control Circuit of Shunt APFs

A three-phase shunt APF compensator is depicted in Fig. 5.1. This circuit corresponds to the circuit from Fig. 1.14a, without feedback (with unity gain). The shunt APF injects the compensation current $i_C(t)$, into the power network and offers a notable compensation for harmonics and reactive power. The compensation current can be determined by

$$i_C(t) = i_L(t) - I_{H1}\sin(2\pi f_M t) \quad , \tag{5.1}$$

where: I_{H1}—amplitude of first harmonic, f_{M}—frequency of first harmonic.

In the case when harmonics compensation is perfect, the line current $i_{\text{M}}(t)$ consists of only the first harmonic line current

$$i_M(t) = I_{\text{H1}}\sin(2\pi f_M t) \quad . \tag{5.2}$$

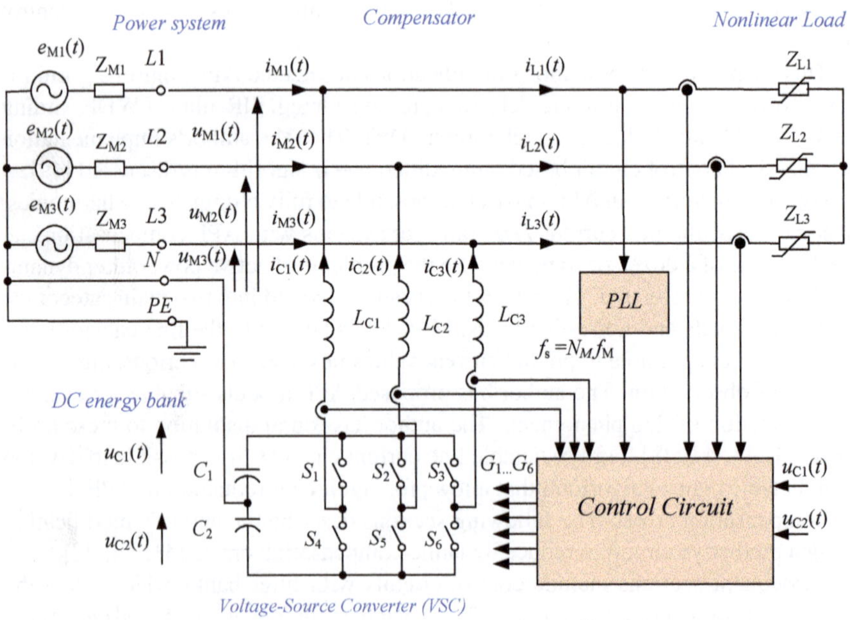

Fig. 5.1 Three phase shunt APF compensator

When the phase angle between line voltage $u_1(t)$ and line current $i_M(t)$ is equal to zero, the reactive power is compensated too.

In the APF (Fig. 5.1) three load currents $i_{L1}(t), i_{L2}(t), i_{L3}(t)$ are measured, and then used to determine the instantaneous values of compensation currents $i_{C1}(t)$, $i_{C2}(t)$, $i_{C3}(t)$. To generate compensation currents, a three-phase inverter is applied. Due to the fact that the inverter is like a voltage source, it is necessary to use output inductors L_{C1}, L_{C2}, L_{C3} and a current controller with feedback $(i_{C1}(t), i_{C2}(t), i_{C3}(t))$ to obtain the qualities of a current source. The three-phase inverter is supplied from a DC energy bank consisting of two electrolytic capacitors C_1 and C_2. The capacitors are charged from power lines through a three-phase inverter using an additional voltage controller implemented in the APF's main control algorithm.

The shunt APF multirate control circuit is depicted in Fig. 5.2, but it should be noted that in this figure the controller of the capacitor voltage and the circuits for sample rate conversion are omitted.

There are many APF control methods, among which the following can be cited: Gyugyi and Strycula [22], the instantaneous reactive power (IPT) theory by Akagi [2–6], by Fryze [18, 19] and by Czarnecki [15–17], $p - q - r$ algorithm by Kim et al. [29], a review of first harmonic detection by Asiminoaei et al. [10], Aredes [7], Singh et al. [38], closed loop harmonic detection by Mattavelli [13, 33], Ghosh and Ledwich [21] and sliding DFT by the author [39, 41, 42, 48, 52, 53]. An interesting set of methods for improved power quality is presented by Benysek et al. [11]. A review of the principles of electrical power control is also described by Pasko and Maciazek in [36].

5.2.1 *Synchronization*

Figure 5.3 presents an analog synchronization circuit used by the author in the APF control circuits. In this circuit, the voltage of the three-phases is supplied through isolation transformers to the inputs of the active low-pass filters (LPF). 4-order Butterworth with passband frequency 0–50 Hz is used here. For frequency f_{cr} = 50 Hz the phase response is equal to $-180°$ and it can be easily compensated by the inverting amplifier. Sinusoidal signals from the filters are converted to the square wave by comparators. Then, the signal from one phase is connected to the phase detector input of an analog PLL. The PLL generates sampling signal frequency

$$f_s = N_M f_M \quad , \tag{5.3}$$

where: f_M—power line frequency, N_M—number of samples per power line period $T_M = 1/f_M$. Another function of the circuit is detecting the presence of all phases and the phase sequence.

In order to avoid beat frequency between the power line frequency and the compensation current modulation frequency, a fully synchronized control system should be applied. Preferably, the modulation frequency should be a multiple of power line

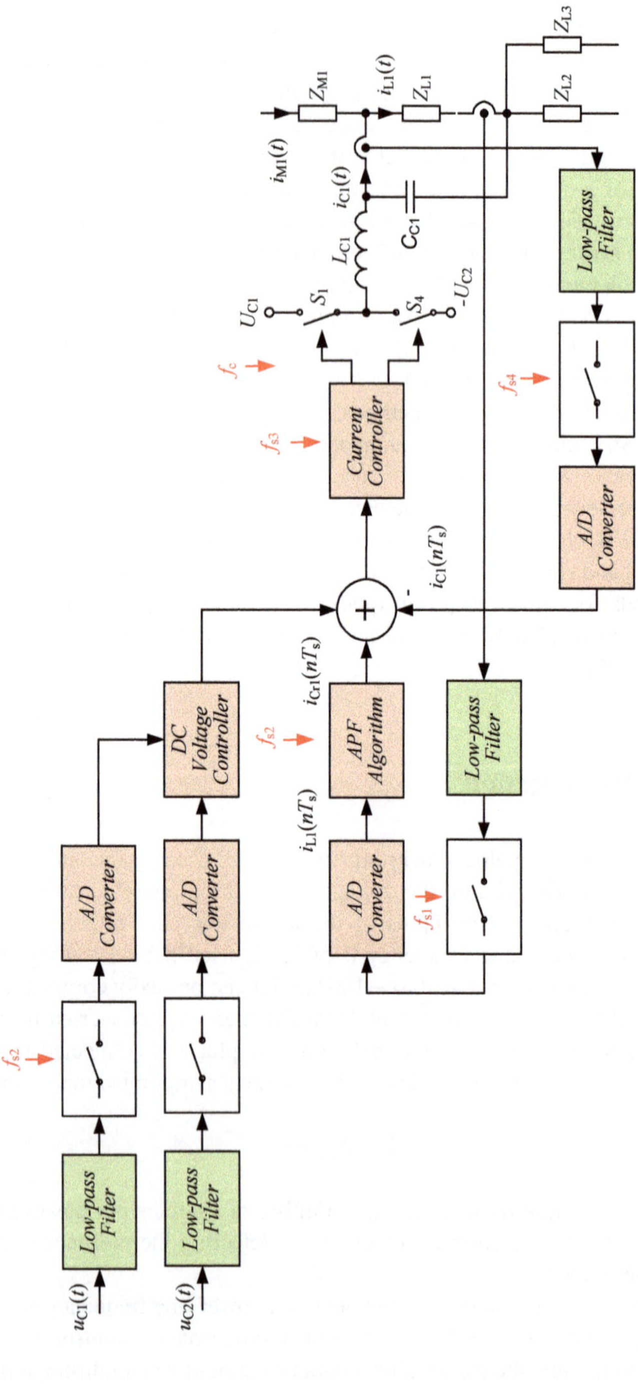

Fig. 5.2 The shunt APF multirate control circuit

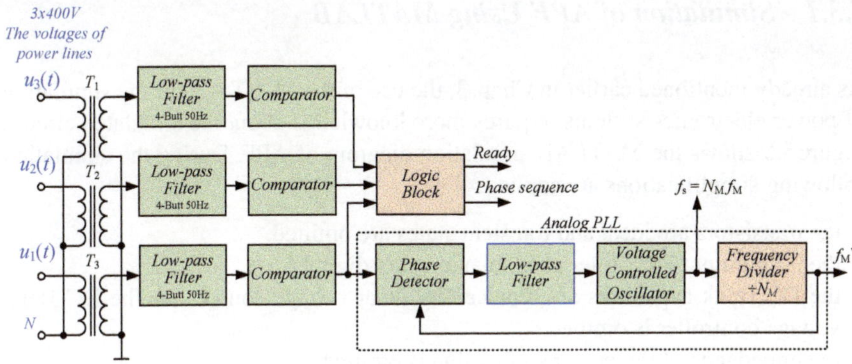

Fig. 5.3 APF analog synchronization circuit with PLL

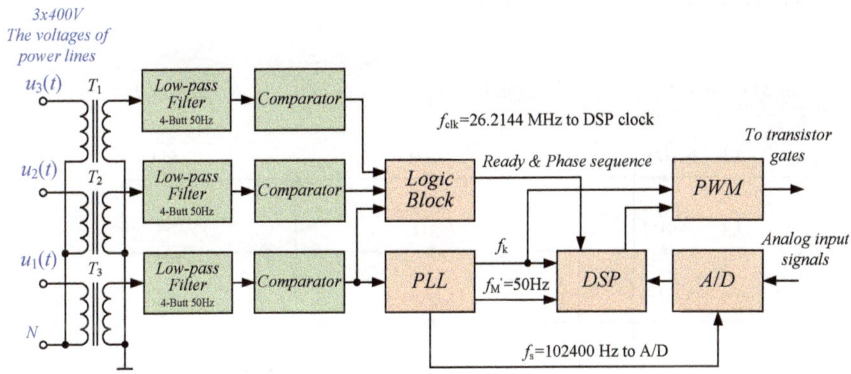

Fig. 5.4 Digital control system with full synchronization

voltage frequency, because to avoid beat frequency. The block diagram of such a solution is depicted in Fig. 5.4. In this circuit, the DSP, PWM and A/D converter are synchronized with the power line by a common PLL circuit.

5.3 Simulation of APF

Simulation problems of digital APF control algorithms and whole APF compensation circuit are not easy to solve. The author uses two solutions. In the first MATLAB with a digital version of the APF output RLC filter is used; in the second solution PSIM is used, wherein the APF control algorithm is implemented using C code.

5.3.1 Simulation of APF Using MATLAB

As already mentioned earlier in Chap. 3, the use of the MATLAB for the simulation of power electronics systems requires more knowledge about the simulated circuit. Figure 5.5 shows the MATLAB simulation diagram of APF. During the simulation following simplifications are applied:

- the transistors are ideal and transient states are omitted.
- the APF output RLC filter is simulated using digital form.
- the DC Bank capacitors are replaced by ideal voltage sources, so the DC Bank voltage controller is omitted,
- the impedance of the power supply line is omitted,
- sample rate signal is changed using only upsampler and downsampler.

A model simulating typical power load is used. However, it is also possible to use recorded waveform of load current, when available.

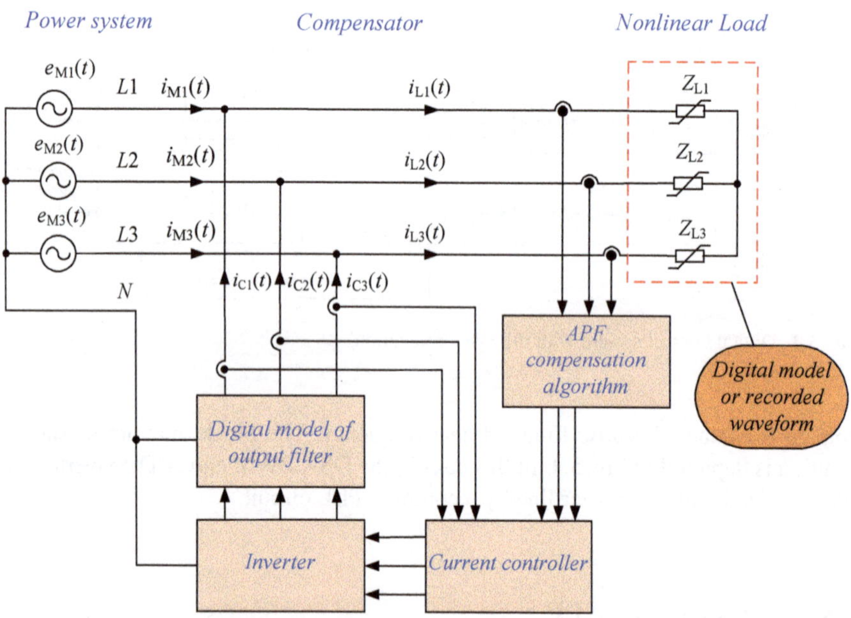

Fig. 5.5 The MATLAB simulation diagram of APF

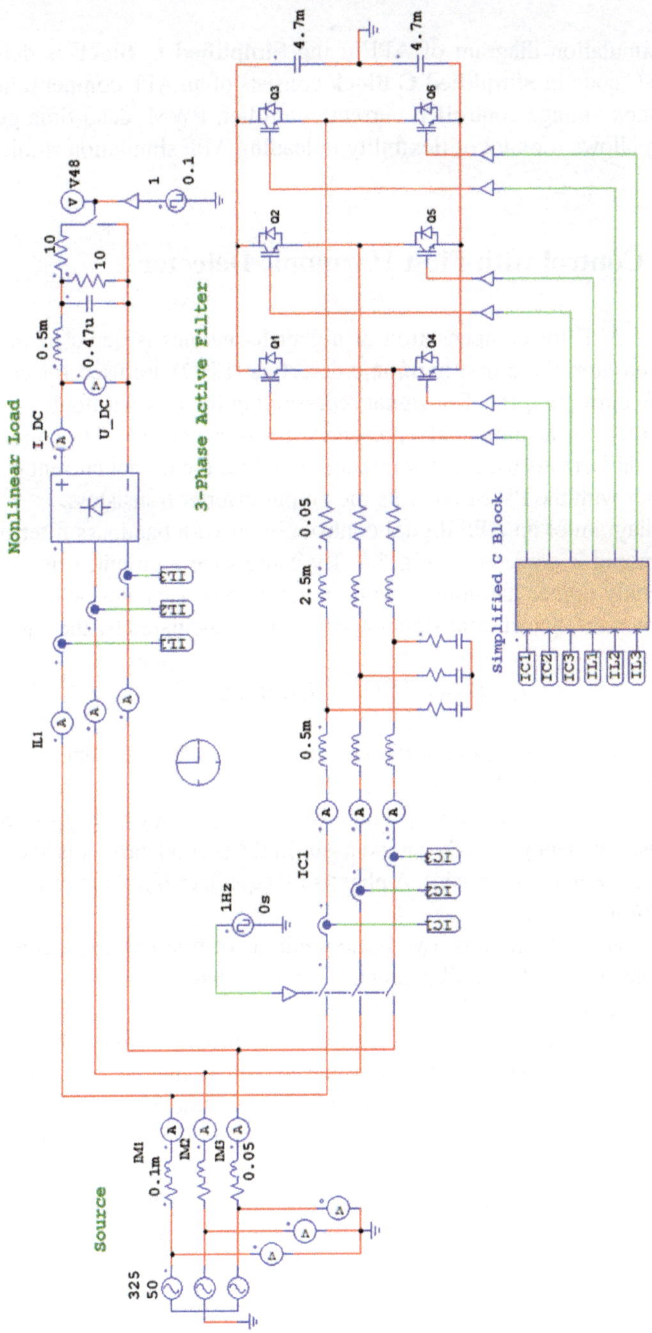

Fig. 5.6 The PSIM simulation diagram of APF using simplified C block

5.3.2 *Simulation of APF Using PSIM*

The PSIM simulation diagram of APF using **Simplified C Block** is depicted in Fig. 5.6. The C code in **Simplified C Block** consists of an APF compensation algorithm, DC Bank voltage controller, current controller, PWM, dead-time generator. This solution allows for a lot of flexibility in leading APF simulation studies.

5.4 APF Control with First Harmonic Detector

A three-phase APF for compensation of higher harmonics is depicted in Fig. 5.7. In this compensator, three first harmonic detectors (FHD) are used for calculating compensation currents $i_C(t)$. The signal representing the first harmonic component $i_{H1}(n)$ is subtracted from the signal representing load current $i_L(n)$. The compensation reference signal $i_{Cr}(n)$ is used as a reference signal for the output current controller, which together with the PWM controls the output inverter transistors.

A block diagram of an APF digital control circuit with bandpass filter tuned for the first harmonic is depicted in Fig. 5.8. By using such a circuit, it is possible to compensate only higher harmonics, while reactive power cannot be compensated. Compensation reference current signal $I_{Cr}(z)$ can be calculated by the equation

$$I_{Cr}(z) = I_L(z)(1 - H_1(z)H_p(z)) \quad , \tag{5.4}$$

where: $H_1(z)$—transfer function of digital filter, $H_p(z)$—transfer function of digital filter for phase correction, $I_L(z)$—signal of load current.

In such a circuit, a low-pass filter with crossover frequency $f_{cr} = f_M$ or bandpass filter with pass frequency $f_p = f_M$ can be used. In the second case, it is also possible to compensate lower subharmonics. A phase shift equalizer $H_p(z)$ is applied to obtain phase correction.

In the following subsections, two basic methods of first harmonic detection are considered: one using a digital filter and the other one based on a DFT. The common drawbacks of the DFT Fourier-based harmonic detection methods are their imprecise results in transient conditions and high requirements: proper design of the antialiasing filter, synchronization between the sampling and fundamental frequency, careful application of the windowing function, proper usage of the zero-padding to achieve the power of two series of samples, large memory requirements to store the achieved samples, large computational power required for the DSP.

In the following sections, the author presents his own solution for the first harmonic detection circuits. The author has examined these circuits using simulation and experimental tests.

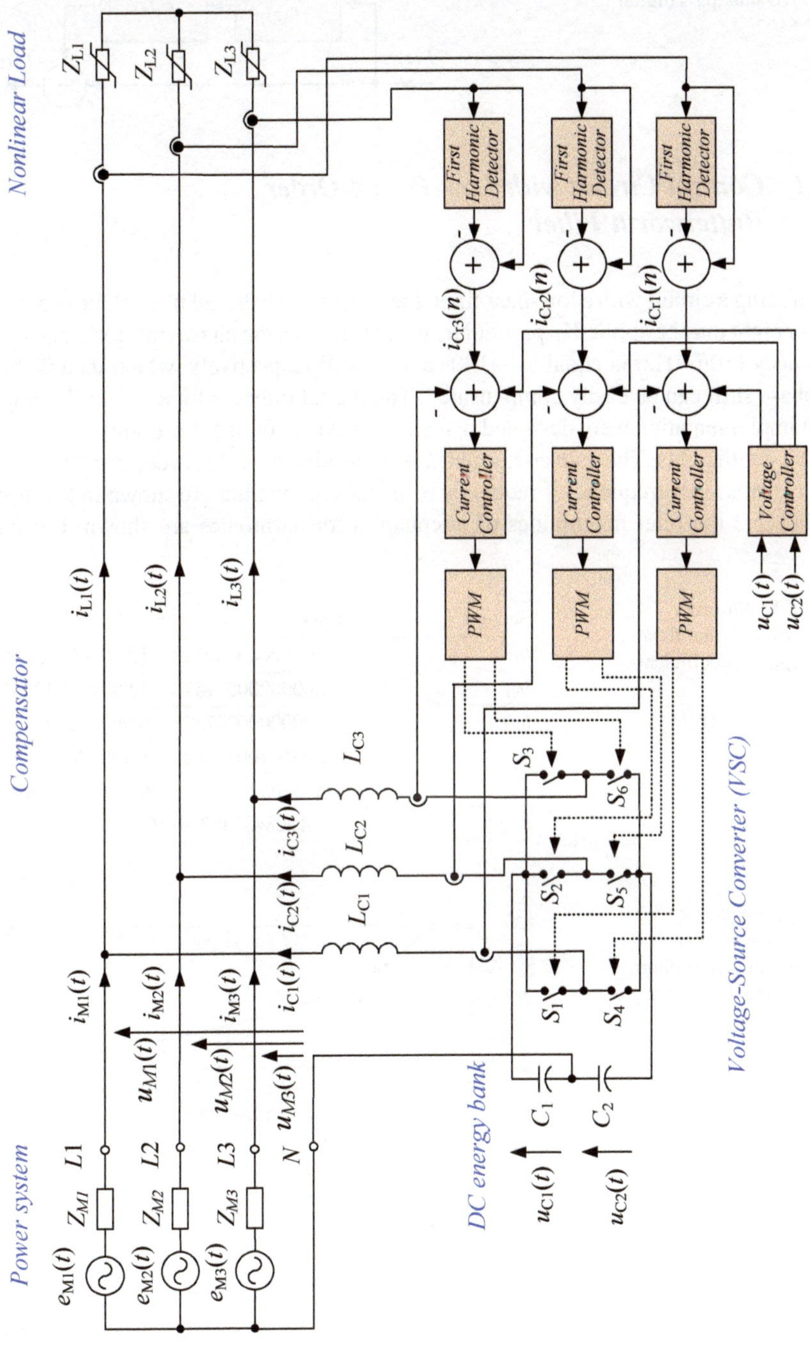

Fig. 5.7 Three-phase APF with FHD control circuit

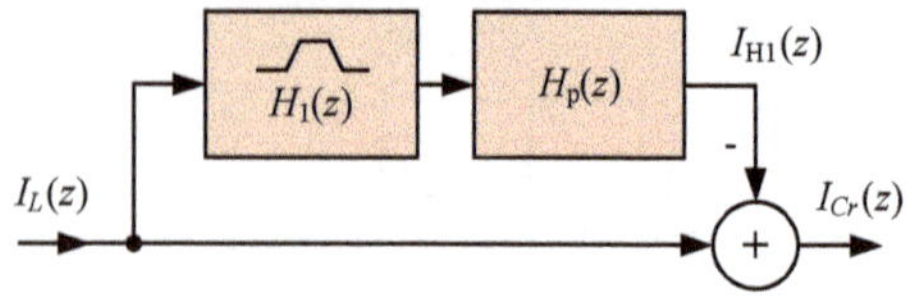

Fig. 5.8 APF control circuit with FHD bandpass digital filter

5.4.1 *Control Circuit with Low-Pass 4-Order Butterworth Filter*

When using a circuit with a low-pass filter, the Butterworth digital filter of the order of four or eight may be applied in particular, and in this case, the phase shift for crossover frequency (50/60 Hz) is equal to $-180°$ and $-360°$ respectively, which means that the phase shift can be easily compensated. The digital filter coefficients for floating-point implementation were designed using MATLAB®. Table 5.1 presents the values of filter coefficients. The gain coefficient k is multiplied by $\sqrt{2}$ in order to get the unit gain for crossover frequency. Frequency responses of the filter are shown in Fig. 5.9. In Table 5.2 the filter magnitudes of attenuation for harmonics are shown. For the

Table 5.1 4th-order Butterworth digital filter floating-point coefficients

Section	$n = 0$	$n = 1$
b_{n0}	1.000000000000	1.000000000000
b_{n1}	2.000000057757	1.999999942243
b_{n2}	1.000000007739	0.999999992261
a_{n1}	-1.955070062590	-1.98079495886
a_{n2}	0.955659070541	0.981391716995
k	$2.196845545754 \cdot 10^{-8}\sqrt{2}$	

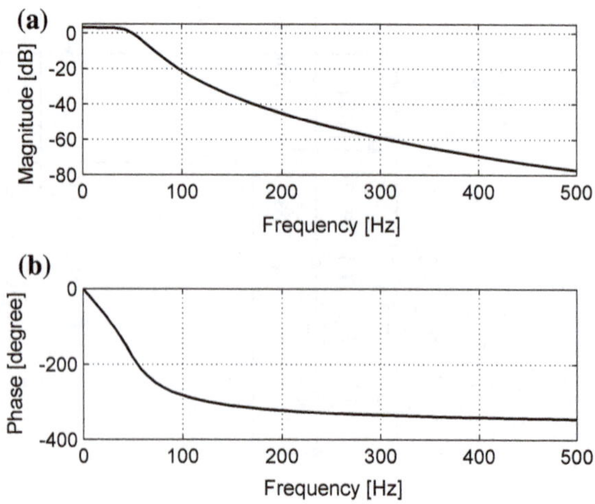

Fig. 5.9 Frequency responses of 4-order Butterworth digital filter: **a** magnitude, **b** phase angle

Table 5.2 Magnitude attenuation for the harmonics of 4th-order Butterworth digital filter

Frequency [Hz]	50	100	150	200	250	300	350
Magnitude [dB]	0	−21.09	−35.17	−45.18	−52.95	−59.30	−64.68

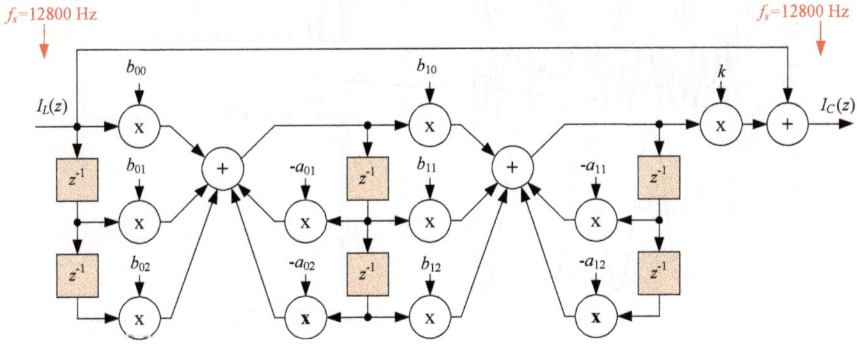

Fig. 5.10 APF control circuit with 4-order Butterworth digital filter

second harmonic (100 Hz) the magnitude of attenuation is equal to −21.09 dB, so the second harmonic is suppressed 11.3 times. If this is not enough, it possible to apply an 8th-order Butterworth filter, which has the magnitude of attenuation equal to −45.39 dB for the second harmonic. In this solution the second harmonic is suppressed 186 times. The APF control circuit with 4th-order filter is shown in Fig. 5.10. A simulation of three-phase compensation APF was made for this circuit. The results of such a simulation are presented in Fig. 5.11. Waveforms of currents $i_L(t)$, $i_C(t)$, $i_M(t)$ are depicted in Fig. 5.11a and their spectra in Fig. 5.11b. The same line current parameters are presented in Table 5.3, which confirms the effectiveness of such harmonic compensation.

5.4.2 *Control Circuit with Low-Pass 5-Order Butterworth LWDF*

Due to the numerous advantages of the LWDF (**L**attice **W**ave **D**digital **F**ilter), which are described in Chap. 3, it would be beneficial to use it in an APF control circuit. However, low-pass LWDF filters can only be realized for odd orders, so it is not possible to use the idea from the previous section. Therefore, a low-pass 5-order LWDF with a crossover frequency shifted to $f_{cr} = 59.1$ Hz is applied, chosen to obtain phase shift equal to −180° for the first harmonic. If the sampling frequency f_s is synchronized with the line voltage frequency f_M, the filter maintains its attenuation and phase shift despite the changing frequency f_M. To find the correct compensation, the filter gain should also be corrected by a factor of $k = 1.0898708$. A block

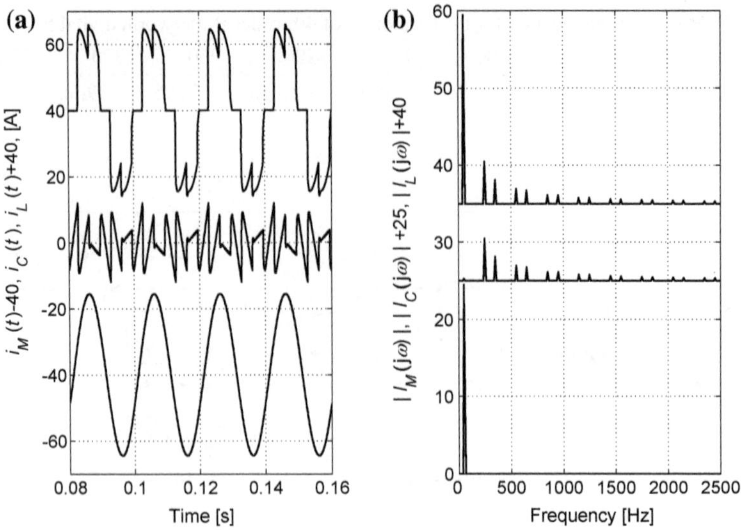

Fig. 5.11 Simulation result of APF with control circuit with 4th-order Butterworth digital filter: **a** waveforms, **b** spectra

Table 5.3 Effects of APF compensation

Current i_M	I_M(rms) [A]	*THD* [%]	*SINAD* [dB]	*THD*50 [%]
Without compensation	18.0	29.7	−10.9	29.6
Algorithm with 4-order Butterworth	17.3	0.7	−43.0	0.1

diagram of an APF control circuit with a 5-order Butterworth LWDF is depicted in Fig. 5.12. The filter coefficient values were calculated using the (L)WDF Toolbox for MATLAB [8, 9] and the results are shown in Tables 5.4 and 5.5. The coefficient values of this filter in comparison with the filter from the previous section are more suitable for low precision arithmetics, especially fixed-point arithmetics.

5.4.3 *Control Circuit with Sliding DFT*

The sliding DFT algorithm is described in Chap. 3. In the author's opinion the sliding DFT is highly suitable for APF control circuits [39–41, 43, 44]. The principle of the algorithm is described in Sect. 3.7.3. In the considered solution only one signal-bin sliding DFT filter structure for detecting the first harmonic of load current is used [48]. The first harmonic spectral component signal of load current is thus calculated

$$S_1(nT_s) = e^{j2\pi k/N_M}(S_1((n-1)T_s) - i_L((n-N_M)T_s) + i_L(nT_s)) \quad , \tag{5.5}$$

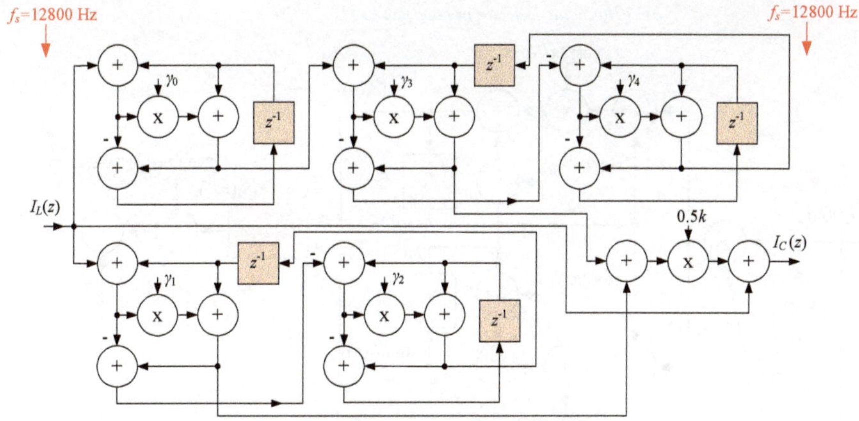

Fig. 5.12 APF control circuit with 5th-order Butterworth LWDF

Table 5.4 5th-order Butterworth lattice wave digital filter coefficients

γ	Value
γ_0	0.9714041474
γ_1	−0.9541456454
γ_2	0.9995792798
γ_3	−0.9822334470
γ_4	0.9995792798

Table 5.5 Magnitude attenuation for the harmonics of 5-order Butterworth LWDF

Frequency [Hz]	50	100	150	200	250	300	350
Magnitude [dB]	0	−22.14	−39.71	−52.23	−61.95	−69.89	−76.60

where: $i_L(nT_s)$—discrete signal representing load current, $S_1(nT_s)$—discrete signal representing first harmonic complex spectral component of first phase load current, N_M—number samples per line period. The discrete signal representing the first harmonic signal of a load current with zero phase angle between line voltage $u_1(t)$ and line current $i_{H1}(t)$ can be described by the equation

$$i_{H1}(nT_s) = 2/N_M |S_1(nT_s)| \sin(2\pi 50 nT_s) \quad . \tag{5.6}$$

Compensation current signal is the result of a difference between the load current signal and the first harmonic reference sinusoidal signal

$$i_C(nT_s) = i_L(nT_s) - 2/N_M |S_1(nT_s)| \sin(2\pi 50 nT_s) \quad . \tag{5.7}$$

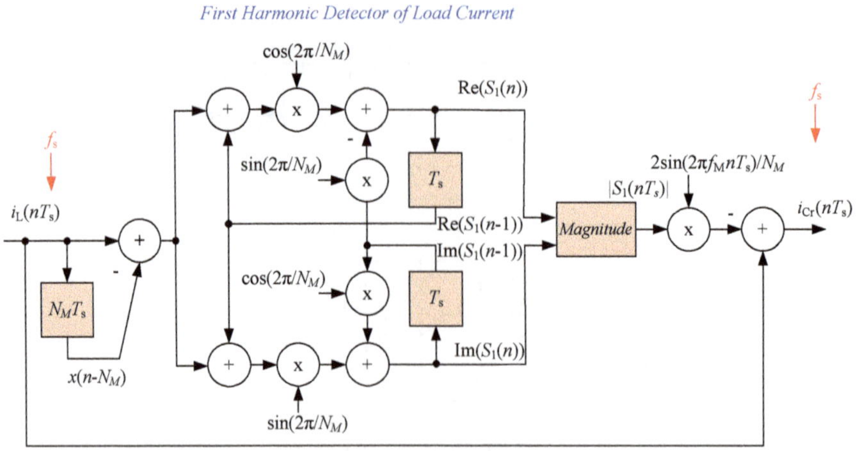

Fig. 5.13 Control algorithm for three-phase APF with harmonics and reactive power compensation

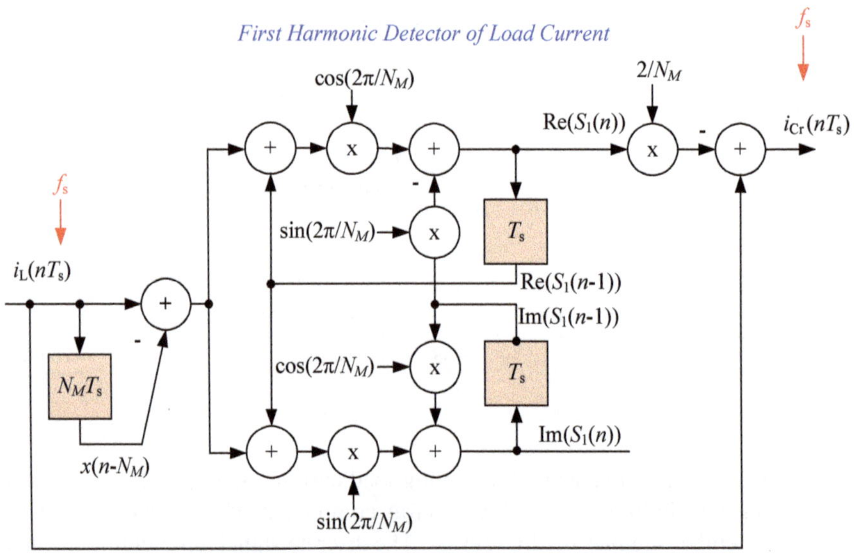

Fig. 5.14 Control algorithm for three-phase APF with harmonic compensation

The block diagram of this type of control circuit is shown in Fig. 5.13. If a shunt APF with compensation of harmonics is required, compensation current can be determined by the equation

$$i_C(nT_s) = i_L(nT_s) - 2/N_M Re(S_1(nT_s)) \sin(2\pi 50 nT_s) \quad . \tag{5.8}$$

A block diAgram of such a solution is depicted in Fig. 5.14.

In the case when current imbalance must be compensated, compensation current for the first phase $i_{C1}(nT_s)$ has to be calculated by the formula

$$i_{C1}(nT_s) = i_{L1}(nT_s) - 2/N_M \frac{|S_1(nT_s)| + |S_2(nT_s)| + |S_2(nT_s)|}{3} \sin(2\pi 50 nT_s) \ , \tag{5.9}$$

where: $S_1(nT_s)$, $S_2(nT_s)$, $S_3(nT_s)$—discrete signal representing first harmonic complex spectral component signal of first, second and third phase load current. The block diagram of the APF control algorithm for a three-phase APF with harmonic, reactive power and asymmetry compensation is depicted in Fig. 5.15. In the summing block the resultant magnitude of three phase current is calculated. The magnitude is used to modulate the amplitude of each phase's first harmonic reference signal. By subtracting these signals from appropriate input signals, the output compensation signals $i_{C1}(nT_s)$, $i_{C2}(nT_s)$, $i_{C3}(nT_s)$ are calculated.

One of the most difficult tasks for the control algorithm is calculation of the magnitude using fixed point arithmetic, which is a potential source of big errors, especially in the calculation of a square root. In the proposed algorithm the square root is calculated using the following formula

$$\sqrt{x} \approxeq -0.2831102x^2 + 1.0063284x + 0.272661 \quad \text{for} \quad 0.25 < x \leqslant 1 \ . \tag{5.10}$$

To ensure sufficient accuracy, the numbers ranging from 0 to 1 should be divided into at least three ranges. An example of implementing square root calculation, the program written in C language is shown in Listing 5.1

Listing 5.1 Square root calculation

```
float x,u,y;
...
if x < 0.0625 {
    u=16*x;
    y=(-0.2831102*u^2+2*0.5031642*u+0.272661)/4; }
else if (x >= 0.0625)&&(x<0.25) {
    u=4*x;
    y=(-0.2831102*u^2+2*0.5031642*u+0.272661)/2; }
else
    y=-0.2831102*x^2+2*0.5031642*x+0.272661;
```

The results of using such a sqrt algorithm are shown in Fig. 5.16. This program can be easily modified for arithmetic Q15.

The block diagram of three-phase APF with harmonic and reactive power compensation is depicted in Fig. 5.17. In the figure are shown DC bank voltage proportional controllers one for voltage level and second for voltage balancing. The amplitude of the control signals $u_s(nT_s)$ and $u_r(nT_s)$ is limited by signal limiters. They are used to reduce the inrush current when the DC bank capacitors are discharged. There are also current hysteresis controllers, used on the output of the circuit.

Based on the system shown in Fig. 5.17, the shunt APF (in laboratory version) was built with the following parameters $L_{C1,2,3} = 2\,\text{mH}$, $C_{1,2} = 4.7\,\text{mF}$, $U_{C1,2} = 400\,\text{V}$. In Fig. 5.18 4 experimental results of the compensation circuit with such shunt APF are shown. As the load, the three-phase transformer with an uncontrolled three-phase

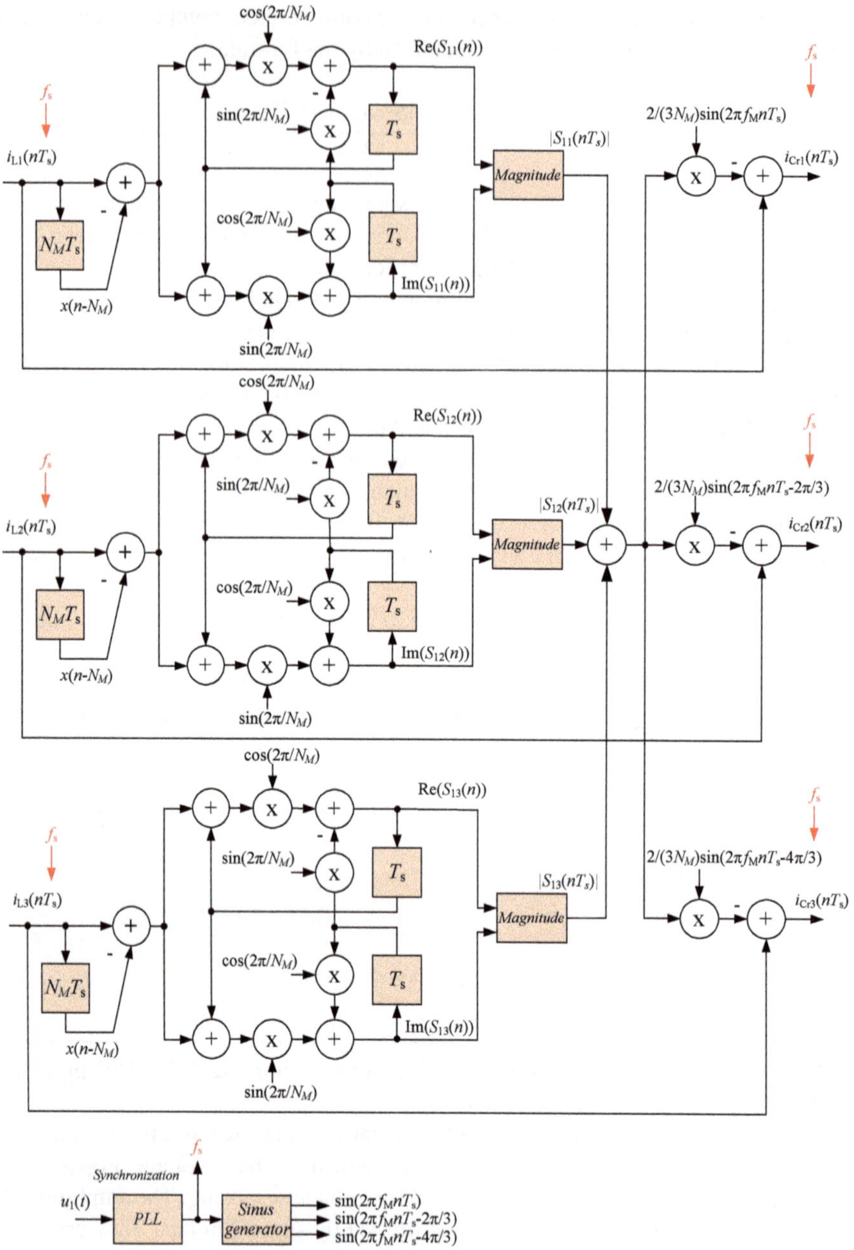

Fig. 5.15 Control algorithm for three-phase APF with harmonic, reactive power and asymmetry compensation

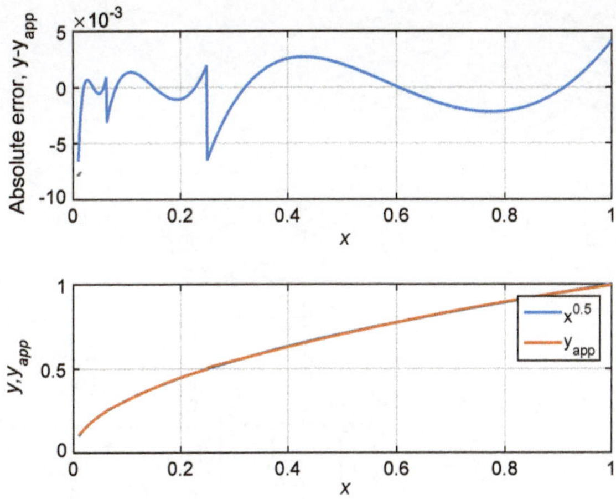

Fig. 5.16 Approximation of square root calculation

bridge rectifier is applied. The waveforms of the currents: $i_L(t), i_C(t), i_M(t)$ are shown in Fig. 5.18a and their spectra are shown in Fig. 5.18b.

5.4.4 *Control Circuit with Sliding Goertzel*

The principles of the sliding Goertzel algorithm are described in Chap. 3. The range of applications for the SGDFT algorithm is similar to the SDFT. A block diagram of the shunt APF control algorithm with harmonics and reactive power compensation based on SGDFT is depicted in Fig. 5.19. The SGDF algorithm, like the SDFT algorithm, requires the determination of the magnitude for the first harmonic of the load current signal. It also requires a smaller number of arithmetic operations in comparison with the SDFT algorithm.

5.4.5 *Control Circuit with Moving DFT*

A detailed description of the moving DFT algorithm is presented in Chap. 3. The algorithm output result is in the time domain, because it does not require determination of the signal magnitude. A block diagram of the MDFT control circuit for the shunt APF is shown in Fig. 5.20. The computational workload for the MDFT control circuit is less than for SDFT and SGDFT.

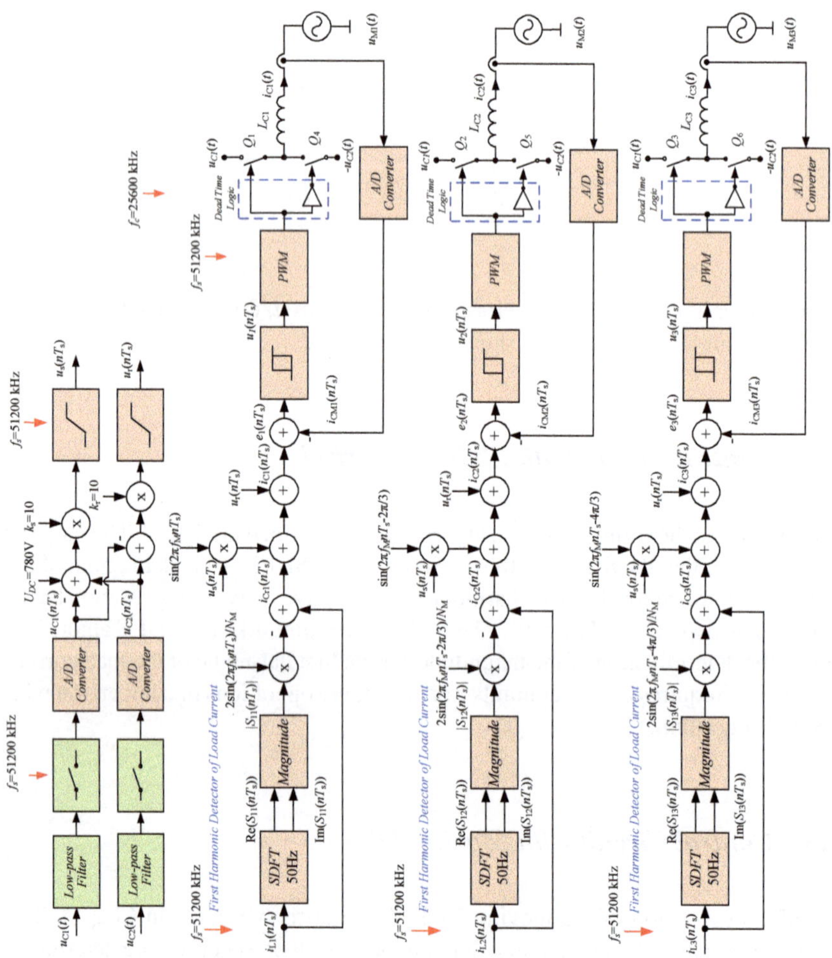

Fig. 5.17 Block diagram of three-phase APF with harmonic and reactive power compensation with the DC bank voltage controller

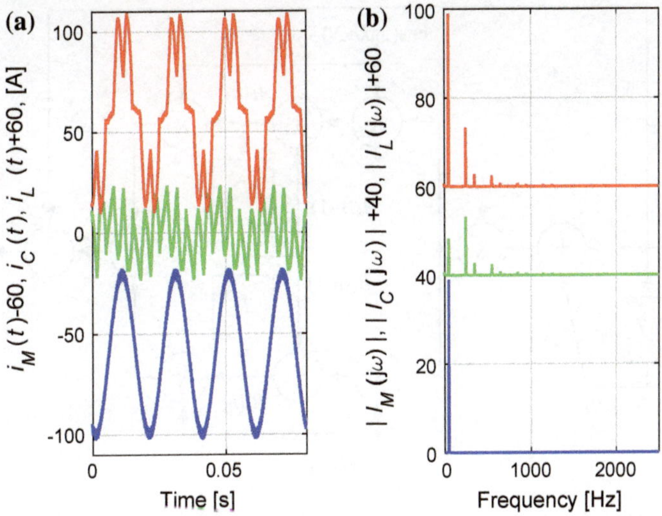

Fig. 5.18 Exemplary experimental waveforms of currents: $i_L(t)$, $i_C(t)$, $i_M(t)$ in a compensation circuit with shunt APF: **a** waveforms, **b** spectra

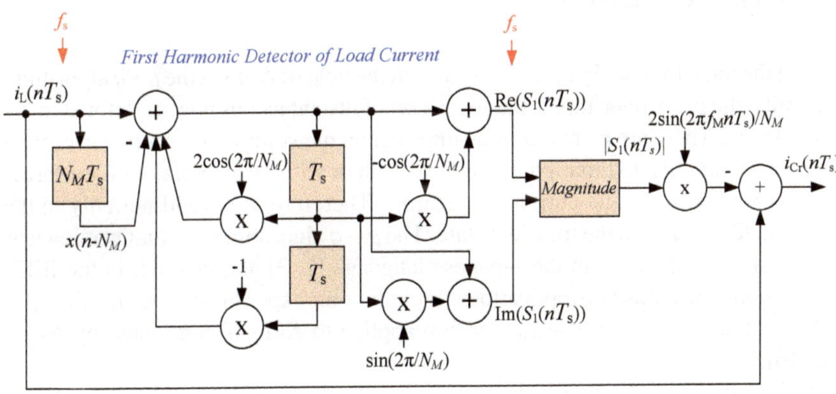

Fig. 5.19 Shunt APF control algorithm with harmonics and reactive power compensation based on SGDFT

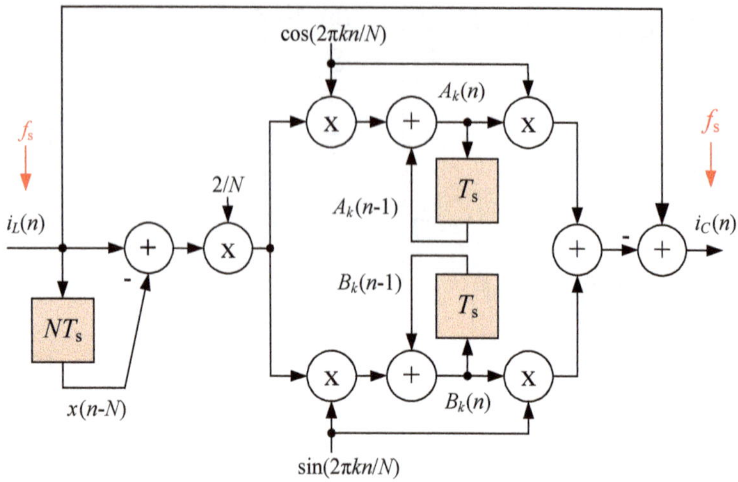

Fig. 5.20 Control algorithm with moving DFT

5.5 The Control Circuit for the Shunt APF Based on $p - q$ Algorithm

One of the most frequently used algorithms in the field of APFs is the $p-q$ algorithm. Generally, the $p-q$ algorithm is based on a set of instantaneous powers defined in the time domain. There are no restrictions imposed on the voltage or current waveforms and it can be applied to three-phase systems with or without a neutral wire for three-phase generic voltage and current waveforms. Therefore, it is valid not only in the steady-state, but also in the transient state. The $p-q$ algorithm in its first version was published by Akagi et al. in the Japanese language [2, 3] and in 1984, in the IEEE Transactions on Industry Applications, including an experimental verification [4]. A detailed description of $p - q$ algorithm applied to APFs is considered by Akagi et al. [6].

A block diagram of a shunt APF $p-q$ algorithm control algorithm test circuit for a balanced three-phase and three-wire system is shown in Fig. 5.21.

The waveforms which are a result of the test circuit simulation in steady-state are shown in Figs. 5.22 and 5.23. Three-phase orthogonal input signals $i_{L1}(t)$, $i_{L2}(t)$, and $i_{L3}(t)$ are used as test signals, which are shown in Fig. 5.22. The input signals are converted into $\alpha - \beta$ system using Clark transformation (Fig. 5.22)

$$\begin{bmatrix} i_{L\alpha}(t) \\ i_{L\beta}(t) \end{bmatrix} = \sqrt{\frac{2}{3}} \begin{bmatrix} 1 & -\frac{1}{2} & -\frac{1}{2} \\ 0 & \frac{\sqrt{3}}{2} & -\frac{\sqrt{3}}{2} \end{bmatrix} \begin{bmatrix} i_{L1}(t) \\ i_{L2}(t) \\ i_{L3}(t) \end{bmatrix} , \tag{5.11}$$

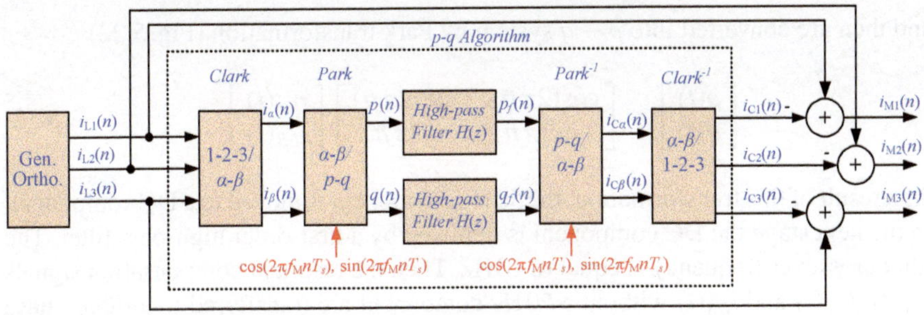

Fig. 5.21 A block diagram of $p-q$ algorithm test circuit

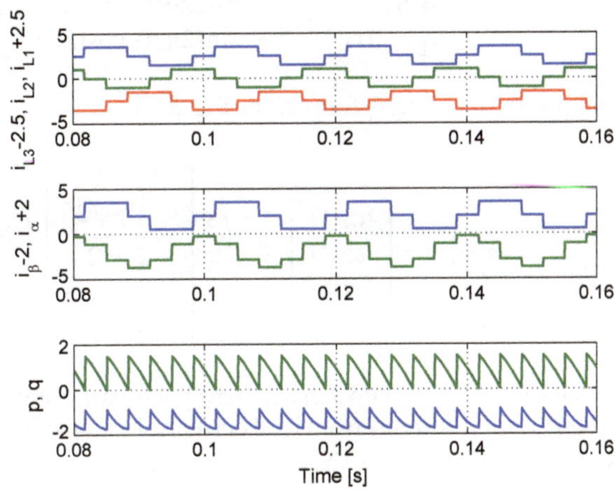

Fig. 5.22 Simulation waveforms of the $p-q$ control algorithm: orthogonal load currents $i_{L1}(n)$, $i_{L2}(n)$, $i_{L3}(n)$, components $i_\alpha(n)$, $i_\beta(n)$, $p(n)$, $q(n)$

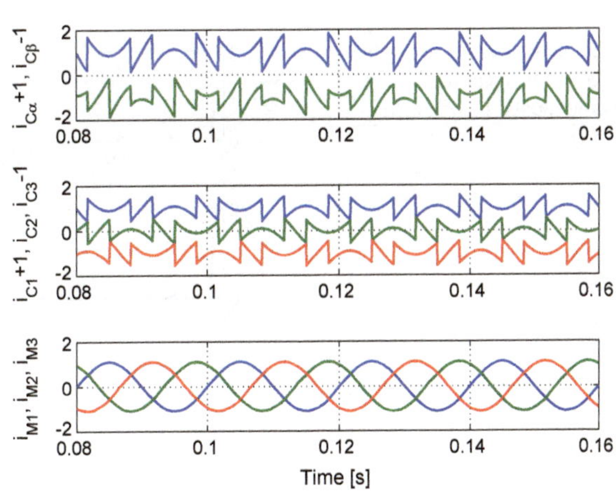

Fig. 5.23 Simulation waveforms of the $p-q$ control algorithm, components: $p_f(n)$, $q_f(n)$, $i_{C\alpha}(n)$, $i_{C\beta}(n)$, compensation currents $i_{C1}(n)$, $i_{C2}(n)$, $i_{C3}(n)$, line currents $i_{M1}(n)$, $i_{M2}(n)$, $i_{M3}(n)$

and then are converted into $p-q$ system by Park transformation (Fig. 5.22)

$$\begin{bmatrix} p(t) \\ q(t) \end{bmatrix} = \begin{bmatrix} \cos(2\pi f t) & -\sin(2\pi f t) \\ \sin(2\pi f t) & \cos(2\pi f t) \end{bmatrix} \begin{bmatrix} i_{L\alpha}(t) \\ i_{L\beta}(t) \end{bmatrix} . \tag{5.12}$$

As a result of this transformation, the 50 Hz component is moved to a DC component. In the next stage the DC component is removed by a first-order high-pass filter. The filter crossover frequency is equal to 10 Hz. Then the resultant compensation signals $i_{C1}(t)$, $i_{C2}(t)$ and $i_{C3}(t)$ without a 50 Hz component are transferred to a three-phase system by inverse Park transformation (Fig. 5.23)

$$\begin{bmatrix} i_{C\alpha}(t) \\ i_{C\beta}(t) \end{bmatrix} = \begin{bmatrix} \cos(2\pi f t) & \sin(2\pi f t) \\ -\sin(2\pi f t) & \cos(2\pi f t) \end{bmatrix} \begin{bmatrix} p_f(t) \\ q_f(t) \end{bmatrix} , \tag{5.13}$$

and inverse Clark transformation (Fig. 5.23)

$$\begin{bmatrix} i_{C1}(t) \\ i_{C2}(t) \\ i_{C3}(t) \end{bmatrix} = \sqrt{\frac{2}{3}} \begin{bmatrix} 1 & 0 \\ -\frac{1}{2} & \frac{\sqrt{3}}{2} \\ -\frac{1}{2} & -\frac{\sqrt{3}}{2} \end{bmatrix} \begin{bmatrix} i_{C\alpha}(t) \\ i_{C\beta}(t) \end{bmatrix} . \tag{5.14}$$

Finally the line currents $i_{M1}(t)$, $i_{M2}(t)$ and $i_{M3}(t)$ are determined by subtraction of compensation currents from load currents. Listing of the MATLAB program for $p-q$ simulation algorithm is shown in Listing 5.2.

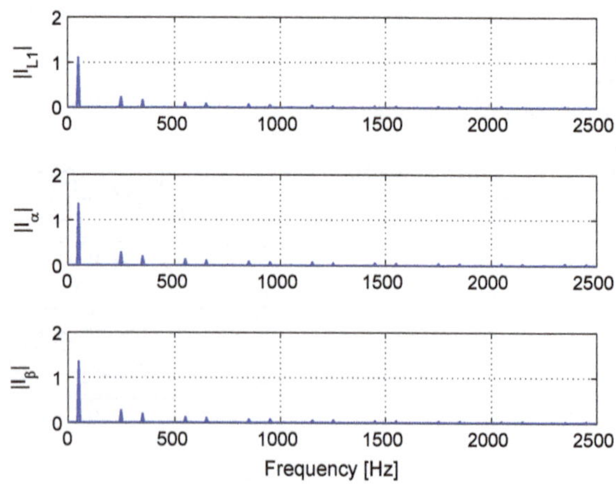

Fig. 5.24 The $p-q$ control algorithm, spectra of components: $i_{L1}(n)$, $i_{\alpha}(n)$, $i_{\beta}(n)$

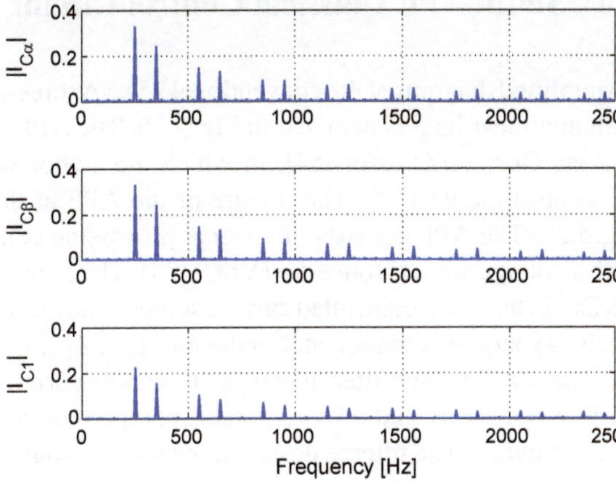

Fig. 5.25 The $p-q$ control algorithm, spectra of components: $i_{C\alpha}(n)$, $i_{C\beta}(n)$, $i_{C1}(n)$

Listing 5.2 $p-q$ simulation algorithm

```
clear all; close all;
NM=2^8; % number of samples per line period
fs=50*NM; % sampling frequency
fM=50; % line frequency
N_period=4; % number of periods
N=N_period*NM; % number of samples
fi=0; %phase shift
t=(0:N-1)/fs; % time
s1=sin(2*pi*fM*t+fi); c1=cos(2*pi*fM*t+fi);
% ------- Orthogonal signals -------
fi2=-5*pi/6;
inA=sin(2*pi*fM*t+0+fi2);
inB=sin(2*pi*fM*t-2*pi/3+fi2);
inC=sin(2*pi*fM*t-4*pi/3+fi2);
in1=sign(inA)/2; in2=sign(inB)/2; in3=sign(inC)/2;
iL3=in1-in2; iL2=in2-in3; iL1=in3-in1;
% ------- Clark transformation -------
ialfa=(iL1 - 0.5*iL2 - 0.5*iL3)*(2/3)^0.5;
ibeta =( (3^0.5)/2*(iL2-iL3))*(2/3)^0.5;
% ------- Park transformation -------
p=c1.*ialfa-ibeta.*s1;
q=s1.*ialfa+c1.*ibeta;
% ----- Butterworth first-order high-pass filter
fcr=10; % crossover frequency 10\,Hz
Fcr=fcr/fs*2; % relative crossover frequency
[b a]=butter(1,Fcr,'high');
pf=filter(b,a,p); qf=filter(b,a,q);
% ------- Park inverse transformation -------
iCalfa = pf.*c1 + qf.*s1;
iCbeta = -pf.*s1 + qf.*c1;
% ------- Clark inverse transformation -------
iC1 = (iCalfa)*(2/3)^0.5;
iC2 = (-1/2*iCalfa + (3^0.5)/2*iCbeta)*(2/3)^0.5;
iC3 = (-1/2*iCalfa - (3^0.5)/2*iCbeta)*(2/3)^0.5;
% ------- Line current calculation -------
iM1=iL1-iC1; iM2=iL2-iC2; iM3=iL3-iC3;
```

Spectra of selected signals are shown in Figs. 5.24 and 5.25.

5.6 Shunt APF Classical Control Circuit

A simplified diagram of the conventional 75 kVA three-phase shunt active power filter with nonlinear load is depicted in Fig. 5.26. The APF was built by the University of Zielona Gora (UZ) team [54], in which the author was involved in the design of the control circuit [49]. The picture of the APF at the UZ laboratory is shown in Fig. 5.27. The APF consists of a signal processing control circuit and output circuit with a voltage-source converter (VSC) [34]. The APF control circuit should force the VSC to behave as a controlled current source. The output circuit consists of two kinds of energy storage components: inductors L_{C1}, L_{C2}, L_{C3} and two DC capacitors C_1, C_2. The active power filter injects the harmonic currents $i_{C1}(t)$, $i_{C2}(t)$ and $i_{C3}(t)$ into the power network and offers a notable compensation for harmonics, reactive power and unbalance. The filter is designed for three or four wire loads. The nonlinear load consists of a thyristor power controller with a resistor load. Experimental waveforms of the compensation circuit in steady-state with the resistor load, from the top to the bottom: load current $i_L(t)$, compensation current $i_C(t)$, line current $i_M(t)$ are shown in Fig. 5.28. The control algorithm for the proposed APF is based $p-q$ control algorithm. A simplified block diagram of the APF control algorithm is depicted in Fig. 5.29, based on the circuit designed by Strzelecki and Sozanski et al. [49]. The algorithm

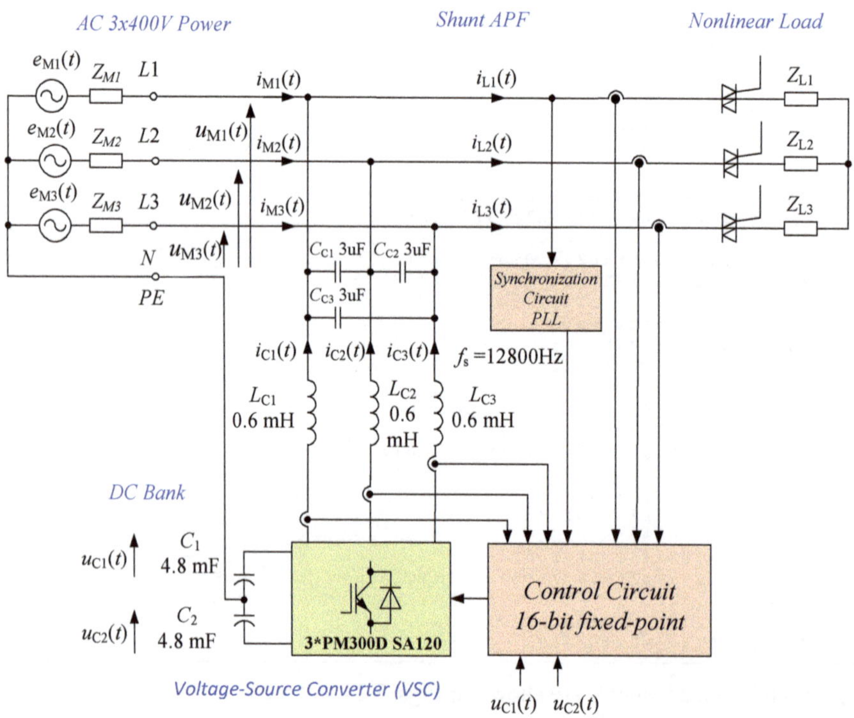

Fig. 5.26 Tests circuit of classical three-phase shunt active power filter

Fig. 5.27 The three-phase shunt APF at the UZ laboratory

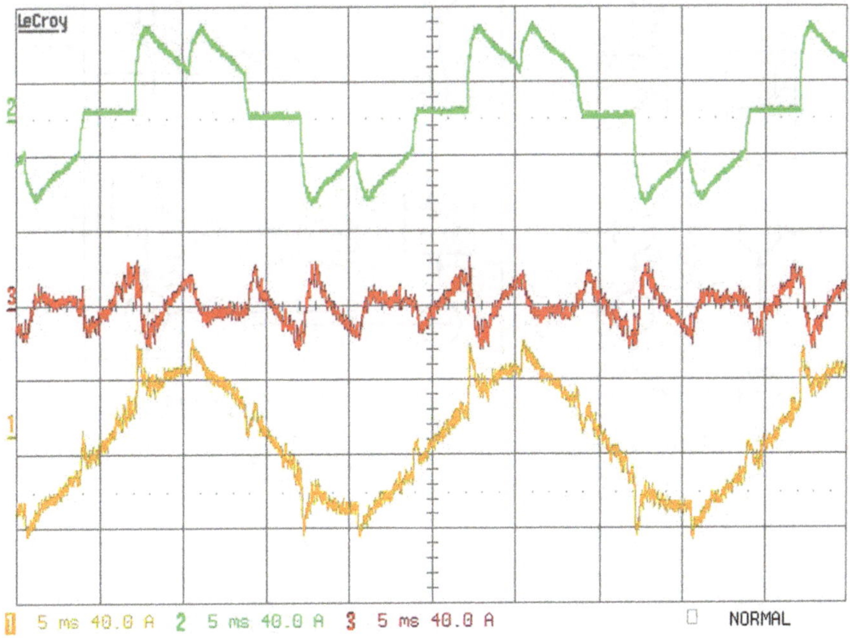

Fig. 5.28 Experimental waveforms (shown for one phase) of the three-phase compensation circuit with the shunt APF, $i_L(t)$—*green*, $i_C(t)$—*red*, $i_M(t)$—*yellow*

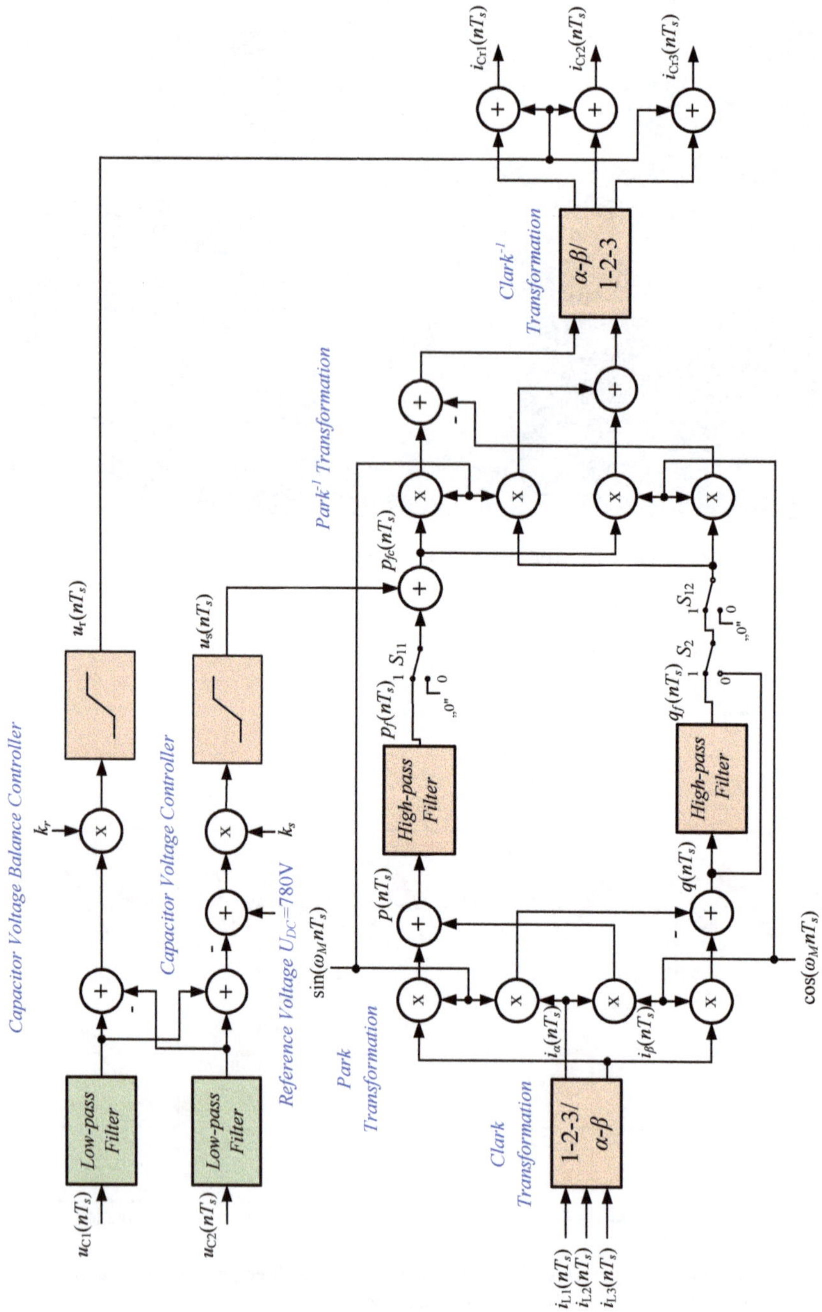

Fig. 5.29 Block diagram of control algorithm

is realized using the fixed-point 16-bit digital signal processor TMS320C50 which has the sampling rate f_s. The digital signal processor is synchronized using the PLL circuit with the line voltage U_1 and the algorithm is performed N_M times per line period. The sampling periods can be calculated using the formula

$$T_s = \frac{T_M}{N_M}, \tag{5.15}$$

where: T_M—period of the line voltage, $f_M = T_M^{-1}$—frequency of the line voltage, N_M—total number of samples per line voltage period. For the line voltage frequency of $f_M = 50\,\text{Hz}$ and the number of samples chosen to $N_M = 256$, the sampling period is equal to $T_s = 78.125\,\mu\text{s}$ and the sampling rate is equal to $f_s = 12800$ samples/s.

Three-phase current signals can be transformed into the equivalent two-phase representation. The transformation $(1-2-3 \rightarrow \alpha-\beta-0)$ from the three-phase current signals $i_{L1}(nT_s)$, $i_{L2}(nT_s)$, $i_{L3}(nT_s)$, to the two-phase $i_{L\alpha}(nT_s)$, $i_{L\beta}(nT_s)$ with an additional neutral signal $i_{L0}(nT_s)$ can be written into a matrix form as

$$\begin{bmatrix} i_{L\alpha}(nT_s) \\ i_{L\beta}(nT_s) \\ i_{L0}(nT_s) \end{bmatrix} = \sqrt{\frac{2}{3}} \begin{bmatrix} 1 & -\frac{1}{2} & -\frac{1}{2} \\ 0 & \frac{\sqrt{3}}{2} & -\frac{\sqrt{3}}{2} \\ \frac{1}{\sqrt{2}} & \frac{1}{\sqrt{2}} & \frac{1}{\sqrt{2}} \end{bmatrix} \begin{bmatrix} i_{L1}(nT_s) \\ i_{L2}(nT_s) \\ i_{L3}(nT_s) \end{bmatrix}, \tag{5.16}$$

where: $i_{L\alpha}(nT_s)$—digital representation of signal $i_{L\alpha}$ for sampling period T_s, n—index of the current sample.

In the next step, the two-phase signals are transformed from the rotating to the stationary reference frame. This transformation is commonly called the reverse Park transformation and can be digitally calculated by equations

$$\begin{cases} p(nT_s) = i_{L\alpha}(nT_s)\sin\left(\frac{2\pi n}{N_M}\right) + i_{L\beta}(nT_s)\cos\left(\frac{2\pi n}{N_M}\right) \\ q(nT_s) = i_{L\alpha}(nT_s)\cos\left(\frac{2\pi n}{N_M}\right) - i_{L\beta}(nT_s)\sin\left(\frac{2\pi n}{N_M}\right) \end{cases}, \tag{5.17}$$

and the digital sinusoidal reference signal is given by the formula

$$\sin(2\pi f_M n T_s) = \sin\left(2\pi f_M n \frac{T_M}{N_M}\right) = \sin\left(\frac{2\pi n}{N_M}\right). \tag{5.18}$$

5.6.1 High-Pass IIR Filter

In order to generate reference sinusoidal and cosinusoidal signals, a table containing sinus function values is allocated in the digital signal processor program memory. Signal $p(nT_s)$ represents instantaneous active power and signal $q(nT_s)$ represents instantaneous reactive power. The DC components of signals $p(nT_s)$ and $q(nT_s)$ are removed by first order high-pass digital IIR filters. The filter design is based on the analog reference prototype using a bilinear transform. The high-pass filter transfer function is described by equations

$$H(z) = \frac{b_0 + b_1 z^{-1}}{1 + a_1 z^{-1}} , \tag{5.19}$$

and

$$b_0 = -b_1 = \frac{2\frac{T_1}{T_s}}{1 + 2\frac{T_1}{T_s}} , \quad a_1 = \frac{1 - 2\frac{T_1}{T_s}}{1 + 2\frac{T_1}{T_s}} \tag{5.20}$$

where: T_1—the reference (analog) filter time constant. Assuming the values, $T_1 = 0.016\,\mathrm{s}$ and $f_s = 12.8\,\mathrm{kHz}$, the filter transfer function is determined by the equation

$$H(z) = \frac{0.9975645 - 0.9975645 z^{-1}}{1 - 0.995129 z^{-1}} . \tag{5.21}$$

Frequency response of the high-pass filter is shown in Fig. 5.30, cutoff frequency of such a filter is equal to about 10 Hz. This filter, as well as the entire control algorithm, was implemented using a 16-bit digital signal processor with Q15 arithmetic. A schematic realization of this filter is shown in Fig. 5.31.

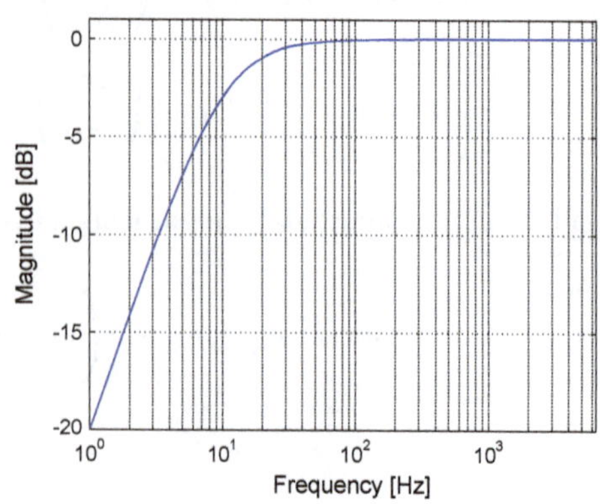

Fig. 5.30 Frequency response of high-pass digital filter

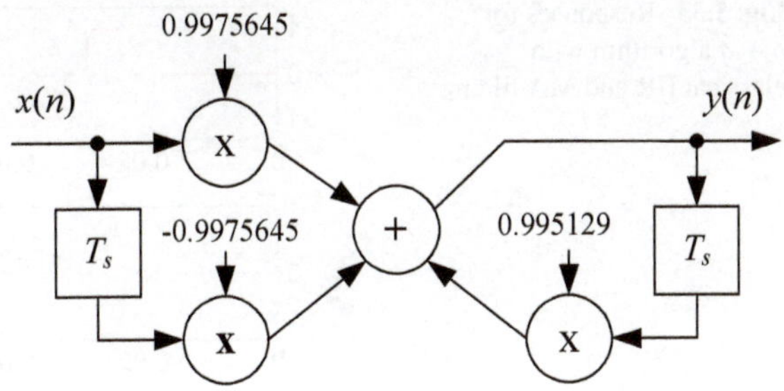

Fig. 5.31 Realization block diagram of the high-pass digital filter

5.6.2 *Improved High-Pass Filter*

In the algorithm for removal of the first harmonic typically two high-pass IIR filters $H(z)$ with cutoff frequency about 10 Hz are used. These filters introduce additional delays equal to three periods of line voltage. Therefore, to avoid this, the author suggests an improved solution using of high-pass filters based on moving average (MA). This filter has a shorter impulse response than IIR [37, 59], the response time is equal to one period of line voltage. Transfer function of such a filter is described by equation

$$H(z) = 1 - \frac{1}{N_M}\sum_{k=0}^{N_M-1} z^{-k} = 1 - \frac{1}{N_M}\frac{1-z^{-N_M}}{1-z^{-1}} \quad , \tag{5.22}$$

where: N_M—number of samples per period of line voltage.

Block diagram of such filter is depicted in Fig. 5.32. The response of that filter is calculated by

$$\begin{aligned} y(n) &= x(n) - ((x(n) - x(n-N_M))\frac{1}{N_M} + w(n-1)) \\ &= x(n)\left(1 - \frac{1}{N_M}\right) - x(n-1) + \frac{1}{N_M}x(n-N_M) + y(n-1) \quad . \end{aligned} \tag{5.23}$$

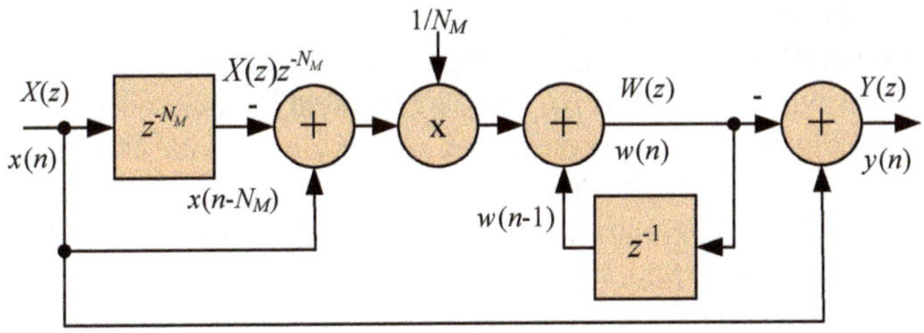

Fig. 5.32 Block diagram of a high-pass filter with MA

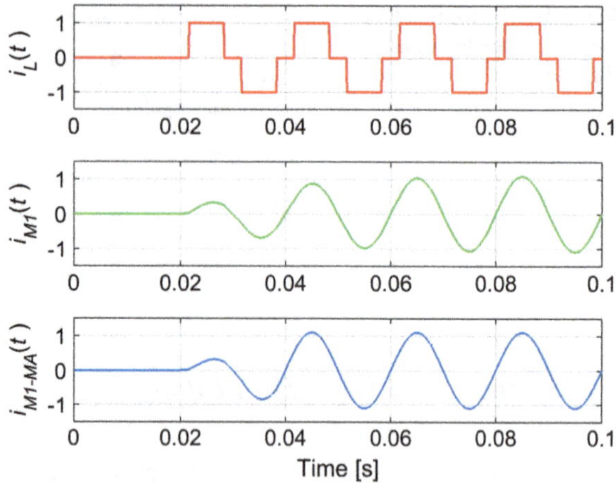

Fig. 5.33 Responses for $p-q$ algorithm with classical IIR and MA filters

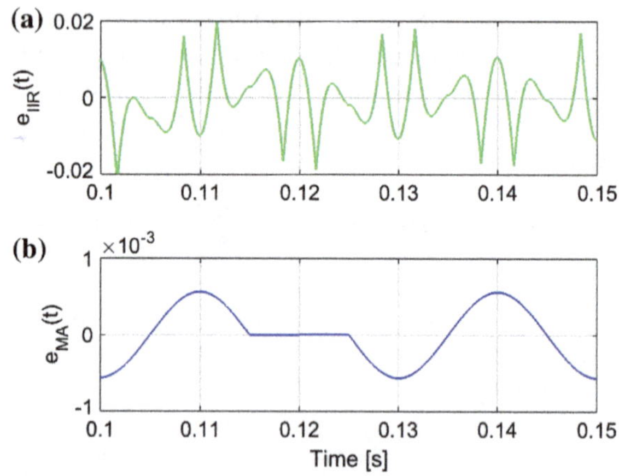

Fig. 5.34 Compensation error for $p-q$ algorithm with classical IIR (**a**) and MA (**b**) filters

To demonstrate this, simulation studies have been conducted using compensation system such as that shown in Fig. 5.21. As an input test signals $i_{L1}(t)$, $i_{L2}(t)$, $i_{L3}(t)$, unit orthogonal three-phase signals are used (Fig. 5.22). The results of simulation for one phase are shown in Fig. 5.33.

Waveform $i_{M1}(t)$ presents compensating signal generated by a circuit with traditional IIR filters while waveform $i_{M1_{MA}}(t)$ is generated by a circuit with MA filters. As seen in the Fig. 5.33, the response time of the circuit with MA filters is three times shorter than with the IIR filter. Therefore, such simple modification of $p-q$ algorithm allows decreasing the dynamic response of the whole APF. Additionally, $p-q$ algorithm with MA filters has smaller compensation error than that with IIR filters. Compensation error waveforms for both filters are shown in Fig. 5.34 (please note the difference in scale between the graphs).

5.6.3 DC Bank Voltage Controller

In addition to the compensation of harmonics and compensation of reactive power, the control algorithm must also control the voltage on the DC capacitor bank (C_1, C_2). In order to stabilize the DC voltage, a proportional controller is used and its response is calculated using the equation

$$u_s(nT_s) = k_s(U_{DC} - (u_{C_1}(nT_s) + u_{C_2}(nT_s))) \quad , \tag{5.24}$$

where: $u_{C_1}(nT_s)$, $u_{C_2}(nT_s)$—voltage on capacitors C_1 and C_2, respectively, k_s—gain of voltage controller, U_{DC}—DC bank reference voltage.

Signal $u_s(nTs)$ is added from to the component $p_f(nT_s)$

$$p_{fc}(nT_s) = p_f(nT_s) + u_s(nT_s) \quad . \tag{5.25}$$

5.6.4 The Remaining Part of the $p - q$ Algorithm

In the next step components p_{fc} and q_f are transformed by Park transformation into the two-phase representation

$$\begin{cases} i_{Cr\alpha}(nT_s) = p_{fc}(nT_s)\sin\left(\dfrac{2\pi n}{N_M}\right) - q_f(nT_s)\cos\left(\dfrac{2\pi n}{N_M}\right) \\ i_{Cr\beta}(nT_s) = -p_{fc}(nT_s)\cos\left(\dfrac{2\pi n}{N_M}\right) - q_f(nT_s)\sin\left(\dfrac{2\pi n}{N_M}\right) \end{cases} , \tag{5.26}$$

and then transformed back to the three-phase reference current signals

$$\begin{bmatrix} i_{Cr1}(nT_s) \\ i_{Cr2}(nT_s) \\ i_{Cr3}(nT_s) \end{bmatrix} = \sqrt{\frac{3}{2}} \begin{bmatrix} 1 & 0 & \frac{1}{\sqrt{2}} \\ -\frac{1}{2} & \frac{\sqrt{3}}{2} & \frac{1}{\sqrt{2}} \\ -\frac{1}{2} & -\frac{\sqrt{3}}{2} & \frac{1}{\sqrt{2}} \end{bmatrix} \begin{bmatrix} i_{Cr\alpha}(nT_s) \\ i_{Cr\beta}(nT_s) \\ i_{L0}(nT_s) \end{bmatrix} , \tag{5.27}$$

Signal $s_{df}(nT_s)$ from the capacitor voltage balance controller is then added to the reference compensation signals $i_{Cr1}(nT_s)$, $i_{Cr2}(nT_s)$ and $i_{Cr3}(nT_s)$. Thanks to such control, the voltages on both capacitors are equal.

The control algorithm for the proposed APF is based on the $p-q$ control algorithm which was first developed by Akagi et al. [4]. Additional control switches, whose functions are described in Table 5.6, are added to the classical $p - q$ algorithm.

Table 5.6 APF control algorithm switch functions

Switch	Position	Function
S_2	1	Harmonics compensation
S_2	0	Full compensation of reactive power
S_{11} and S_{12}	1	APF compensator is switched on
S_{11} and S_{12}	0	APF compensator is switched off, working only capacitors voltage controller, capacitors C_1 and C_2 are charged to nominal working voltage

5.6.5 *Output Current Controller*

In the next step the output compensation reference current signals $i_{Cr1}(nT_s)$, $i_{Cr2}(nT_s)$ and $i_{Cr3}(nT_s)$ are converted to analog form by a 12-bit D/A converter. The block diagram of the control circuit is shown in Fig. 5.35. Finally, the output compensation reference current signals are transformed to the transistor controlling pulses by the current controller. Initially, the proposed circuit employs a hysteresis current controller algorithm realized using analog comparators and additional glue logic. The hysteresis control algorithm is based on a nonlinear feedback loop with two-level hysteresis comparators. The inverter switching speed depends largely on the load parameters. In the proposed APF an advanced hysteresis current controller with a variable width hysteresis is applied. A block diagram of this controller is depicted in Fig. 5.36. It has additional improvements:

- maximum switching speed is limited,
- switching speed depends on the "speed history",
- switching speed is dependent on the reference compensation current signals $i_{cr_1}(t)$; the higher the signal level, the lower is the switching speed.

Figure 5.37 shows a dependence of the level of reference compensation current signals $i_{cr_1}(t)$ on the switching speed. For a lower signal level the switching speed is around 25 kHz, and for a higher level of signal the speed is around 10 kHz. The shunt

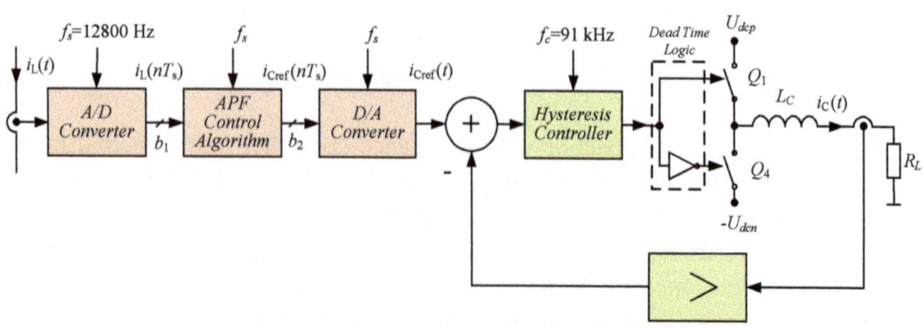

Fig. 5.35 Block diagram of control circuit for one phase

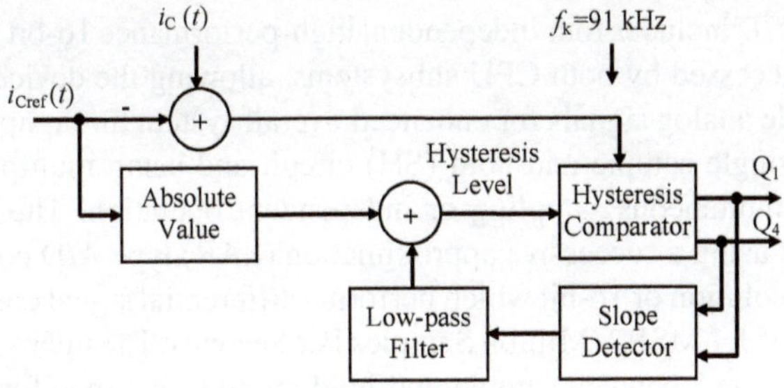

Fig. 5.36 Block diagram of output modulator for one phase

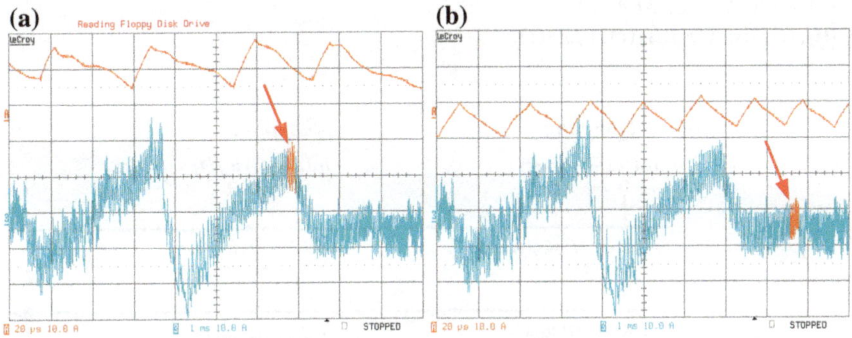

Fig. 5.37 Experimental waveforms, illustration of the dependence of reference compensation current signals $i_{CR1}(t)$ level on the switching speed: **a** for high level of signal, **b** for low level of signal

APF should not be connected to the mains without a boot routine. Before switching on compensation, it is necessary to charge the capacitors to the operating voltage of 700 V. The starting procedure includes the following steps:

- inverter transistors are switched off, capacitors C_1 and C_2 are charged to peak line voltage up to 560 V,
- APF compensator is switched off, only capacitor voltage regulator is working, capacitors C_1 and C_2 are charged to nominal working voltage ($S_{11} = 0$, $S_{12} = 0$),
- APF compensator is switched on ($S_{11} = 1$, $S_{12} = 1$).

5.6.6 *Modernized Digital Controller for the APF*

In the last year control circuit for the three-phase shunt active power filter was modernized. The old circuit control based on the TMS320C50 digital signal processor was replaced by modern TMS320F28377D digital signal microcontroller [57]. The

TMS320F28377D includes four independent high-performance 16-bit ADC modules which can be accessed by both CPU subsystems, allowing the device to efficiently manage multiple analog signals for enhanced overall system throughput. Each ADC module has a single sample-and-hold (SH) circuit and using multiple ADC modules enables simultaneous sampling or independent operation. The ADC module is implemented using a successive approximation (SAR) type A/D converter with a configurable resolution of 16-bit which performs differential signal conversions with a performance of 1.1 MSPS (Million Samples Per Second). The microcontroller consists of only four independent sample-and-hold circuits, and therefore the required twelve analog input signals are divided into three groups:

- $i_{L1}(t)$, $i_{L2}(t)$, $i_{L3}(t)$, $i_{L0}(t)$,
- $i_{C1}(t)$, $i_{C2}(t)$, $i_{C3}(t)$, $u_{DC1}(t)$,
- $u_{M1}(t)$, $u_{M2}(t)$, $u_{M3}(t)$, $u_{DC2}(t)$.

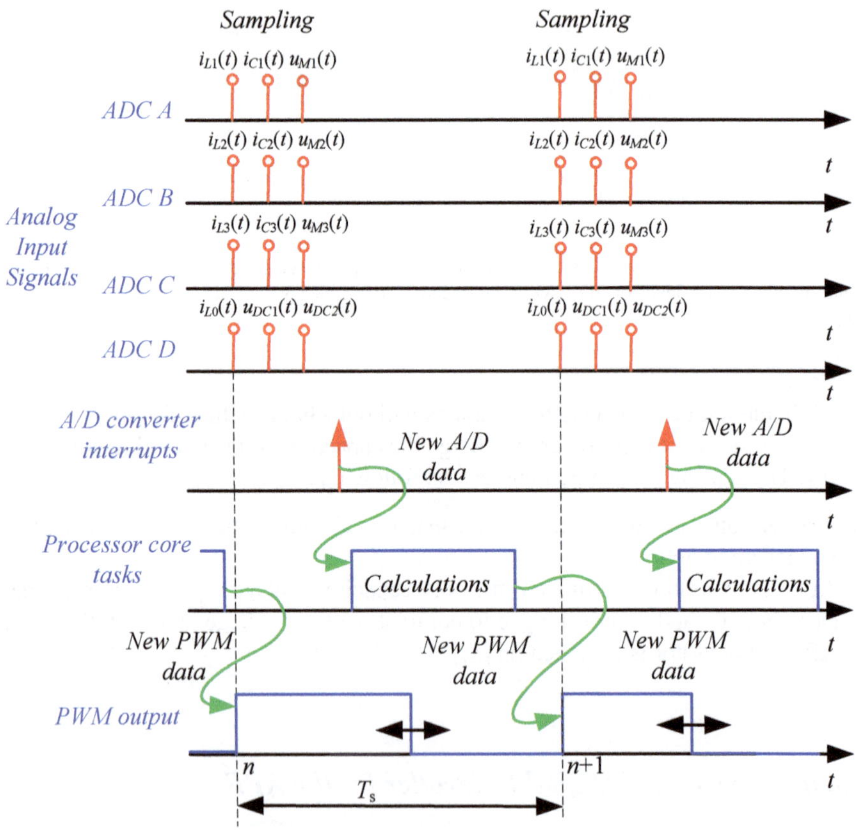

Fig. 5.38 Timing diagram of the modernized control circuit for the shunt APF

The signals of each group are sampled simultaneously. Individual groups are sampled sequentially. This solution minimizes the error associated with non-simultaneously sampling of all the signals. Timing diagram of presented sampling methods is depicted in Fig. 5.38. Such solution allows a considerable simplification of the APF control circuit.

5.7 Dynamics of Shunt APF

Examples of waveforms of line $i_M(t)$, compensating $i_C(t)$ and load $i_L(t)$ currents in a compensation circuit for a resistive load with thyristor controller are depicted in Fig. 5.39. When the value of load current changes rapidly, as in current $i_L(t)$ in Fig. 5.39b, the APF transient response is too slow and the line current $i_M(t)$ suffers from dynamic distortion. The amplitude of these distortions is dependent on the value of changes in load current. This distortion causes an increase in harmonic content in the line current, which is mainly dependent on the time constant of the output circuit. The value of the total harmonic distortion (*THD*) ratio for the presented waveform of line current is equal to over 10%. Such a large value of *THD* is associated with the presence of dynamic distortion resulting from too small slew rate of the compensating current. The biggest influence on the time constant of the output system has an inductance value of the output inductors L_{C1}, L_{C2}, L_{C3}. It is worth noting, however, that the curves from Fig. 5.39b represent one of the worst-case-scenarios.

The output circuit of a shunt active power filter acts as a controllable current source. The output circuit is typically realized using three-phase transistor inverter (Fig. 5.39a). Such an output circuit is a voltage source inverter, and therefore it must be converted into a controlled current source. To achieve this, the current controllers and the output inductors L_{C1}, L_{C2}, L_{C3} should be used.

5.7.1 *Methods of Reducing APF Dynamic Distortion*

The dynamics of shunt APF control current are mainly dependent on the inverter output time constant, consisting of the APF output impedance and resultant impedance of load and line. Problems related to the dynamics of the APF are not very widely described in the literature, but it is possible to find a few publications: e.g., Mariethoz and Rufer [31], Marks and Green [32], Wojciechowski and Strzelecki [61] and the author [41, 44, 45, 48]. Possible solutions for the improvement of APF dynamics are shown in Table 5.7. Proposed circuits for reducing APF dynamic distortion are presented in Fig. 5.40. In the simplest solution, the speed of load current changes should be decreased. In order to reduce the impact of this phenomenon in a practical compensation circuit, the rate of change of the load current is reduced by adding a serial inductor L_k (Fig. 5.40a). However, this solution increases the weight and cost of the compensation system and, with respect to additional power loses, it is not

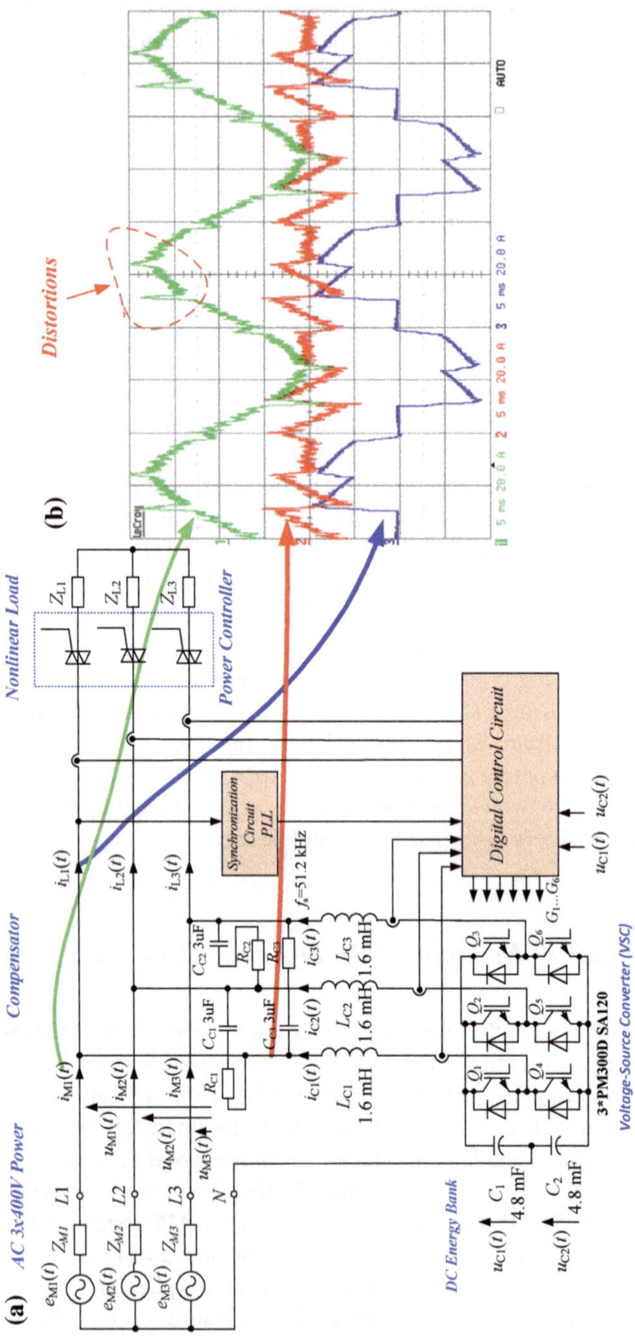

Fig. 5.39 Illustration of the shunt APF dynamic distortions: **a** the shunt APF circuit, **b** waveforms

Table 5.7 Methods to reduce the APF dynamic distortion

Predictable loads	Unpredictable loads
• Load with additional serial inductor	• Load with additional serial inductor
• High speed APF	• High speed APF
• Set of two APFs: high power low speed APF, low power high speed APF	• Set of two APFs: high power low speed APF, low power high speed APF
• APF with multirate output inverter	• APF with multirate output inverter
• APF with prediction control algorithm	

acceptable in every case. Another simple solution is a high speed APF with high frequency switching output transistors. This allows for a reduction in the value of the output inductor inductance L_C and thus decreases the output time constant. The disadvantage of this solution are the large power losses in output transistors. A compromise solution is to use two APF filters: one high power slow APF and a second low power high speed APF (Fig. 5.40b). Then the compensation signal should be divided into two subbands, low-pass and high-pass. However, to ensure a good compensation result, the high speed APF output current should have the same value as in the "slow" APF, which reduces the benefits of such a solution in comparison to the single fast filter. However, it is possible to turn on the "fast" APF only in transient states—when the value of load current rapidly changes. This idea was applied in a mutlirate APF with modified output inverter [44], which is presented later in this Chapter.

The loads can be divided into two categories, with predictable and unpredictable (noise-like) load currents [23, 30, 37, 48, 61, 62]. A current can be considered predictable if the root mean square (*RMS*) of the load current changes relatively slowly compared to the period of the line voltage. This is the case in e.g. controlled rectifiers, power controllers, impulse power pulse power supplies, inverters etc. The unpredictable load changes, on the other hand, often occur e.g. in arc furnaces, electrical discharge machining etc.

Most loads belong to the first category. For predictable load currents, from the observations of the current in the previous periods, it is possible to predict the waveform of current in the ongoing period with a high degree of certitude [31, 41]. The waveform presented in Fig. 5.39b is an example where a load undergoes predictable current changes. A block diagram of circuit with predictive control is shown in Fig. 5.40c. In the case of unpredictable loads, a mutlirate APF with modified output inverter [44] is proposed. Both the circuit and the inverter are described later in this chapter.

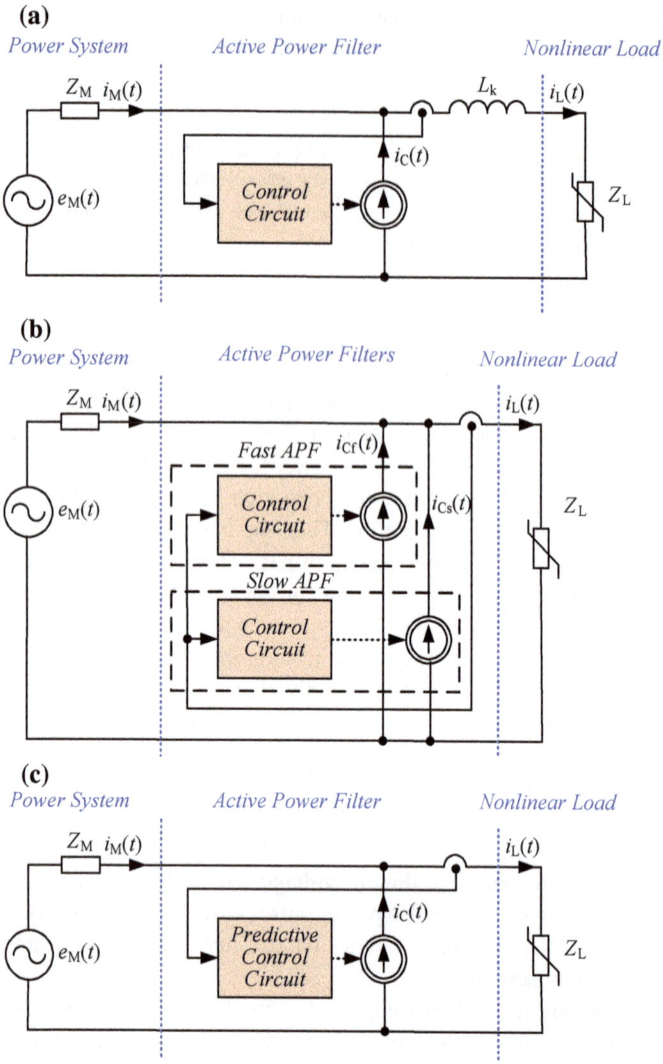

Fig. 5.40 Proposed circuits for decreasing APF dynamic distortion: **a** by reducing the rate of change of load current, **b** by using two APFs, "fast" and "slow", **c** by using predictive control circuit

5.7.2 *Control Circuit*

A schematic diagram of shunt active power filter control circuit for one phase is depicted in Fig. 5.41. Its primary aim is to generate the output current $i_C(t)$ according to the compensating reference signal $i_{Cr1}(nTs)$. Since switches S_1, S_5 work in an impulse mode at the output, it is necessary to use an inductor L_{C1}. The value of inductance of inductor L_{C1} needs to be a balance between the amplitude of ripple

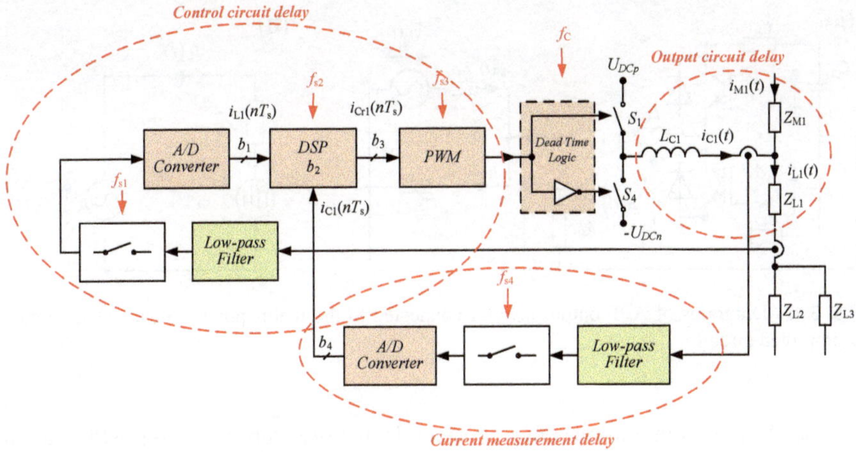

Fig. 5.41 Delays in a shunt APF

of output current and the speed of APF reaction. For typical values of switching frequency of power transistor, this inductance is the major factor determining the time constant of the output of APF circuit. As shown in Fig. 5.41, the typical delays in the APF circuit include those caused by:

- output circuit: S_1, S_4, L_{C1},
- the main APF algorithm,
- control circuit: digital signal processor, output pulse width modulator (PWM), A/D converter for $i_{L1}(t)$,
- A/D converter for $i_{C1}(t)$.

In this figure the main sources of signal delay occurring in the shunt APF are indicated: control circuit delay, output circuit delay, compensation current measurement delay. These delays generate the dynamic distortion of power line current visible in Figs. 5.28 and 5.39b. An equivalent APF output circuit (for one phase) is depicted in Fig. 5.43a, where Z_M represents resultant power line impedance and Z_L represents load impedance.

The block diagram of control circuit model is depicted in Fig. 5.42. The model consists of three blocks: $H_{alg}(z)$—resultant transfer function for APF algorithm, z^{-Nalg}—resultant "pure" delay of control algorithm, $H_{out}(z)$—resultant transfer function of the output circuit.

Fig. 5.42 APF delay model

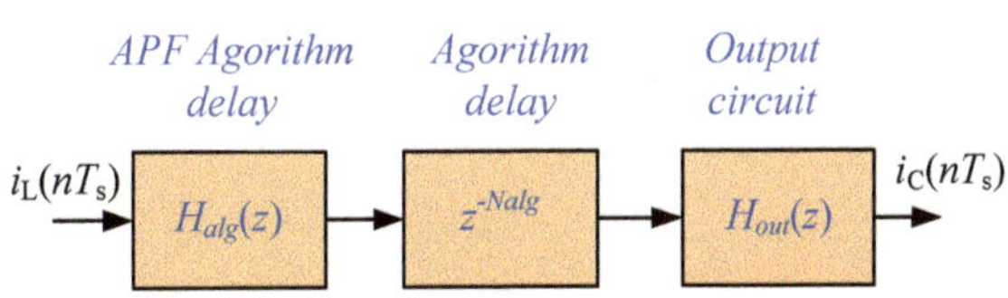

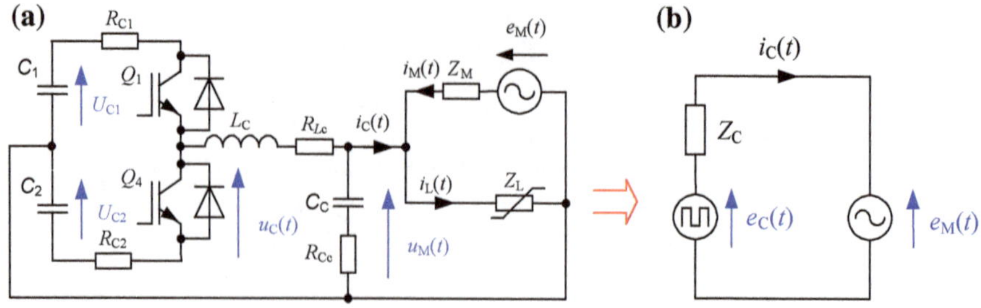

Fig. 5.43 Diagrams of APF output inverter connected to the mains power: **a** equivalent circuit, **b** simplified circuit

The APF output circuit time constant is mainly dependent on APF output impedance and therefore it is possible to simplify the circuit (for one phase), as presented in Fig. 5.43b, where Z_C represents resultant impedance of APF output impedance and power line impedance. This circuit has been proposed by the author [41, 44, 48] and in the author's opinion this simplified circuit allows the simulation of system dynamics with sufficient accuracy. The voltage-to-current transfer function of the simplified circuit is described by the formula

$$H(s) = \frac{I_C(s)}{E_C(s) - E_M(s)} = \frac{1}{R_C + sL_C} \quad . \tag{5.28}$$

The voltage-to-current transfer function of the analog APF output circuit described by Eq. (5.28) is converted into digital domain using bilinear transform. Due to the greater accuracy of simulation, the response of the output circuit is calculated R times more often than the whole algorithm (oversamplig), so its sampling frequency is equal to the Rf_s. The voltage-to-current transfer function of the digital representation of the APF output circuit for the assumed values of the circuit elements $R_C = 0.1\ \Omega$, $L_C =$ 0.6 mH, oversampling ratio $R = 8$ and the sampling frequency of $Rf_s = 102.4$ kHz is described by the equation

$$H(z) = \frac{I_C(z)}{E_C(z) - E_M(z)} = \frac{0.008131 + 0.008131z^{-1}}{1 - 0.9984z^{-1}} \quad . \tag{5.29}$$

Simulation research has been carried out for this circuit. In Listing 5.3 the author's MATLAB program for the realization of this digital circuit is presented.

Listing 5.3 Output circuit simulation program

```
clear all; close all;
line_thickness=2; font_size=18;
R=8; % oversampling ratio
NM=2^8; % number of samples per line period
NP= 100; % number of predios
fs=50*NM*R; % sampling frequency
Ts=1/fs;
N=NM*R*NP;
L_C=0.6e-3; R_C=0.1;
%% ----- Analog transfer function --------
num_s=[1];
den_s=[L_C R_C];
tf(num_s,den_s)          % check the transfer function
%% -------- Bilinear Transform ------------
[num_d den_d]=bilinear(num_s,den_s,fs)
% check the digital transfer function
tf(num_d,den_d,Ts)
tf(num_d,den_d,Ts,'variable','z^-1')
% check the frequency response of RL
figure('Name','Freq.␣resp.','NumberTitle','off');
freqz(num_d,den_d);
title('Freq.␣response');
%% algorithm delay
Nop=R*50; % number of samples for algorithm delay
aop=1;
bop=zeros(1,Nop+1); bop(1,Nop+1)=1;
%% step response
t=(0:N-1)*Ts;
x_step=ones(1,N); % step
i_C_step=filter(bop,aop,x_step);
i_C_step=filter(num_d,den_d,i_C_step);
figure('Name','Step␣response','NumberTitle','off');
  plot(t,i_C_step,'r','LineWidth',line_thickness);
  set(gca,'Xlim',[0 0.05]);
  set(gca,'FontSize',[font_size],'FontWeight','d'),
  ylabel('i_C(t)'); grid on;
  xlabel('Time␣[s]');
  print('APF_out_step.pdf','-dpdf');
%% impulse response
x_imp=[1 zeros(1,(N-1))]; % impulse
i_C_imp=filter(bop,aop,x_imp);
i_C_imp=filter(num_d,den_d,i_C_imp);
% frequency response
i_C_freq=fft(i_C_imp);
%% Frequency response
f=(0:N-1)*fs/N;
i_C_freq_abs_db=20*log10(abs(i_C_freq));
figure('Name','Frequency␣response','NumberTitle','off');
  plot(f,i_C_freq_abs_db,'r','LineWidth',line_thickness);
  set(gca,'Xlim',[0 500]);grid on;
  set(gca,'FontSize',[font_size],'FontWeight','d'),
  ylabel('|I_C(e^{j\omegaT})|');
  xlabel('Frequency␣[Hz]');
  print('APF_out_freq.pdf','-dpdf');
```

Using the program shown in Listing 5.3 the step response of the compensation circuit is determined. The step response of the output RL circuit and algorithm delay is shown in Fig. 5.44. The frequency response is calculated using the same MATLAB program. The magnitude response is shown in Fig. 5.45. It should be noted, however, that these characteristics are for the open loop of the output current controller.

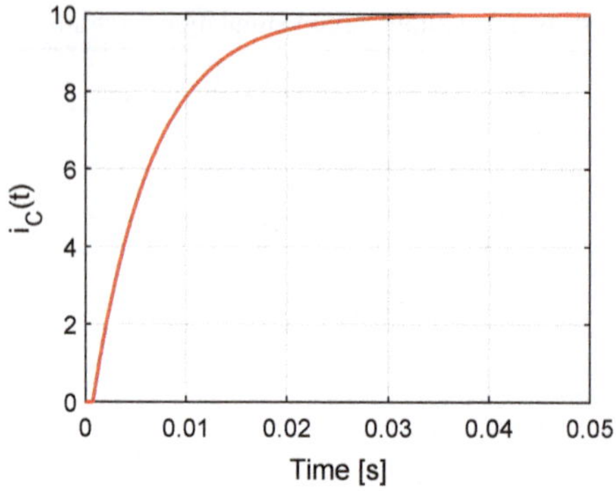

Fig. 5.44 The step response of APF output circuit

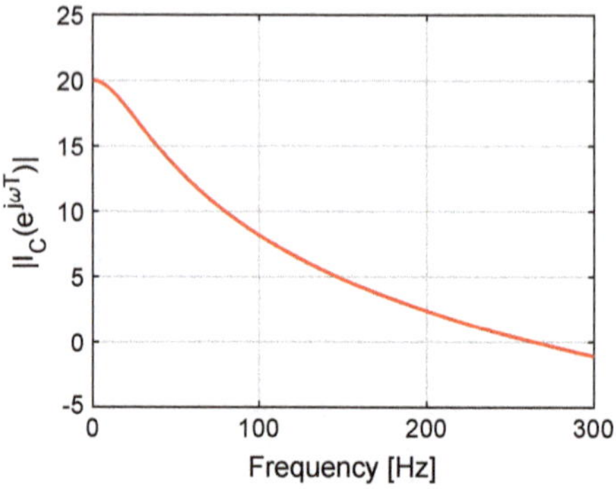

Fig. 5.45 The frequency response of APF output circuit

5.7.3 APF Output Current Ripple Calculation

A diagram of a simple power inverter model connected to the mains power is shown in Fig. 5.43a, where C_C is an output filter capacitance for damping modulation components. In this circuit, the time constant is typically and mainly dependent on inductor L_C value. Therefore when the transistor switching period $T_c = 1/f_c$ is much less than the circuit main time constant τ_C, it is possible to simplify the circuit from Fig. 5.43a to the circuit shown in Fig. 5.46a. It possible to assume that the resulting circuit resistance R_C mainly depends on the resistance of inductor L_C,

$$R_C \cong R_{L_C} \quad . \tag{5.30}$$

Assuming that during the switching period T_c voltages $e_M(t)$, $u_{C1}(t)$ and $u_{C2}(t)$ are constant, for switching state $S_1 = 1, S_4 = 0$ and for time t from 0 to t_1, compensation current $i_C(t)$ can be calculated using the formula

$$i_C(t) = \frac{u_{C1}(t_0) - e_M(t_0)}{R_C}\left(1 - \mathrm{e}^{-t/\tau}\right) + i_C(t_0) \quad , \tag{5.31}$$

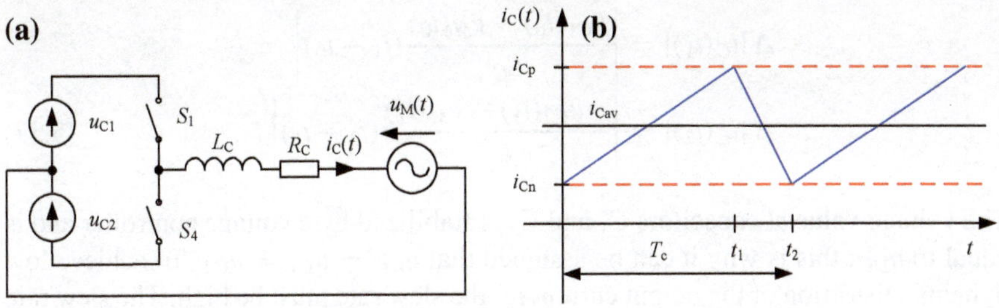

Fig. 5.46 The APF output current ripple: **a** simplified circuit used for current ripple calculation, **b** time diagram of idealized compensation current $i_C(t)$

where:

$$\tau = \frac{L_C}{R_C} \quad , \tag{5.32}$$

for state $S_1 = 0$, $S_4 = 1$ and for time t from t_1 to t_2, $i_C(t)$ can be calculated by

$$i_C(t) = \frac{-u_{C2}(t_0) - e_M(t_0)}{R_C}\left(1 - \mathrm{e}^{-(t-t_1)/\tau}\right) + i_C(t_1) \quad . \tag{5.33}$$

If it is assumed that

$$f_c \gg \frac{1}{\tau} \quad \text{and} \quad f_c \gg \frac{1}{T_M} \quad , \tag{5.34}$$

where: T_M—mains period.

Assuming that average current i_{Cav} is constant too, then the output current can be calculated by simplified equations: for state $S_1 = 1$, $S_4 = 0$

$$i_C(t_1) = \frac{u_{C1}(t_0) - u_s(t_0)}{L_C}(t_1 - t_0) + i_{Cn} \quad , \tag{5.35}$$

and for state $S_1 = 0$, $S_2 = 1$

$$i_C(t_2) = \frac{-\,u_{C2}(t_0) - e_M(t_0)}{L_C}(t_2 - t_1) + i_{Cp} \quad , \tag{5.36}$$

where: $t_1 - t_0$ switch-on time for S_1, $t_2 - t_1$ switch-on time for S_4.

A time diagram of idealized compensation current $i_C(t)$ is shown in Fig. 5.46b. The output ripple can be calculated by the equation

$$\Delta\left|i_C(t_1)\right| = \left|\frac{u_{C1}(t_0) - e_M(t_0)}{L_C}(t_1 - t_0)\right|$$
$$\Delta\left|i_C(t_2)\right| = \left|\frac{-u_{C2}(t_1) - e_M(t_1)}{L_C}(t_2 - t_1)\right| \quad . \tag{5.37}$$

The voltage value at capacitors C_1 and C_2 is stabilized by a voltage controller and is equal to u_{DC}; this is why it can be assumed that $u_{DC} = u_{C1} = u_{C2}$. To achieve low dynamic distortion of the output current i_C, the slew rate must be high. The slew rate can be calculated by the formula

$$\left|\frac{\Delta i_C(t_1)}{\Delta t}\right| = \left|\frac{\pm u_{DC} - e_M(t_0)}{L_C}\right| \quad , \tag{5.38}$$

and the maximum and minimum values of current slew rate

$$\left.\frac{\Delta i_C(t)}{\Delta t}\right|_{max} = \frac{u_{DC} + e_{Mmax}}{L_C} \quad ,$$
$$\left.\frac{\Delta i_C(t)}{\Delta t}\right|_{min} = \frac{u_{DC} - e_{Mmax}}{L_C} \quad . \tag{5.39}$$

Currently IGBT transistors are mostly used as switching elements in the inverters. For the ordinary IGBT the maximum switching frequency is equal to 20 kHz [34] and around 60 kHz [20] for the fast IGBT. The transistor switching power losses can be approximated using the formula

$$P = P_{cond} + f_c E \quad , \tag{5.40}$$

where: E—energy lost in single switching cycle, P_{cond}—power losses in switched-on state.

So it is possible to assume that transistor power losses are directly proportional to switching frequency.

In the light of the above mentioned problem, choosing the right value of inductor L_C is very difficult. Factors to be considered in selecting the right value of APF output inverter inductor are shown in Table 5.8. For a higher L_C value, the time constant is higher and dynamic distortion is bigger, while for a lower L_C value, circuit dynamic distortions are smaller, but the value of compensation current ripple i_C is higher. One of the ways to decrease the dynamic distortion and keep current ripple at a reasonable value is to increase transistor switching frequency, but in this case switching losses and influence from the switching transition are increased.

Table 5.8 Pros and cons of using different values of inductor

Bigger value of inductor	Lower value of inductor
Pros	*Pros*
• Lower current ripple	• Faster transition response
• Lower transistor switching frequency	• Lower cost and lower weight
Cons	*Cons*
• Slower transition response	• Higher value of current ripple
• Bigger cost and weight	• Higher switching frequency
	• Bigger influence from the switching transition

5.7.4 *Simulation of APF Control Circuit*

As the whole APF control circuit is realized using a fixed-point or floating-point DSP, it is sensible to realize a digital simulation of the entire compensation system. For the simulation, the reference compensation signal $I_C(z)$ is calculated by an $p-q$ algorithm, and load current $I_L(z)$ is calculated by the power controller with resistance load. A timing diagram of the digital realization of the APF control circuit is shown in Fig. 1.11. Analog signal is sampled at a frequency of f_s, and is then converted into digital form by an A/D converter (Fig. 5.41). This digital signal is later processed by the main control algorithm of the APF and the output current controller and is then sent to the PWM modulator. The minimum delay of the entire control algorithm is therefore at least one sampling period T_s. This delay must be taken into account during the simulation. For the case under consideration, an algorithm delay time equal to two sample periods was chosen, $N_{alg} = 2$. The block diagram of a simplified digital simulation circuit is depicted in Fig. 5.47. In this case, in the output circuit transfer function $H_{out}(z)$ is also included in the output current controller transfer function (without PWM). The simulated line current $I_M(z)$ results from subtracting both signals $I_L(z)$ and $I_C(z)$. Simulated current signal waveforms of the compensation circuit are shown in Fig. 5.48. The waveforms obtained from the simulation are very close to the experimental waveforms (Fig. 5.28). For simplicity, in the simulation model, the PWM modulator is omitted, so the waveforms do not have modulation components. The spectra of the currents $I_L(z)$, $I_C(z)$, $I_M(z)$ are shown in Fig. 5.48b. In Table 5.9, some parameters of line current $I_M(z)$ are presented. Shown are the

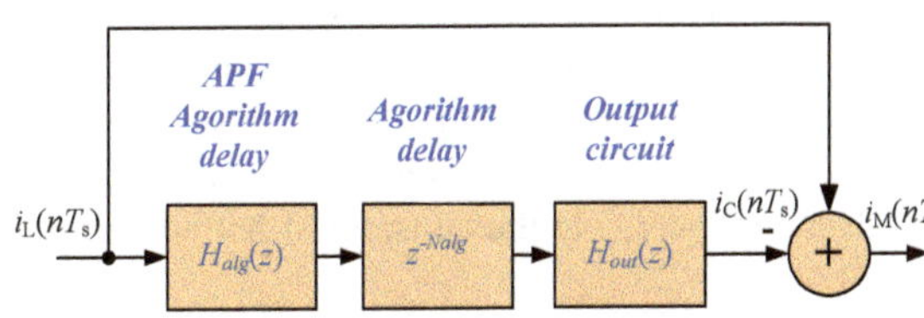

Fig. 5.47 Block diagram of digital simulation of compensation circuit (for one phase)

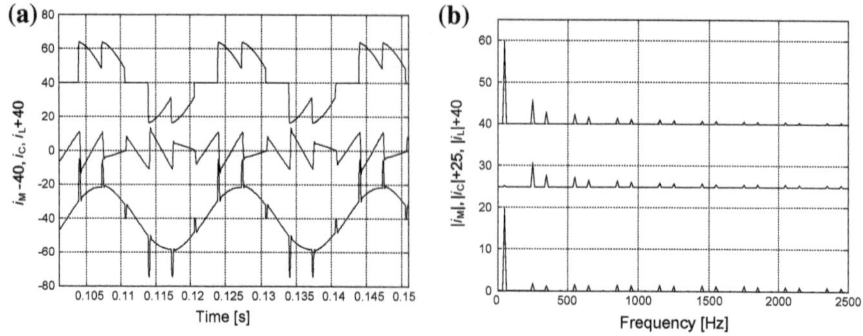

Fig. 5.48 Simulation currents of compensation circuit, load current i_L (*top*), compensation current i_C (*middle*), line current i_M (*bottom*): **a** waveforms, **b** spectra

Table 5.9 Line current parameters

Type of compensation	I_M(rms) [A]	*THD* [%]	*SINAD* [dB]	*THD*50 [%]
Without compensation	15.0	39.3	−8.7	39.0
Classical shunt APF	14.5	27.8	−11.5	27.2
Shunt APF with prediction algorithm $T_A = 214\,\mu s$	14.0	3.2	−29.8	1.3

line parameters for a circuit without compensation and for a circuit with classical compensation. For example, the *THD* ratio of simulated current is equal to 39.3% for a circuit without compensation, and is equal up to 27.8% for a circuit with compensation. This shows that the classical compensation circuit does not work effectively with rapid changes in load current.

5.8 Predictive Control Algorithm for APF

For predictable loads the compensation error depends on the dynamics of the output circuit and occurs periodically, so it can be partially compensated by sending a compensation current in advance. Such a solution would be impossible to implement in an analog control system, but it is easy to implement in a digital control system. In the proposed solution, for predictable loads, it is possible to use a circuit with prediction [31, 41, 44, 48], as shown in Fig. 5.49. Previous load current signal samples are stored in memory and are sent to present output in advance. This compensation is dependent on the inverter output time constant. Because the time constant is mainly dependent on the output inverter inductor value, it is possible to set a constant value for advance time T_A. In the considered APF the discrete advance time T_A is

$$T_A = N_A T_s \quad , \tag{5.41}$$

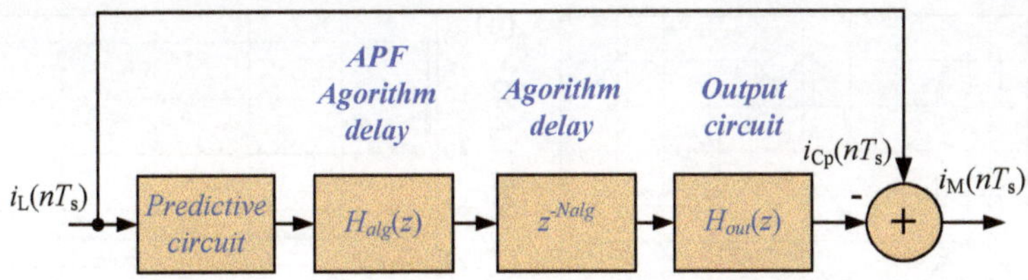

Fig. 5.49 Block diagram of compensation circuit for digital simulation with predictive circuit (for one phase)

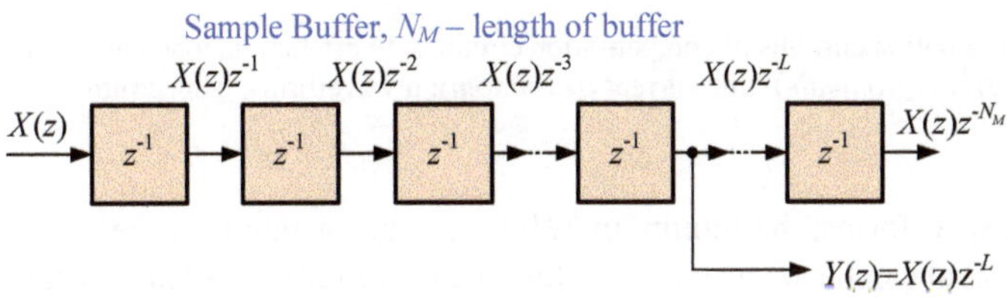

Fig. 5.50 Block diagram of a simple predictive circuit

where: $T_s = 1/(f_M N_M)$—sampling period, N_M—number of samples per power line voltage period, N_A—number of samples sent ahead.

The block diagram of a predictive circuit is depicted in Fig. 5.50. The circuit consists of sample buffer of length N_M, where samples of current period are stored. In the next period of power line voltage samples are send in advance from a buffer tap to the output ahead in time, which decreases the dynamic distortion. The length of sample buffer tap can be calculated by the formula

$$L = N_M - N_A \quad . \tag{5.42}$$

In the simulated compensation circuit the digital advance time is equal to $T_A = 214\,\mu s$. Simulated current signal waveforms for the compensation circuit with prediction are shown in Fig. 5.51a. The waveforms of line current $i_M(t)$ show that the goal of compensation has been reached and that the shape of the current is very close to sinusoidal. This is also supported by the spectrum shown in Fig. 5.51b. Similarly, the same is true with the signal parameters listed in Table 5.9. As a result of the prediction algorithm, the *THD* ratio was reduced from 27.8 to 3.2%.

5.8.1 Advance Time T_A

The main question is what should be the value of advance time T_A in order to achieve the goal of achieving a minimum value of *THD* of line current? Assuming that, for

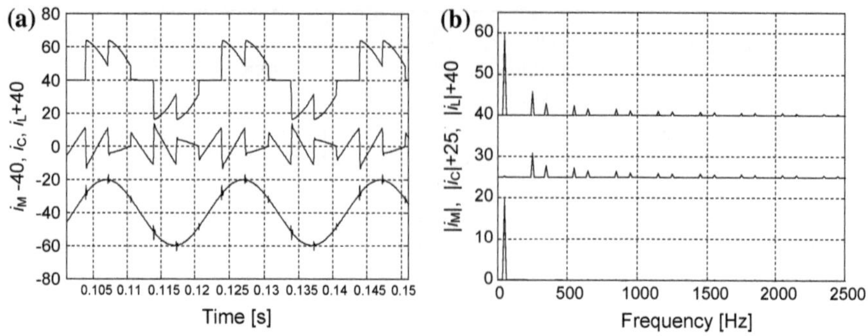

Fig. 5.51 Simulation currents of compensation circuit with prediction, load current i_L (*top*), compensation current i_C (*middle*), line current i_M (*bottom*): **a** waveforms, **b** spectrum

sinusoidal waveforms, the minimum *THD* value corresponds to the minimum *RMS* value, this goal can be achieved by finding the minimum *RMS* value, which is easier to determine. Such approach is used in this book

$$THD_{I_M}\Big|_{min} \longrightarrow I_{Mrms}|_{min} = f(T_A)|_{min} \quad . \tag{5.43}$$

Due to the complexity of the function $f(T_A)$, finding the minimum using analytical method is difficult. Therefore, numerical methods have been used for that purpose. The simulation studies were performed, investigating the effect of advance time T_A on the *RMS* current line value. A block diagram of the simulation is shown in Fig. 5.52. The minimum *RMS* value of line current is present for the time T_A equal to the time in which the APF output of the step response has reached 50% of the final value. The same relationship exists for line current *THD* ratio. Thus, the minimum value of *THD* current line is for

$$T_A = T_{alg} + T_{step(50\%)} \quad , \tag{5.44}$$

where: T_{alg}—"pure" delay of whole APF algorithm, $T_{step(50\%)}$—50% of step response time of whole APF control algorithm.

For predictive APF, T_A time for minimum *RMS* value of line current is used. Using the new control algorithm with predictive harmonic compensation, it is possible to decrease the harmonic contents in line currents from *THD* ratio from around 20% to below 10%. The current waveforms of compensation circuits with classical APF and with predictive APF for optimal value of T_A are shown in Fig. 5.53.

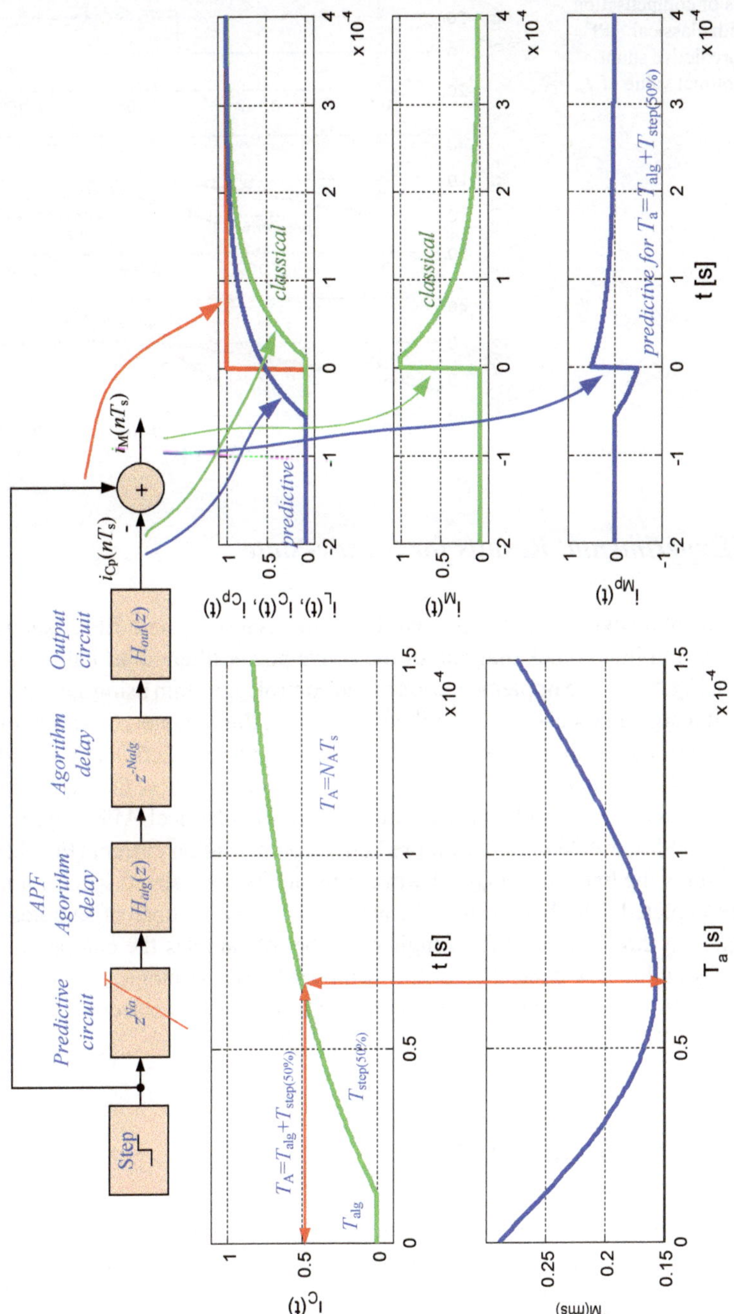

Fig. 5.52 The dependency of line current value I_{Mrms} versus T_A time for compensating step response

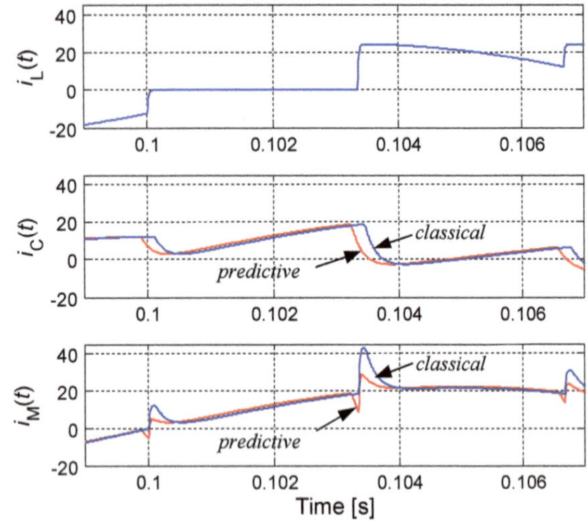

Fig. 5.53 Current waveforms of compensation circuits with classical APF and with predictive shunt APF for optimal value of T_A

5.8.2 Experimental Results for Steady-State

For experimental studies the author used the compensation system EFA1 shown in Fig. 5.26. A simplified block diagram of the active power filter control algorithm is depicted in Fig. 5.29. The implementation of the control algorithm using a fixed-point 16-bit digital signal processor is described in detail in this Chapter. Modification of the $p - q$ control algorithm is shown in Fig. 5.54, with two prediction circuits added to the classical control circuit [47, 48]. Figures 5.55 and 5.56 show the experimental waveforms in the same steady-state conditions, for the classical APF (Fig. 5.55a), and for the modified APF circuit with predictive compensation current (Fig. 5.56a). In these figures the line current $i_{M1}(t)$ waveform and its normalized spectrum magnitude are depicted. The best results are achieved for the number of samples sent ahead equaling three $N_A = 3$; for a higher number of samples the compensator is unstable. Using the new control algorithm with predictive harmonic compensation, it is possible to decrease the harmonic contents in power line currents from *THD* ratio around 22% to near 5% for $N_A = 3$.

5.8.3 Step Response of APF

The first issue when using the APF with a predictive circuit is determining how it will work when the value of load current changes rapidly. For this reason the author investigated a step response of load currents for the considered shunt APF [48]. Figure 5.58a shows the waveforms of a load current regulated by a power controller with resistive load without the APF compensation. The load current value is adjusted

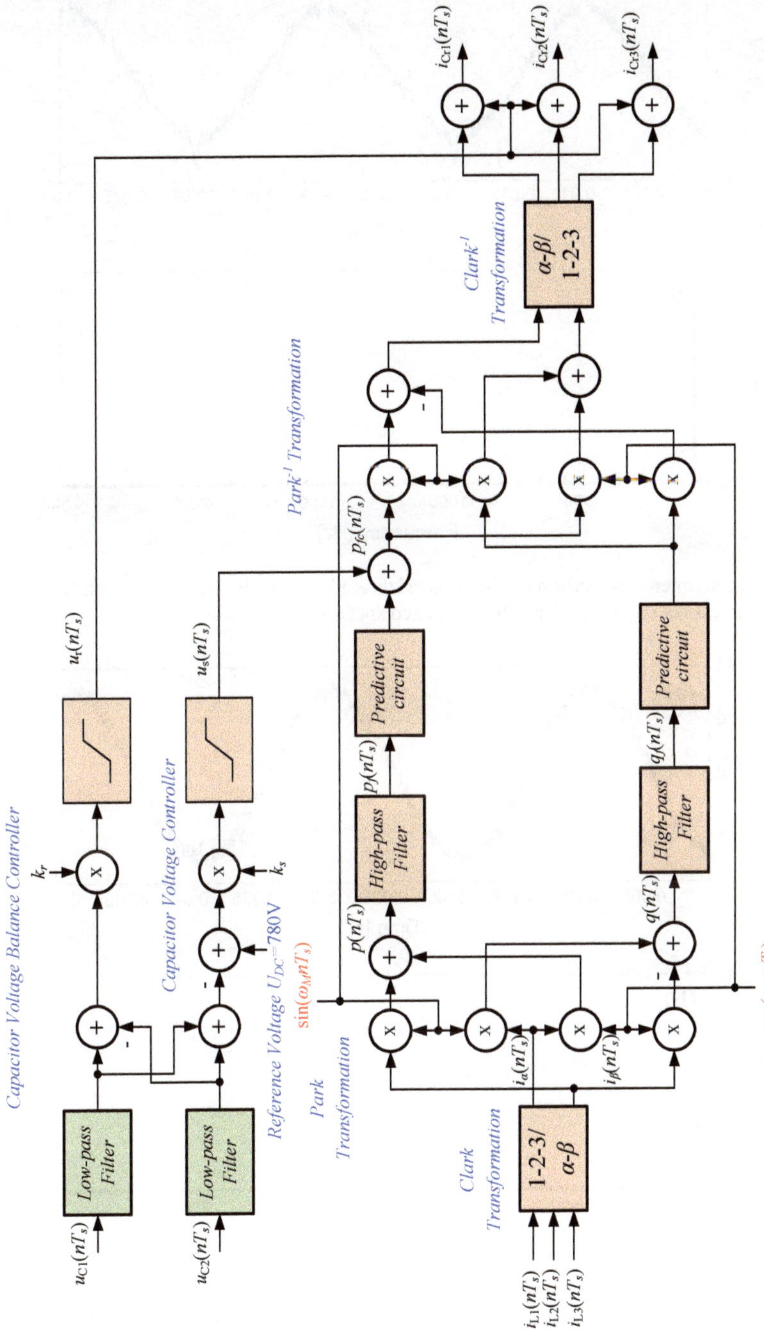

Fig. 5.54 Block diagram of control algorithm with predictive circuit

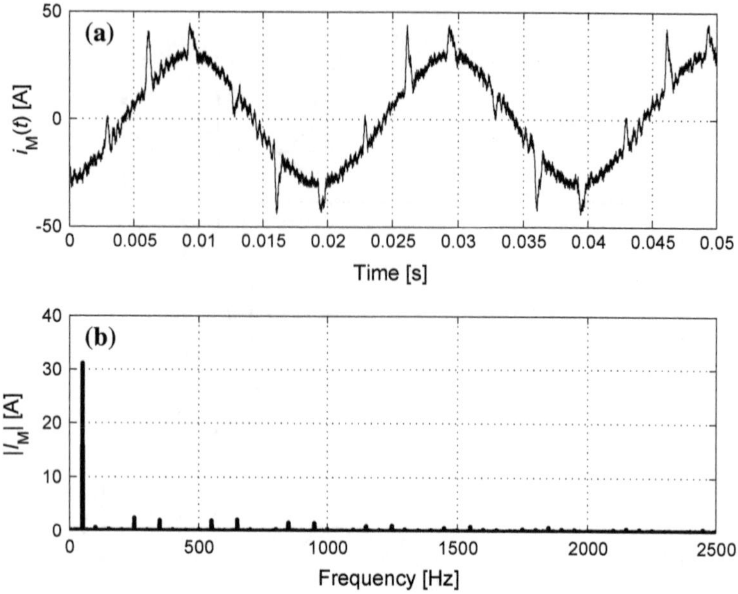

Fig. 5.55 Experimental waveforms of the classical three phase APF in steady-state with the resistive load line current $i_{M1}(t)$: **a** waveform, **b** normalized spectrum magnitude

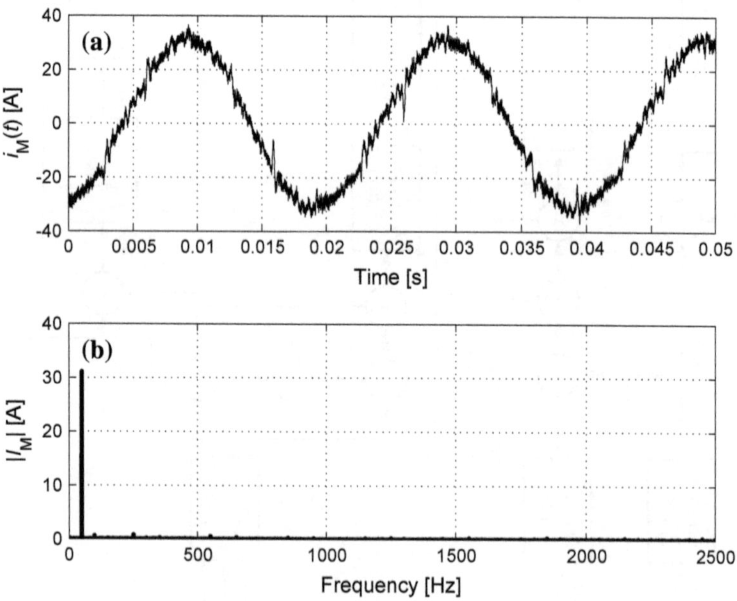

Fig. 5.56 Experimental waveforms of the modified three-phase APF in steady-state with the resistive load, line current $i_{M1}(t)$ for $N_A = 3$: **a** waveform, **b** normalized spectrum magnitude

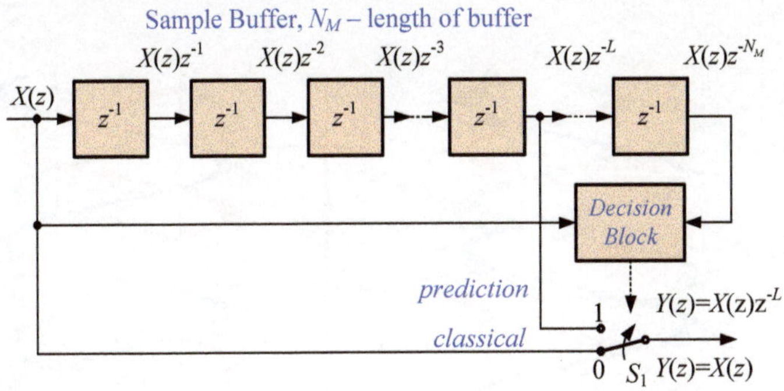

Fig. 5.57 Block diagram of a predictive circuit

by a power controller regulated by thyristor phase control, by changing trigger angle. The fire angle is controlled by a signal shown in Fig. 5.58a–d (the waveform at the very top). The step response of the load currents for the considered shunt APF with classical control algorithm is shown in Fig. 5.58b, for the same load current values as in Fig. 5.58a. The APF step response setting time is equal to one period of line voltage. Figure 5.58c shows step response of the APF with modified control algorithm with predictive circuits constantly switched on. In the first line voltage period, after the value of load currents is changed, the APF sends compensation current $i_C(t)$ adequate to previous load currents and hence the resultant currents are non-sinusoidal (Fig. 5.58c). Therefore the predictive circuit should be modified to the circuit presented in Fig. 5.57. In this circuit current samples $X(z)$ are stored in DSP memory—sample buffer of length N_M. In the next period of line current they are compared to present samples and if the absolute difference of the present sample value and respective sample stored in memory is less than an assumed value, the predictive current compensation algorithm is switched on (switch S_1 in position 1). As a result of this comparison, we can assume that the current waveform will be the same as in the previous period and therefore samples $X(z)z^{-L}$ can be sent from the buffer tap L to the output ahead in time, which decreases the dynamic distortion. If load current is changed, the current predictive algorithm is switched off (switch S_1 in position 0), the algorithm waits for a steady-state and if the current waveform does not change in subsequent periods, it switches on again (switch S_1 in position 1). The output signal is described by the formula

$$\begin{cases} Y(z) = X(z)z^{-L} & \text{for} \quad S_1 = 1 \\ Y(z) = X(z) & \text{for} \quad S_1 = 0 \end{cases} \tag{5.45}$$

Decision block (Fig. 5.57) operates according to the relation described in Listing 5.4.

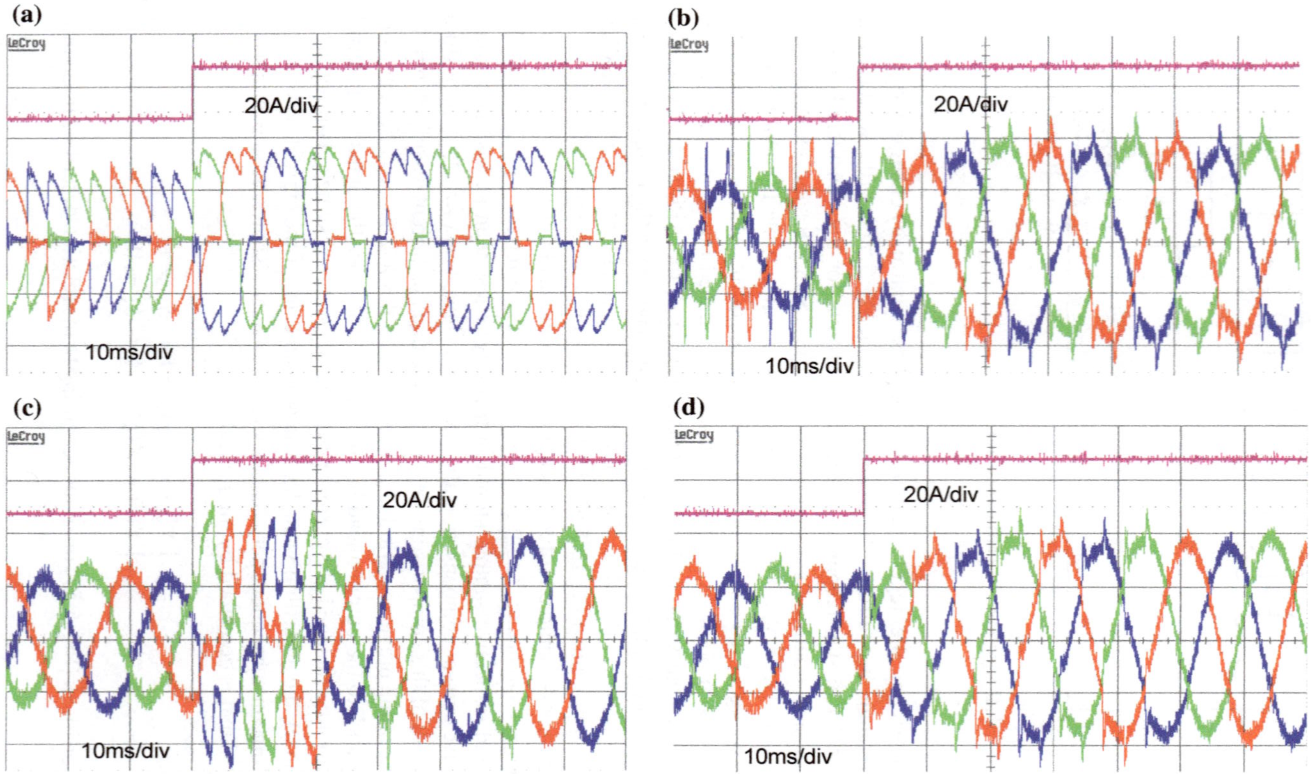

Fig. 5.58 Experimental waveforms of line currents: $i_{M1}(t)$, $i_{M2}(t)$, $i_{M3}(t)$ for step response of the load currents for the power controller with the resistive load: **a** APF is switched off, **b** APF with classical control algorithm, **c** APF with predictive algorithm constantly switched on, $N_A = 3$, **d** APF with adaptive predictive algorithm, $N_A = 3$

Listing 5.4 C program for decision block

```
if abs(x(n) - x(n - NM)) > Emax {
          S1 = 0;          // classical
          y(n) = x(n);
          }
else {
          S1 = 1;          // predictive
          y(n) = x(n - L);
          }
```

The constant E_{max} is maximum difference between the current sample and the sample from the last period of the power line voltage.

Waveforms with the result of such modification are shown in Fig. 5.58d. After the rapid change in value of load current, the predictive (non-causal) circuits are switched off. In the considered case (Fig. 5.58d), they are switched on again after two and a half line voltage periods. It should be noted that interesting changes in the spectrum of line current occur for rapid changes of load current. For this reason a test with square changes of load current was performed. Its result is shown in Fig. 5.59. A waveform of line current $i_{M1}(t)$ (for $N_A = 3$) is depicted for square changes of load current $i_{M1}(t)$. However, a waterfall diagram of the spectrum of line current is shown in Fig. 5.60.

Using adaptive predictive compensation current it is possible to decrease harmonic content for predictable loads. This modification for an $p - q$ control algorithm is very simple and additional computational workload is very small. Therefore it can be easily realized in an existing APF digital control circuit. The considered current prediction circuit may also be useful for other APF control algorithms. The presented circuit may also be applied in other areas of power electronics.

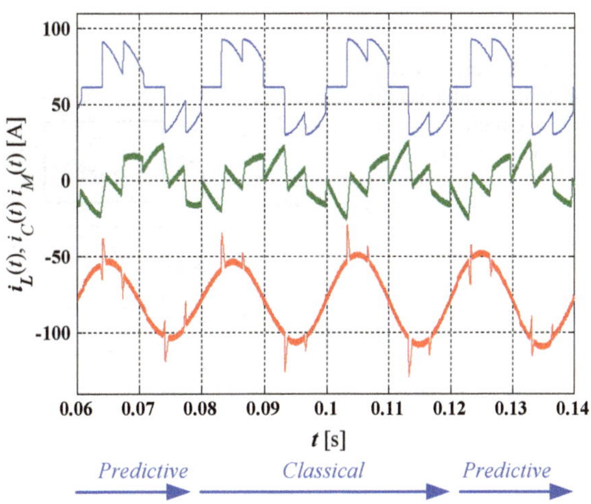

Fig. 5.59 Experimental waveform of line current $i_{M1}(t)$, for $N_A = 3$

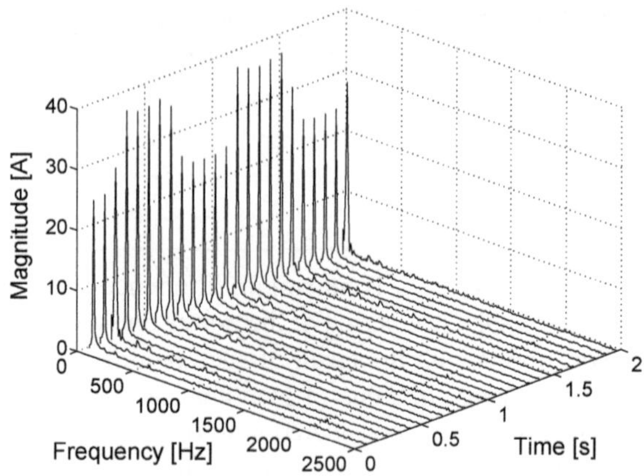

Fig. 5.60 Experimental waterfall diagram of line current $i_{M1}(t)$, for $N_A = 3$

5.9 Selected Harmonics Separation Methods Suitable for APF

In the proposed solution the user can select which harmonics are most important in the active power filtration process. This is very important, especially when several APFs are working in parallel or cascaded connection. Another application of such circuits are power line harmonics resonance dampers. A block diagram of a selective harmonics APF control circuit is shown in Fig. 5.61.

In this circuit K band-pass filters tuned for selected harmonics are used. For such a circuit, the signal representing compensation current is calculated by the equation

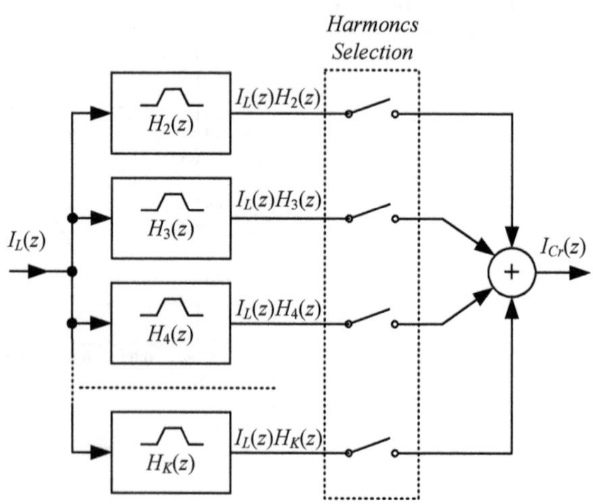

Fig. 5.61 Selective K harmonics compensation control circuit for $k = \{5, 7, 11, 13\}$

$$I_{Cr}(z) = I_L(z) \sum_{k=0}^{K} H_k(z) \quad , \tag{5.46}$$

where: $H_k(z)$—transfer function of band-pass filter for k-th harmonic.

A circuit giving the user the choice of harmonics compensated by the APF has been added to the control system, thus allowing the user to adjust the APF compensation to the local conditions. The set of harmonics selected for elimination (second, third, ...) is set and in the synthesis filter banks harmonic signals are synthesized to the appropriate current compensation signal. In this case the synthesis filter bank is very simple and consists of a summing block only. This kind of solution can also be used for harmonic prediction to cancel permanent dynamic errors (for predictable loads), as described by the author [41].

5.9.1 Control Circuit with MDFT

A moving DFT algorithm [35] works as an analysis filter bank. This is described in Chap. 3. The block diagram of such an analysis filter bank is shown in Fig. 3.68. Simplified characteristics of the amplitude of such a filter bank are shown in Fig. 3.59. The APF control circuit using MDFT filter banks for four harmonics ($k = \{5, 7, 11, 13\}$) is shown in Fig. 5.62. The discrete representation in Z-domain of load current signal $i_L(z)$ is divided into $N = 256$ uniform subbands by the analysis filter banks, with, in this case, only four being selected. An analysis filter bank consists of a common comb filter on the input and the four branches. The frequency response of such a circuit is depicted in Fig. 5.63. It should be noted that for the four selected harmonics phase shift is 0°. For the control system shown in Fig. 5.62 simulation studies were carried out for a three-phase APF compensation circuit. The results of such a simulation are shown in Fig. 5.64.

5.9.2 Control Circuit with $p - q$ Algorithm

The $p-q$ algorithm can be also used in an analysis filter bank. A block diagram of this circuit for the compensation of 5th, 7th and 11th harmonics is shown in Fig. 5.65. The circuit consists of three blocks: the first works with a frequency of $5f_M$, the second of $7f_M$ and the third, of $11f_M$. Resultant compensation signal $i_{Cr}(n)$ for each phase is a result of the sum of output signals for the selected harmonics.

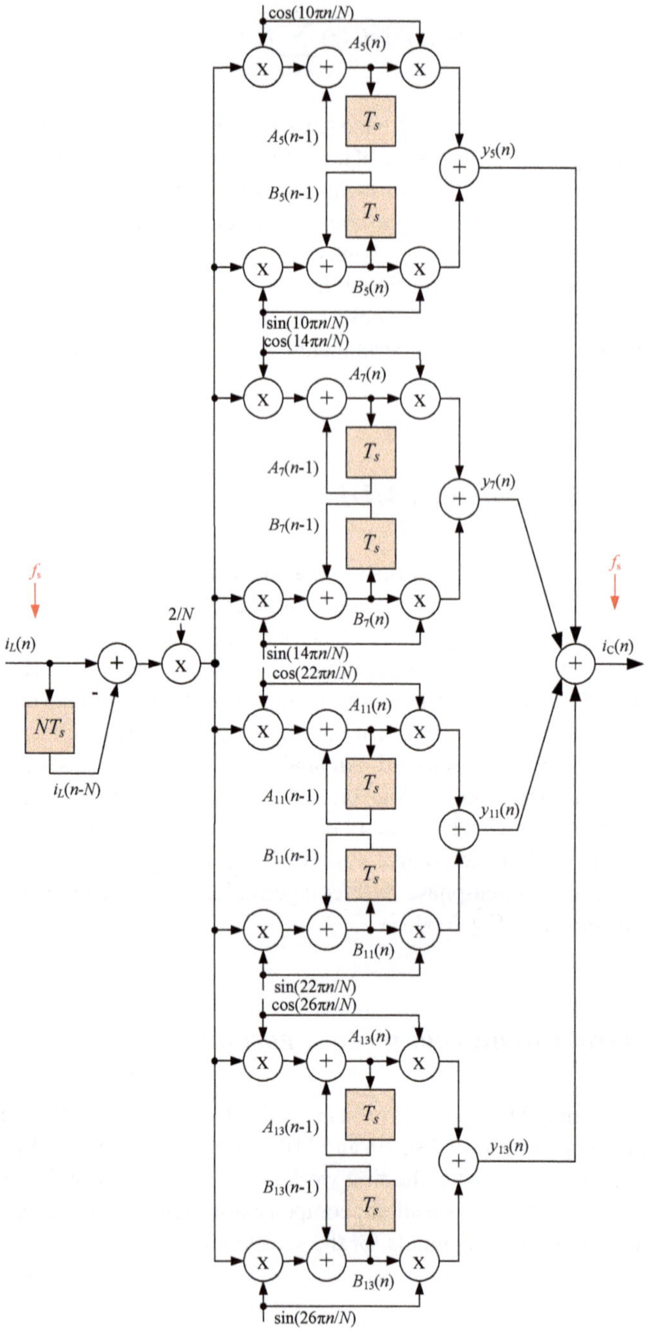

Fig. 5.62 A block diagram of MDFT APF control circuit

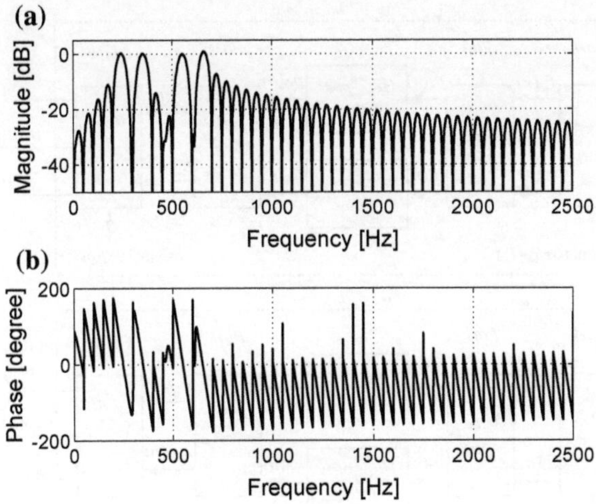

Fig. 5.63 Frequency response of MDFT filter bank $N = 256$, $k = \{5, 7, 11, 13\}$: **a** magnitude, **b** phase angle

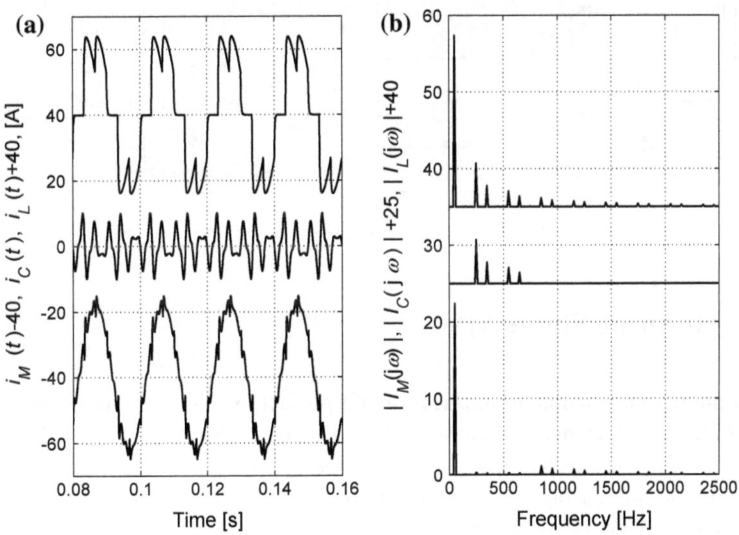

Fig. 5.64 APF MDFT with selective harmonics compensation for $k = \{5, 7, 11, 13\}$: **a** waveforms, **b** spectra

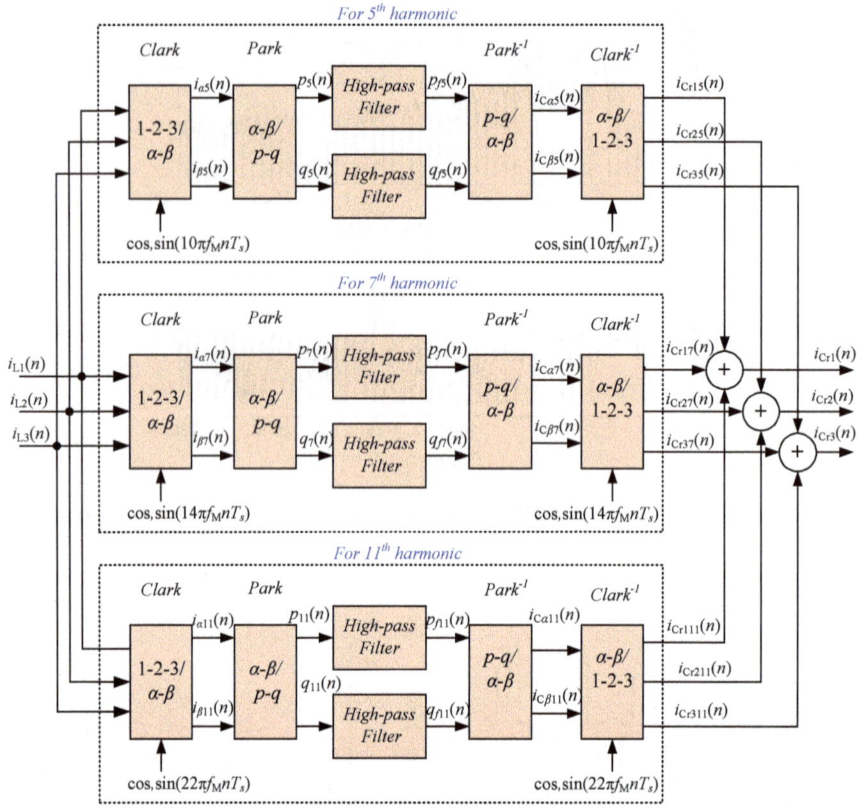

Fig. 5.65 A block diagram of APF $p - q$ algorithm control circuit for $k = \{5, 7, 11\}$

5.10 Multirate Shunt APF

It is commonly known that the application of a prediction algorithm for unpredictable loads is inefficient. Hence, the author suggests using multirate APF. Given that high dynamic performance is necessary only for approximately 10% of the time in the line voltage period, increasing the switching frequency to 60 kHz seems to be unreasonable. Therefore, the author proposes an inverter output stage with two sets of output circuits: "fast" and "slow" [44–46].

The compensation circuit for a one phase active power filter with a modified inverter is depicted in Fig. 5.66. The circuit has a common supply voltage U_{C1}, U_{C2}, made using common DC bank (C_1, C_2), for both parts of the inverter: "slow" (S_{s1}, S_{s2}, L_{Cs}) and "fast" (S_{f1}, S_{f2}, L_{Cf}). The value of inductor L_{Cs} is designed to achieve a low $i_C(t)$ current ripple and the value of inductor L_{cf} is designed to achieve a fast response to the output current. In the considered circuit, the following transistor

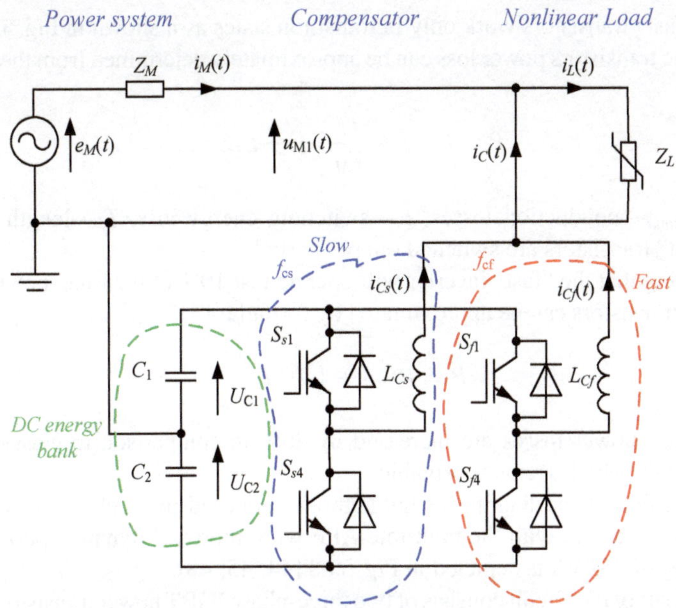

Fig. 5.66 One phase active power filter with modified inverter

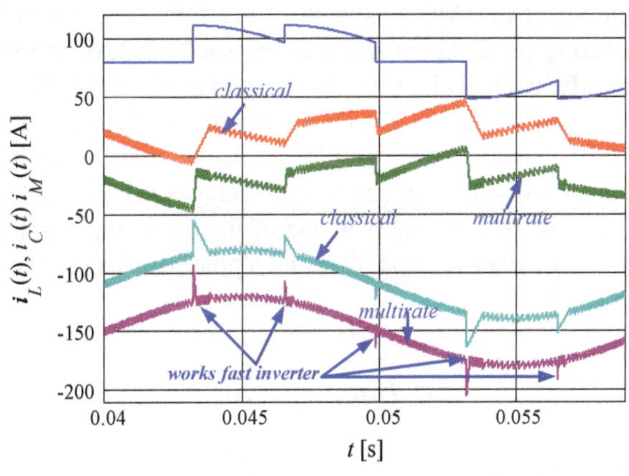

Fig. 5.67 Comparison of current waveforms of currents (for one phase): $i_L(t)$, $i_C(t)$, $i_M(t)$ for classical APF and multirate APF

switching frequencies are adopted: $f_{cs} = 12,800$ Hz, $f_{cf} = 51,200$ Hz and values of inductance: $L_{Cs} = 2.5$ mH, $L_{Cf} = 0.5$ mH.

The big difference between well-known solution, where "fast" and "slow" inverters work continuously, e.g. [14], and author's solution is the working mode of the "fast" inverter. In the proposed circuit "fast" inverter works only in the transition state. In Fig. 5.67 the comparison of current waveforms of currents (for one phase): $i_L(t)$, $i_C(t)$, $i_M(t)$ for classical APF and multirate APF is shown. It should be noted that the "fast" transistors work only in transition states as it shown in Fig. 5.67. The value of the transistors power loss can be approximately determined from the relation

$$P = P_{cond} + \frac{T_f}{T_M} f_{cf} E + f_{cs} E \quad , \tag{5.47}$$

where: P_{cond}—conduction losses, E—switching energy loss, T_f—length of time when "fast" transistors are switched on, per period.

Assuming that the "fast" inverter will operate just 10% of the time, power losses of output transistors can be approximated by formula

$$P = P_{cond} + 1.4 f_{cs} E \quad . \tag{5.48}$$

In such case power losses are increased by 40% in comparison to classical APF output circuit, which seems acceptable.

A simplified block diagram of the author's proposed three-phase active power compensation circuit with the multirate APF with improved dynamic performance for a power of 75 kVA is depicted in Fig. 5.68 [44, 45, 48].

Power part of the circuit consists of two three-phase IGBT power transistor bridges connected to the AC lines through an inductive filtering system composed of inductors L_{Cs1}, L_{Cs2}, L_{Cs3}, L_{Cf1}, L_{Cf2} and L_{Cf3}. The APF circuit contains common DC energy storage, provided by two electrolytic capacitors C_1 and C_2. A view of the output circuit of the APF is shown in Fig. 5.69. The control circuit is realized using the floating-point digital signal processor TMS320F28335 [56, 58]. The block diagram of APF control algorithm is depicted in Fig. 5.70. The control algorithm uses sliding Goertzel DFT [25, 26, 41, 43] for the load current first harmonic detection. Due to sliding Goertzel DFT (SGDFT) characteristics (shown in Chap. 3), control circuits have to be synchronized to line voltages $u_1(t)$, $u_2(t)$ *and* $u_3(t)$ by a synchronization unit, and this is why it is one of the most important parts of the control circuit. It consists of a low-pass filter and a phase-locked loop circuit (PLL).

5.10.1 Analog Input Circuit

The APF has eleven analog input signals: three from load currents $i_{L1}(t)$, $i_{L2}(t)$ and $i_{L3}(t)$, six from inverter output currents $i_{Cs1}(t)$ $i_{Cs2}(t)$, $i_{Cs3}(t)$, $i_{Cf1}(t)$, $i_{Cf2}(t)$, $i_{Cf3}(t)$ and $i_{Cf1}(t)$, and two from DC capacitor voltages U_{DCp} and U_{DCn}. All signals are sampled with a sampling rate equal to $f_s = 102{,}400$ Hz. For the electronic measurement of currents, with a galvanic isolation between the primary circuit (high power) and the secondary circuit (electronic circuit), current transducers LEM LA 125-P are used. Primary current is transformed to secondary current according to conversion ratio K_N. For primary current $i_{pCT} = 125$ A, the secondary is equal to $i_{sCT} = 125$ mA. In a similar way to the currents, the voltages are also transformed. Primary voltage $u_{DC}(t)$ is transformed to current signal and then to voltage secondary current $i_{sVT}(t)$. In the

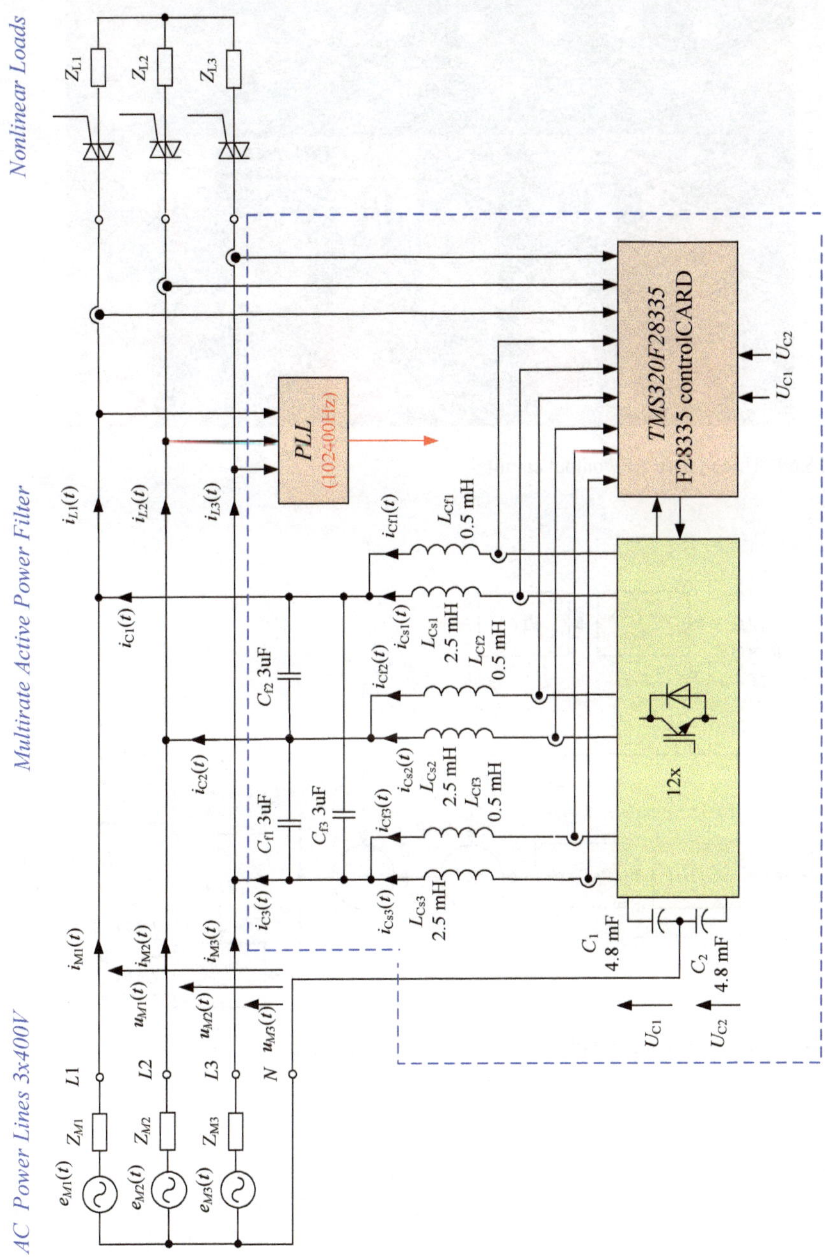

Fig. 5.68 Modified three-phase shunt active power filter

Fig. 5.69 Three-phase APF output circuit

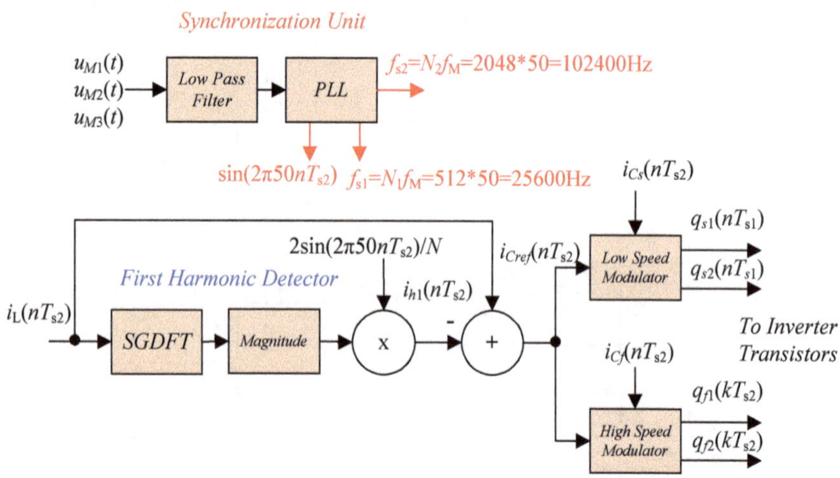

Fig. 5.70 Block diagram of APF control algorithm for one phase

considered circuit LEM voltage transducers LV25-P are used. For primary current $i_{pVT} = 10\,\text{mA}$, the secondary is equal to $i_{sVT} = 25\,\text{mA}$. The TMS320F28335 has a 12-bit A/D converter with voltage input range from 0 to 3$\cdot$ V [56, 58]. Unfortunately, these inputs are unipolar and therefore all analog input signals have to be transformed to this range. Therefore, for analog signals a virtual ground at +1.5 V is made. It is made by a 1.5 V reference voltage diode D_1. Figure 5.71 shows a simplified diagram

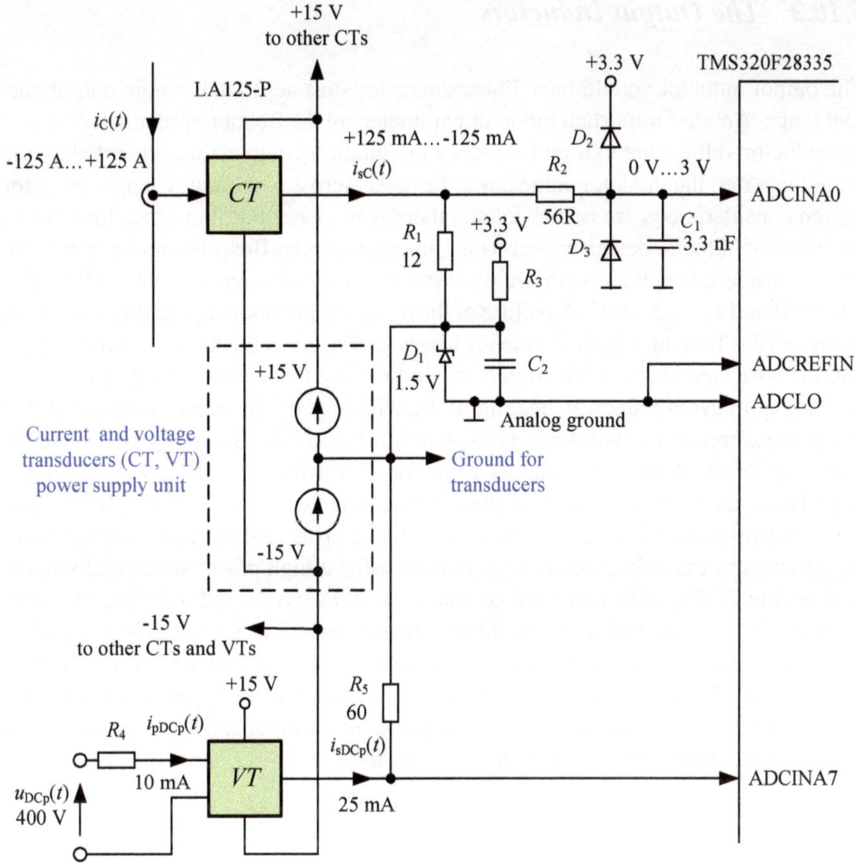

Fig. 5.71 Simplified diagram of galvanically isolated analog input circuit

of the analog input circuit. A/D converter input voltage for current transducer can be calculated by the formula

$$U_{ADCIN0} = \frac{I_C}{K_{NCT}} R_1 - U_{D1} \quad , \tag{5.49}$$

where: K_{NCT}—current transfer ratio of current transducer, and for voltage transducer

$$U_{ADCIN7} = \frac{U_{DCp}}{R_4} \frac{1}{K_{NVT}} R_5 - U_{D1} \quad . \tag{5.50}$$

Such a solution, designed by the author, allows significant simplification of the input circuit, using only one supply voltage 3.3 V.

5.10.2 The Output Inductors

The output inductor should have linear characteristics across the whole output current range. Another important inductor parameter are the frequency characteristics of the inductor value. Due to high frequency modulation components generated by the output inverter, the inductor value has to be linear across a frequency range. Inductor design considerations are not trivial and there is no perfect design procedure. Some problems of inductor design are described among others by Bossche and Valchev [12]. In the considered application the value of inductors was chosen equal [43, 44] to: L_{Cf} = 0.5 mH and L_{Cs} = 2.5 mH. A picture of the designed inductors L_{Cf} and L_{Cs} is shown in Fig. 5.72a. To reduce high-frequency losses, ferrite cores U100/50/25 with air gaps and litz wire were chosen. The inductor windings are located at some distance from the air gap to avoid induction heating of the windings by the transverse field at that place. Location of the windings is depicted in Fig. 5.72b. Small signal frequency responses of the designed inductors were checked using Agilent 4294 A precision impedance analyzer. The result of these measurements is shown in Fig. 5.73a. The frequency response of inductor impedance is linear up to 1 MHz. Additionally, inductor parameters were checked for high current using a high power sinusoidal voltage source (Fig. 5.73b) with regulated output of 0–500 V/50 A, and a frequency range of 500–8,000 Hz. This electric machine voltage source was designed by the author using an old synchronous frequency converter, an electric machine formerly used for induction heating. This electrical machine frequency converter consists of two synchronous electric machines: one three-phase on the input and a second one phase on the output, both fixed on the one common shaft.

5.10.3 APF Simulation Results

The simplified output circuit of the proposed APF is shown in Fig. 5.74. The circuit consists of two output stages: one with switches S_{s1}, S_{s4} and inductor L_{Cs}, and a second with switches S_{f1}, S_{f4} and inductor L_{Cf}. The first output stage works continuously with the slowest switching frequency f_{cs}. In the second output stage, switches S_{f1}, S_{f4} work with a several-times higher frequency f_{cf} only in the case when output current changes very quickly (typically 10% of line voltage power period). Some interesting analyses of a similar circuit were presented by Watanabe et al. [60]. At the beginning, a digital hysteresis modulator was designed for controlling the modified inverter. During the simulation analysis the modified inverter and the classical inverter were taken into consideration. The simulation parameters are: L_{Cf} = 0.5 mH, L_{Cs} = 2.5 mH, U_{C1} = 390 V, U_{C1} = 390 V, f_{cf} = 102,400 Hz, f_{cs} = 25,600 Hz. A simplified diagram of the modified inverter simulation circuit is shown in Fig. 5.74. The control algorithm

(a)

(b)

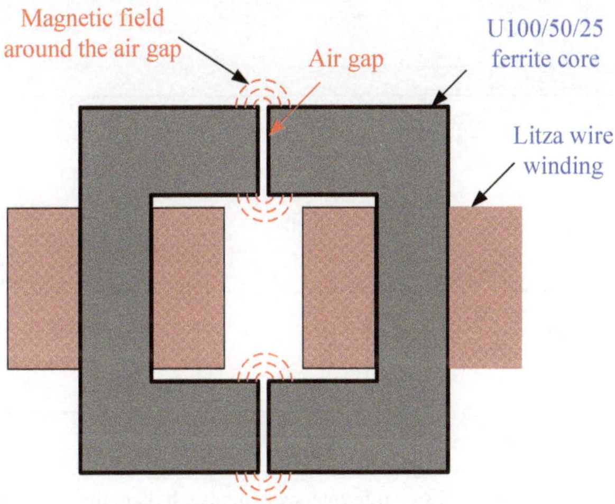

Fig. 5.72 Inductors: **a** the view of inductors **b** location of winding

of the two hysteresis digital modulators with additional conditional control logic implemented in the MATLAB is shown in Fig. 5.75. The MATLAB program of two hysteresis modulators is presented in Listing 5.5.

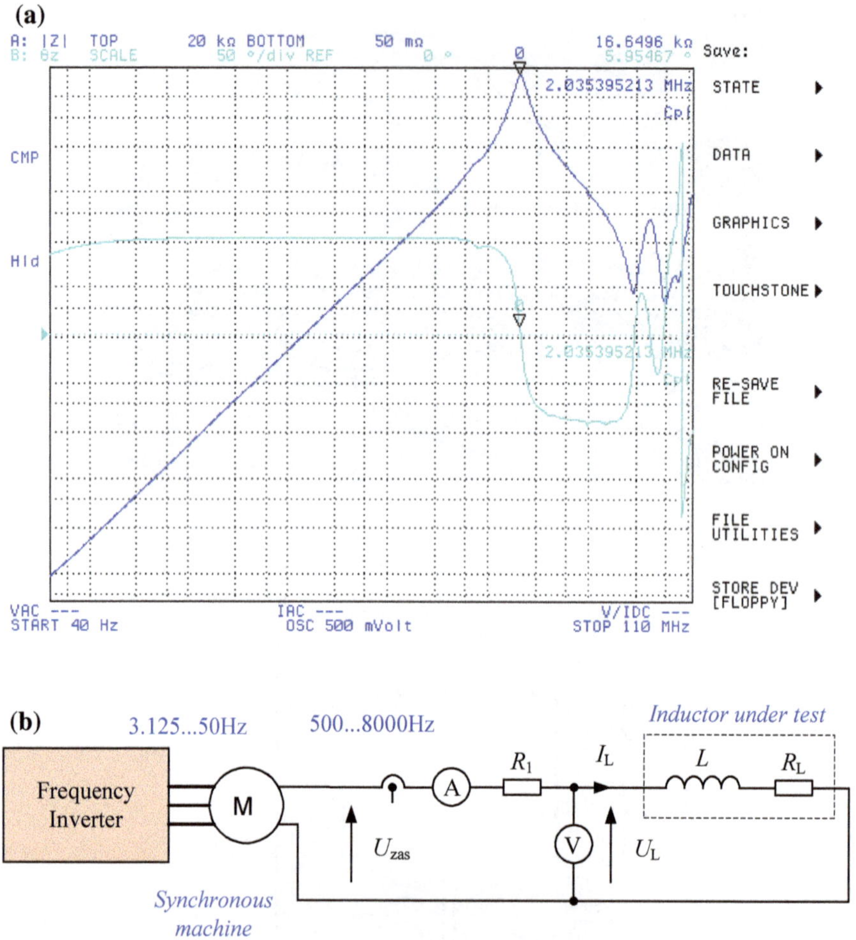

Fig. 5.73 Inductor tests: **a** small signal frequency response, **b** diagram of the inductor test circuit

Fig. 5.74 Simplified diagram of modified inverter model connected to the power line

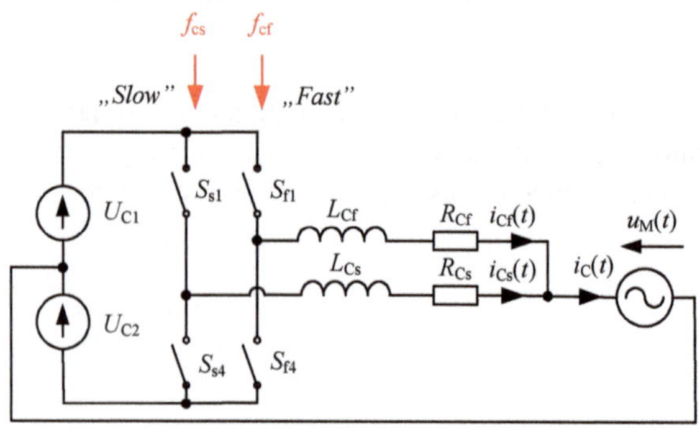

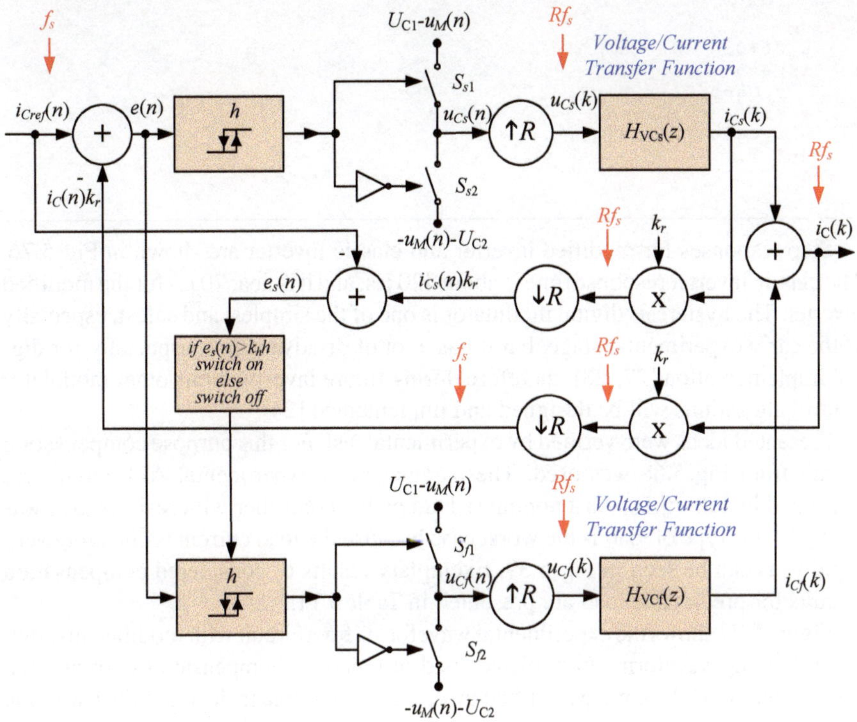

Fig. 5.75 Simplified block diagram of the output inverter simulation circuit

Listing 5.5 Two hysteresis modulators

```
e=i_Cref(n)-i_C(n)*kr;
e_s=i_Cref(n)-i_Cs(n)*kr;
if u_Cs>=0
  if e_s>kh*h
    u_Cs=u_C1-u_M(n);
    if e<-h&&u_Cf>=0
      u_Uf=u_C2-u_M(n);
    elseif e>h&&u_Cf<=0
      u_Cf=u_C2-u_M(n);
    end
  else
    u_Cf=0;
    if blad<-h
      u_Cs=u_C2-u_M(n);
    else
      u_Cs=u_C1-u_M(n);
    end
  end
else %u_Cs <0
  if e_s<-kh*h
    u_Cs=u_C2-u_M(n);
    if e<-h&&u_Cf>=0
      u_Cf=u_C2-u_M(n);
    elseif e>h&&u_Cf<=0
      u_Cf=u_C1-u_M(n);
    end
```

```
    else
      u_Cf=0;
      if e>h
        u_Cs=u_C1-u_M(n);
      else
        u_Cs=u_C2-u_M(n);
      end
    end
end
```

Step responses for modified inverter and classic inverter are shown in Fig. 5.76. The classic inverter response time is about 420 μs, and it is near 70 μs for the modified inverter. The hysteresis digital modulator is one of the simplest and safest, especially at the early experimental stage, but it has a lot of disadvantages, especially for digital implementation [27, 28], therefore during future investigations other modulator control algorithms will be designed and implemented [24].

Presented ideas were verified by experimental test. For this purpose compensation circuit from Fig. 5.68 was used. The parameters of experimental APF circuit are presented in Table 5.10. As a nonlinear load power controller with resistor load was applied. This type of load is the worst case because the load current is changing very rapidly, as can be seen in Fig. 5.39. Exemplary results of considered compensation circuits for predictable load are presented in Table 5.11.

Figure 5.77 shows the experimental waveforms for a circuit with modified inverter. The following waveforms are depicted: load current $i_L(t)$, compensation current $i_C(t)$, line current $i_M(t)$. Using the modified inverter, it is possible to decrease the harmonic

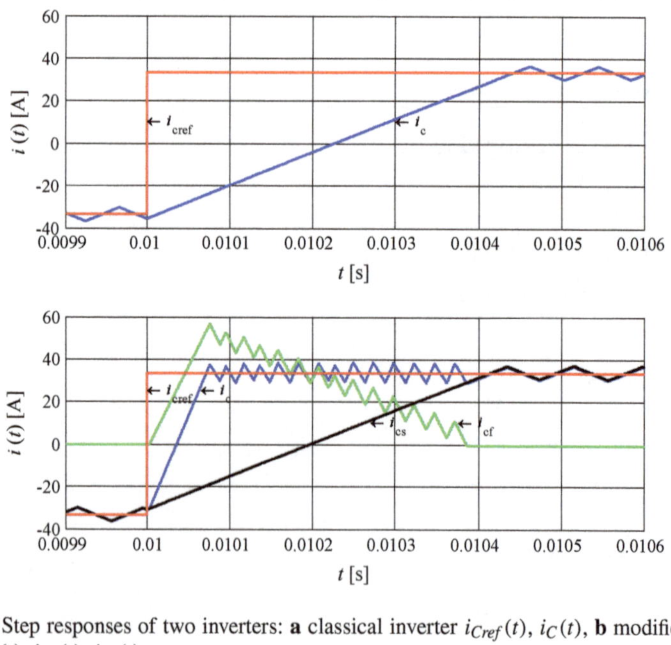

Fig. 5.76 Step responses of two inverters: **a** classical inverter $i_{Cref}(t)$, $i_C(t)$, **b** modified inverter $i_{Cref}(t)$, $i_C(t)$, $i_{Cf}(t)$, $i_{Cs}(t)$

Table 5.10 Parameters of experimental APF circuit

Name	Value
f_{cf}	51,200 Hz
f_{cs}	12,800 Hz
L_{Cs}	2.5 mH
L_{Cf}	0.5 mH

Table 5.11 Comparison of different compensation circuits

Type of compensation	$I_{M(rms)}$ [A]	THD [%]	THD_{50} [%]
Without compensation	29.4	30.1	29.7
Classical shunt APF	29.2	17.8	17.2
Shunt APF with predictive circuit $T_A = 214\,\mu s$	29.0	5.9	5.3
Multirate shunt APF	29.0	6.5	6.3

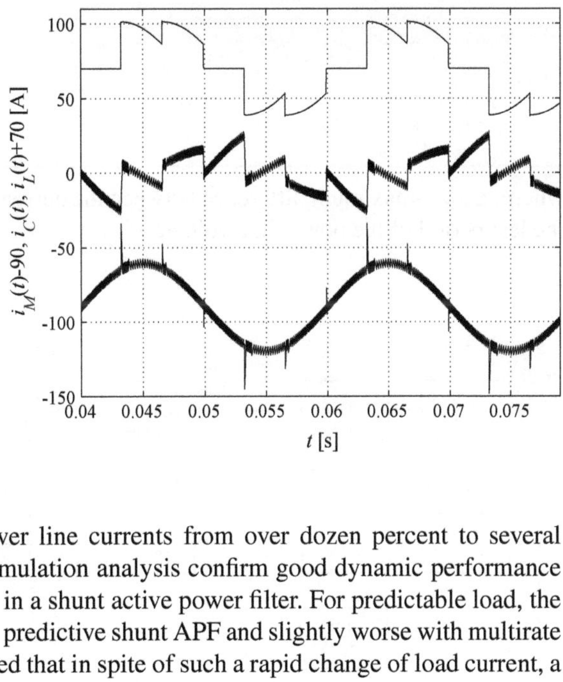

Fig. 5.77 Experimental waveforms of one phase active power filter in steady-state with resistive load, and with modified output inverter: load current $i_L(t)$, compensation current $i_C(t)$, line current $i_M(t)$

contents (THD ratio) in power line currents from over dozen percent to several percent. The results of the simulation analysis confirm good dynamic performance of the modified inverter used in a shunt active power filter. For predictable load, the best results are achieved with predictive shunt APF and slightly worse with multirate shunt APS. It needs to be noted that in spite of such a rapid change of load current, a compensation was achieved at the level of about 7% THD of line current. For typical loads, for which load current changes are usually much slower, the value of THD should be below 5%.

The results of the simulation analysis confirm good dynamic performance of the modified inverter used in a shunt active power filter. For assumed simulation parameters, current ripples are higher when the fastest part of the inverter is switched-on, but the resultant value of *THD* ratio is smaller when compared to the classical inverter.

5.11 Multirate Shunt APF with Prediction

Another possible improvement of the APF can be introduced by combining a predictive circuit with the multirate APF. Such a solution would use the predictive circuit for predictable loads, whilst the multirate circuit would be used for unpredictable loads and for transition states. This new circuit would allow for using the best features of both presented solutions. A diagram of the proposed solution is presented in Fig. 5.78. The APF circuit parameters are the same as shown in Table 5.10. The control algorithm (Fig. 5.78) operates according to the relation described in Listing 5.6.

Listing 5.6 C program for decision block

```
if abs(x(n) - x(n - NM)) > Emax {
    S1 = 0;           // multirate
    y(n) = x(n);
    }
else {
    S1 = 1;           // predictive
    y(n) = x(n - L);
    }
```

where: E_{max}—maximum difference between the current sample and the sample from the last period of the power line voltage.

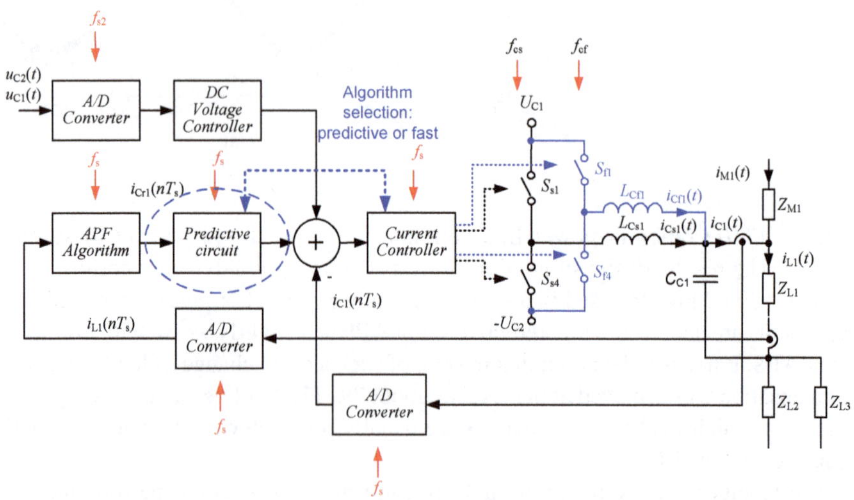

Fig. 5.78 Simplified block diagram of a multirate shunt APF with prediction

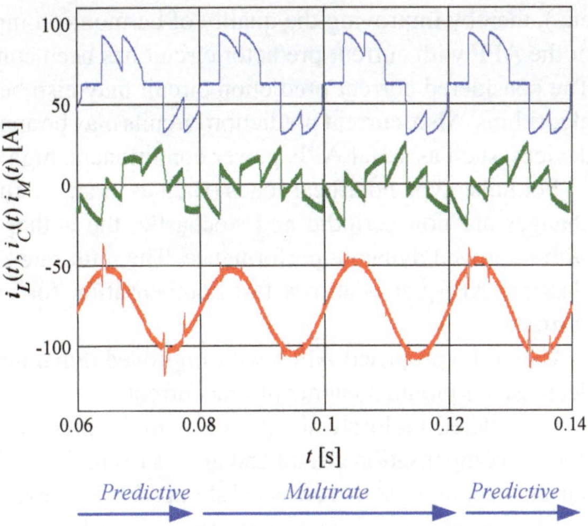

Fig. 5.79 Exemplary experimental waveforms of currents: $i_L(t)$, $i_C(t)$, $i_M(t)$ in a compensation circuit with multirate shunt APF with prediction for a step change in load current

Figure 5.79 shows experimental results of the proposed circuit. The waveforms for a step change in load current are presented for currents $i_L(t)$, $i_C(t)$, $i_M(t)$. As can be seen in the Fig. 5.79, the predictive circuit works during the predictable states and the multirate circuit works during unpredictable states. This allows for a reduction in *THD* for different loads (both predictable and unpredictable).

5.12 Conclusions

The aim of the author's modifications presented in this chapter is to develop control algorithms which allow a reduction in the line current *THD* ratio. The selected digital signal processing algorithms designed for control of active power filters have been designed and investigated. These include algorithms with first harmonics detectors based on: IIR filter, lattice wave digital filter, sliding DFT, sliding Goertzel and moving DFT. Modified classical control circuits based on $p - q$ control algorithm have also been considered. The described APF control circuits with filter banks allows the selection of compensated harmonics or damping of selected resonance in the power line.

Discussion of problems of the active power filter dynamics has been presented. For predictable nonlinear loads which vary slowly compared to line voltage period (rectifiers, motors etc.), it is easier to predict current changes. For such loads, by using a shunt active power filter with a predictive (non-causal) algorithm, it is possible to decrease harmonic contents. This modification a $p - q$ control algorithm is very simple and additional computational workload is very small. Therefore it can be very easy to implement it in an existing APF digital control circuit based on a digital signal processor, microcontroller or programmable digital circuit (FPGA, CPLD, etc.), thereby improving the quality of harmonic compensation. Effective operation of the APF with current predictor circuit has been confirmed by experimental tests. The considered current prediction circuit may also be useful for other APF control algorithms. Also, current prediction circuits may be applied in other power electronics devices, such as serial APF, power conditioners, high quality AC sources, etc.

For noise type nonlinear loads (such as in an arc furnace) where the load current changes are non periodic and stochastic, the author has proposed multirate APF with improved dynamic performance. The multirate APF is more complicated than classical APF but it allows fast compensation for unpredictable changes in load current.

Using the proposed APFs with improved dynamic performance it is possible to decrease harmonic contents of line current.

Considered multirate shunt APF with the predictive circuit is a very good solution for compensation circuit and allows to significantly decrease THD value of line current in comparison to classical shunt APF. However, it is at the expense of increasing the complexity of the circuit and a relatively small increase in transistors losses. Despite the use of high switching frequency of faster transistors, the employed control algorithm allows the switching losses to remain low. The proposed APF works very well both with predictable and unpredictable loads, which makes it a very versatile solution.

References

1. Analog Devices (1994) AC vector processor AD2S100. Analog Devices Inc
2. Akagi H, Kanazawa Y, Nabae A (1982) Principles and compensation effectiveness of a instantaneous reactive power compensator devices. In: Meeting of the power semiconductor converters researchers-IEE-Japan, SPC-82-16. (in Japanese)
3. Akagi H, Kanazawa Y, Nabae A (1983) Generalized theory of instantaneous reactive power and its applications. Trans IEE-Jpn 103(7):483–490
4. Akagi H, Kanazawa Y, Nabae A (1984) Instantaneous reactive power compensators comprising switching devices without energy storage components. IEEE Trans Ind Appl 1A–20(3):625–630
5. Akagi H (1996) New trends in active filters for power conditioning. IEEE Trans Ind Appl 32(6):1312–1322
6. Akagi H, Watanabe EH, Aredes M (2007) Instantaneous power theory and applications to power conditioning, Wiley-Interscience. Wiley, New Jersey
7. Aredes M (1996) Active power line conditioners. PhD thesis, Technische Universitat Berlin
8. Arriens HL (2006) (L)WDF Toolbox for MATLAB reference guide. Technical report, Delft University of Technology, WDF Toolbox RG v1 0.pdf
9. Arriens HL (2006) (L)WDF Toolbox for MATLAB, User's Guide, Technical report, Delft University of Technology, WDF Toolbox UG v1 0.pdf
10. Asiminoaei L, Blaabjerg F, Hansen S (2007) Detection is key—harmonic detection methods for active power filter applications. IEEE Ind Appl Mag 13(4):22–33
11. Benysek G, Pasko M (eds) (2012) Power theories for improved power quality. Springer, London
12. Bossche AV, Valchev VC (2005) Inductors and transformers for power electronics. CRC Press, Boca Raton

13. Buso S, Mattavelli P (2015) Digital control in power electronics, 2nd edn. Morgan & Claypool, San Rafael
14. Cadaval ER, Gonzáilez FB, Montero IM (2005) Active power line conditioner based on two parallel converters topology. In: International conference-workshop compatibility and power electronics. CPE 2005, Gdynia, Poland, pp 134–140
15. Czarnecki LS (2005) Powers in electrical circuits with nonsinusoidal voltages and currents. Publishing Office of the Warsaw University of Technology, Warsaw
16. Czarnecki LS (1984) Interpretation, identification and modification of the energy properties of single-phase circuits with nonsinusoidal waveforms. Silesian University of Technology, Elektryka 19
17. Czarnecki LS (1987) What is wrong with the Budeanu concept of reactive and distortion powers and why is should be abandoned. IEEE Trans Instrum Meas 36(3):673–676
18. Fryze S (1931) Active, reactive and apparent power in non-sinusoidal systems. Przegl Elektrotechniczny (Electr. Rev.) 7:193–203
19. Fryze S (1966) Selected problems of basics of electrical engineering. PWN, Warszawa
20. Fujielectric (2011) 2MBI159HH-120-50 high speed module 1200V/150A. Data sheet, Fujielectric
21. Ghosh A, Ledwich G (2002) Power quality enhancement using custom power devices. Kluwer Academic Publishers, London
22. Gyugyi L, Strycula EC (1976) Active AC power filters. In: Proceedings of IEEE industry applications annual meeting, pp 529–535
23. Hayashi Y, Sato N, Takahashi K (1991) A novel control of a current-source active filter for AC power system harmonic compensation. IEEE Trans Ind Appl 27(2):380–385
24. Holmes DG, Lipo TA (2003) Pulse width modulation for power converters: principles and practice. Institute of Electrical and Electronics Engineers, Inc
25. Jacobsen E, Lyons R (2003) The sliding DFT. IEEE Signal Process Mag 20(2):74–80
26. Jacobsen E, Lyons R (2004) An update to the sliding DFT. IEEE Signal Process Mag 21:110–111
27. Kazimierkowski M, Malesani L (1998) Current control techniques for three-phase voltage-source converters: a survey. IEEE Trans Industr Electron 45(5):691–703
28. Kazmierkowski MP, Kishnan R, Blaabjerg F (2002) Control in power electronics. Academic Press, San Diego
29. Kim H, Blaabjerg F, Bak-Jensen B, Jaeho C (2002) Instantaneous power compensation in three-phase systems by using p-q-r theory. IEEE Trans Power Electron 17(5):701–710
30. Lindgren M, Svensson J (1998) Control of a voltage-source converter connected to the grid through an LCL-filter—application to active filtering. In: Proceedings of the 29th annual power electronics specialists conference (PESC'98), Fukuoka, Japan, vol 1, pp 229–235
31. Mariethoz S, Rufer A (2002) Open loop and closed loop spectral frequency active filtering. IEEE Trans Power Electron 17(4):564–573
32. Marks J, Green T (2002) Predictive transient-following control of shunt and series active power filter. Trans Power Electron 17(4):574–584
33. Mattavelli P (2001) A closed-loop selective harmonic compensation for active filters. IEEE Trans Ind Appl 37(1):81–89
34. Mitsubishi (2000) Mitsubishi intelligent power modules, PM300DSA120. Data Sheet, Mitsubishi
35. Nakajima T, Masada E (1998) An active power filter with monitoring of harmonic spectrum. In: European conference on power electronics and applications EPE, Aachen 1998
36. Pasko M, Maciazek M (2012) Principles of electrical power control. In: Benysek G, Pasko M (eds) Power theories for improved power quality. Springer, London, pp 13–47
37. Routimo M (2008) Developing a voltage-source shunt active power filter for improving power quality PhD thesis, Tampere University of Technology, Tampere
38. Singh B, Al-Haddad K, Chandra A (1999) A review of active filters for power quality improvement. IEEE Trans Industr Electron 46(5):960–971

39. Sozanski (2003) Active power filter control algorithm using the sliding DFT. In: Workshop proceedings, signal processing 2003, Poznan, pp 69–73
40. Sozanski K (2004) Non-causal current predictor for active power filter. In: Conference proceedings: nineteenth annual IEEE applied power electronics conference and exhibition, APEC 2004, Anaheim
41. Sozański K (2004) Harmonic compensation using the sliding DFT algorithm. In: Conference proceedings, 35rd annual IEEE power electronics specialists conference, PESC 2004, Aachen
42. Sozanski K (2006) Harmonic compensation using the sliding DFT algorithm for three-phase active power filter. Electr Power Qual Utilizat J 12(2):15–20
43. Sozanski K (2006) Sliding DFT control algorithm for three-phase active power filter. In: Conference proceedings, 21rd annual IEEE applied power electronics conference, APEC 2006, Dallas
44. Sozanski K, (2007) The shunt active power filter with better dynamic performance. In: Conference proceedings, PowerTech 2007 conference, Lausanne
45. Sozanski K (2008) Improved shunt active power filters. Przegl Elektrotechniczny (Electr. Rev.) 45(11):290–294
46. Sozanski K (2008) Shunt active power filter with improved dynamic performance. In: Conference proceedings: 13th international power electronics and motion control conference EPE-PEMC 2008, Poznan, pp 2018–2022
47. Sozanski K (2011) Control circuit for active power filter with an instantaneous reactive power control algorithm modification. Przegl Elektrotechniczny (Electr Rev) 1:95–113
48. Sozanski K (2012) Realization of a digital control algorithm. In: Benysek G, Pasko M (eds) Power theories for improved power quality. Springer, London, pp 117–168
49. Sozanski K, Strzelecki R, Kempski A (2002) Digital control circuit for active power filter with modified instantaneous reactive power control algorithm, In: Conference proceedings of the IEEE 33rd annual IEEE power electronics specialists conference, PESC 2002, Cairns
50. Sozanski K, Fedyczak Z (2003) Active power filter control algorithm based on filter banks. In: Conference proceedings, Bologna PowerTech—2003 IEEE Bologna, Italy
51. Sozanski K, Fedyczak Z (2003) A filter bank solution for active power filter control algorithms. In: Conference proceedings of the 2003 IEEE 34th annual power electronics specialists conference—PESC '03, Acapulco
52. Sozanski K (2015) Selected problems of digital signal processing in power electronic circuits. In: Conference proceedings SENE 2015, Lodz Poland
53. Sozanski K (2016) Signal-to-noise ratio in power electronic digital control circuits. In: Conference proceedings: signal processing, algorithms, architectures, arrangements and applications - SPA 2016, pp 162–171. Poznan University of Technology
54. Strzelecki R, Fedyczak Z, Sozanski K, Rusinski J (2000) Active power filter EFA1. Technical Report, Instytut Elektrotechniki Przemyslowej, Politechnika Zielonogorska, (in Polish)
55. Strzelecki R, Sozanski K (1996) Control circuit with digital signal processor for hybrid active power filter. In: Conference Proceedings of SENE 1996, Sterowanie w Energoelektronce i Napedzie Elektrycznym, (in Polish)
56. Texas Instruments (2008) TMS320F28335/28334/28332, TMS320F28235/28234/28232 digital signal controllers (DSCs). Data Manual, Texas Instruments Inc,
57. Texas Instruments (2016) TMS320F2837xD Dual-core delfino™ microcontrollers. Data Manual, Texas Instruments Inc
58. Texas Instruments (2010) C2000 Teaching materials, tutorials and applications. SSQC019, Texas Instruments Inc
59. Xie B, Dai K, Xiang D, Fang X, Kang Y (2006) Application of moving average algorithm for shunt active power filter. In: Proceedings of the IEEE international conference on industrial technology (ICIT 2006), Mumbai, India, December 15–17, pp 1043–1047
60. Watanabe S, Boyagoda P, Iwamoto H, Nakaoka M, Takanoet H (1999) Power conversion PWM amplifier with two paralleled four quadrant chopper for MRI gradient coil magnetic field current tracking implementation. In: Conference proceedings, 30th annual IEEE power electronics specialists conference, PESC 1999, Charleston. South Carolina, USA

61. Wojciechowski D, Strzelecki R (2007) Sensorless predictive control of three-phase parallel active filter. In: Conference proceedings, AFRICON 2007, Windhoek
62. Wojciechowski D (2013) High power shunt active compensators. Prace Naukowe Akademii Morskiej w Gdyni, Gdynia. (in Polish)

Chapter 6
Digital Signal Processing Circuits for Digital Class-D Power Amplifiers

6.1 Introduction

Class-D power amplifiers are very similar to typical power inverters. Often, whether the circuit is called a class-D power amplifier or not is determined by output power, application, and precision of operation. The word 'digital' in digital class-D power amplifier indicates that the input signal is in digital form. Output power of such amplifiers ranges from several watts to several kilowatts. Typical applications of digital class-D power amplifiers include:

- high precision DC drives,
- magnetic resonance imaging (MRI) coils,
- high efficiency and high quality audio power amplifiers,
- power signal sources,
- high precision positioning systems.

The most common use of the class-D power amplifier is for amplification of audio signals. Discussion of class-D audio power amplifiers is presented below in this Chapter. Class-D audio power amplifiers are typically around 90% efficient at rated power, versus 40–70% for conventional class-B or class-AB audio amplifiers. This efficiency is most important for battery powered portable devices such as MP3 players, smartphones, laptops, tablets etc. As a result, such devices can have a longer runtime or can be powered by smaller, lighter batteries.

The dynamic range of digital class-D amplifiers reaches 120 dB, which results in high requirements for the algorithm used and its digital realization. The author has proposed a modulator with noise shaping circuit for the class-D amplifier [58, 62] for increasing this D/A conversion quality. In the digital class-D amplifier signal oversampling is required, therefore, considered also are signal interpolators. The interpolators allow for an increased sampling frequency whilst maintaining substantial separation of signal from noise. The author also presents an original analog power supply voltage fluctuation compensation circuit for the open loop digital class-D

K. Sozański, *Digital Signal Processing in Power Electronics Control Circuits*, Power Systems, DOI 10.1007/978-1-4471-7332-8_6 amplifier [58, 62]. The class-D amplifier with digital click modulation is considered as well [60]. Finally, two-way and three-way digital loudspeaker systems, designed by the author, are presented [59, 60].

6.2 Digital Class-D Power Amplifier Circuits

Figure 6.1 shows full bridge and half bridge basic class-D power amplifier circuits. The half bridge circuit has only two switches, but requires dual voltage and does not allow for the implementation of all modes of modulation. The advantage of the half bridge circuit is that the load is connected to ground. The bridge circuit is more complicated as it needs four switches. In this circuit the load is float, which is unacceptable in some applications. Further discussion about the amplifiers can be

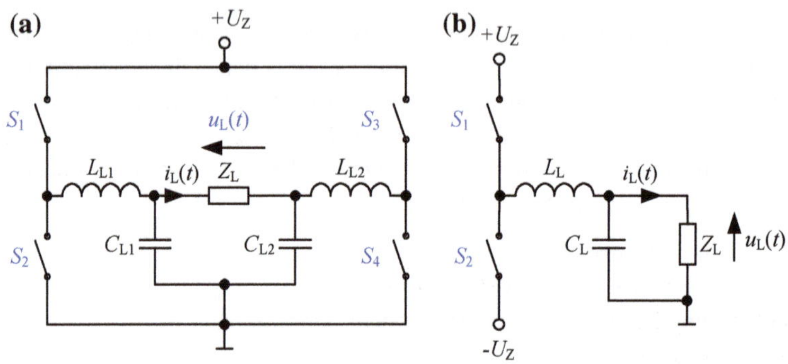

Fig. 6.1 Diagram of class-D power amplifier circuits: **a** full bridge **b** half bridge

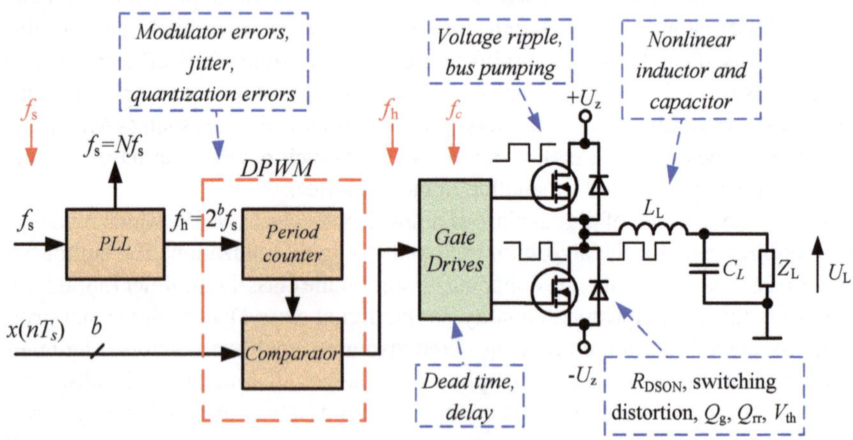

Fig. 6.2 Common sources of errors in the digital class-D power amplifier

found in [41, 46]. During the operation of digital class-D power amplifiers errors will occur that reduce their accuracy. Common sources of errors in the digital class-D power amplifier are shown in Fig. 6.2. The most important error sources include:

- modulation,
- supply voltage,
- switching,
- non-idealities of output filter.

Detailed sources of errors are described in Table 6.1.

Digital PWM as opposed to analog PWM has a finite resolution determined by the number of bits of digital counters. Therefore, digital quantization error occurs in the system. Another source of error is modulator clock jitter. These issues are discussed in Chap. 2. A further problem in class-D amplifiers is bus pumping phenomena, this

Table 6.1 Common sources of errors in the digital class-D power amplifier

Source of error	Type of error
Digital modulator	• Quantization • Counter clock jitter • Modulation components in signal band
Gate drivers	• Dead time • Delay • Timing errors added by the gate drivers • Trigger pulses may be visible in the load current when it has a low value
Supply voltage	• Voltage ripple • Voltage pumping (for half bridge topology) • Voltage source impedance
Switches—MOSFETs	• Switching-on resistance • Current dependent delays of the switching transitions • Embedded diode characteristics, especially high value of diode reverse recovery charge Q_{rr} • Finite switching speed, high value of total gate charge Q_g • Parasitic components that cause ringing on transient edges and generate EMI • Amplitude errors resulting from the non-linear on-state resistance of the MOSFET switches • PCB layout (crucial for both quality of the design and reduction of EMI)
LC output filter	• Non-idealities of inductor • Non-idealities of capacitor

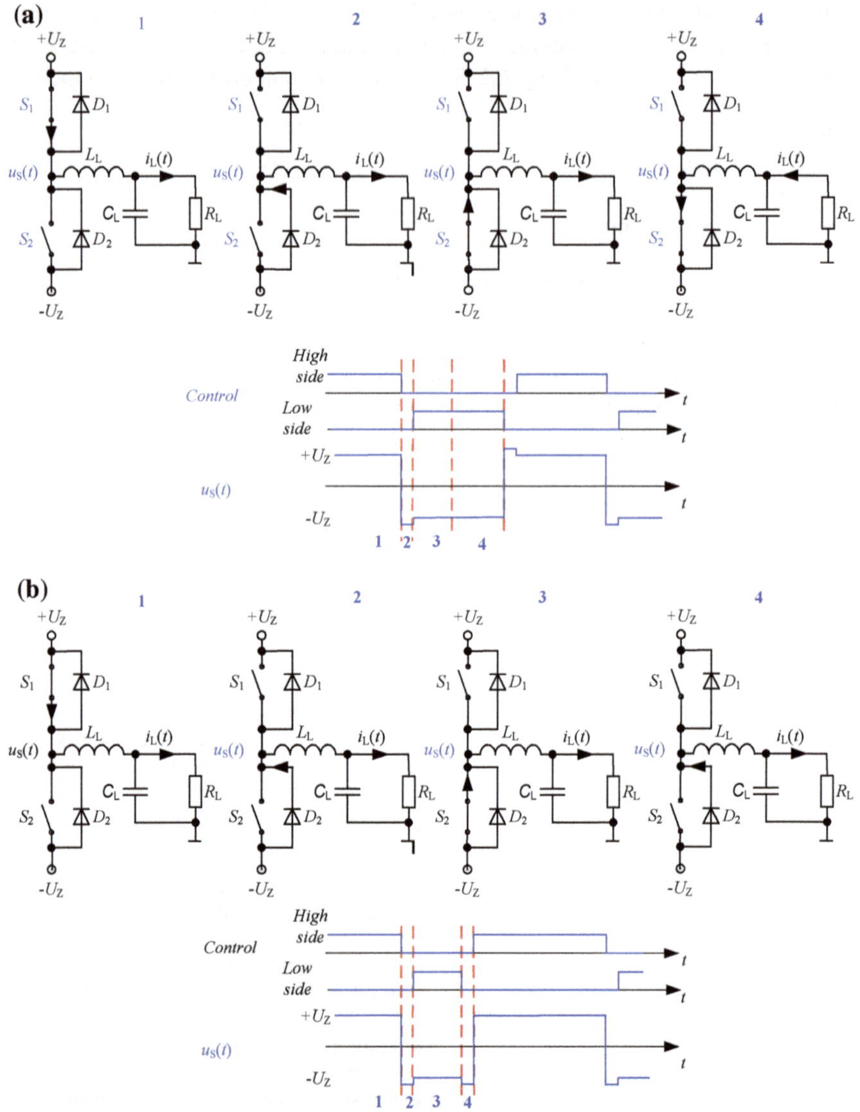

Fig. 6.3 Simplified cycle illustration of class-D power amplifier: **a** for a small signal, **b** for a large signal

occurs in the half bridge topology. In a class-D amplifier output voltage is directly proportional to the bus voltage. Therefore, voltage fluctuation creates distortion. Since the energy flowing in the Class-D switching stage is bi-directional, there is a period where the class-D amplifier feeds energy back to the power supply. Most of the energy flowing back to the supply is from the energy stored in the inductors in

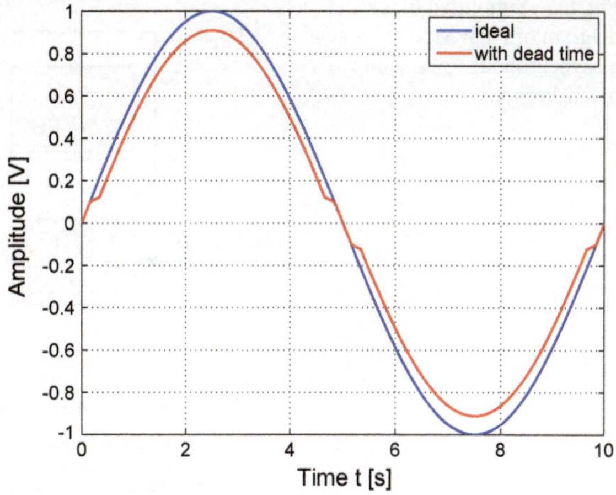

Fig. 6.4 Class-D power amplifier dead time effect illustration

the output low-pass filter. Generally, the power supply has no ability to absorb the energy coming back from the load. So the bus voltage is pumped up, creating voltage fluctuations. The voltage pumping phenomenon occurs mostly at low frequencies, i.e. below 100 Hz.

In order to prevent the condition in which two transistors in one branch are switched on dead time is introduced. This is the time interval in which the control pulses of the transistors are in the off-state. Both transistors are switched off for short period of time to prevent both transistors conducting simultaneously thus causing a short circuit from supply to ground. A simplified switching cycle illustration of class-D power amplifier is depicted in Fig. 6.3. Figure 6.4 depicts a dead time effect in the power class-D amplifier. Illustrated are sinusoidal signal (dashed line) and modulated signal (solid line). For a low level of input signal there is no influence of dead time, but for larger values of signal the output signal is distorted. So the dead time should be as small as possible. An example of discussion of dead time influence on the output signal THD ratio can be found in [39].

6.3 Modulators for Digital Class-D Power Amplifiers

The background of PWM modulators is widely described in the literature, in particular [29] is well recommended. However, the specific problems of modulators for audio class-D amplifiers are described, among others, by [7, 27, 28, 38, 41, 45, 62–64]. The simplified block diagrams of two versions of the digital pulse width modulator (DPWM) are depicted in Fig. 6.5. The first one has an asynchronous clock signal f_h generator and in the second one the clock signal frequency is an integer multiple of the input signal sampling ratio. The second one is better, and the advantages of the synchronous version are described in Chap. 2.

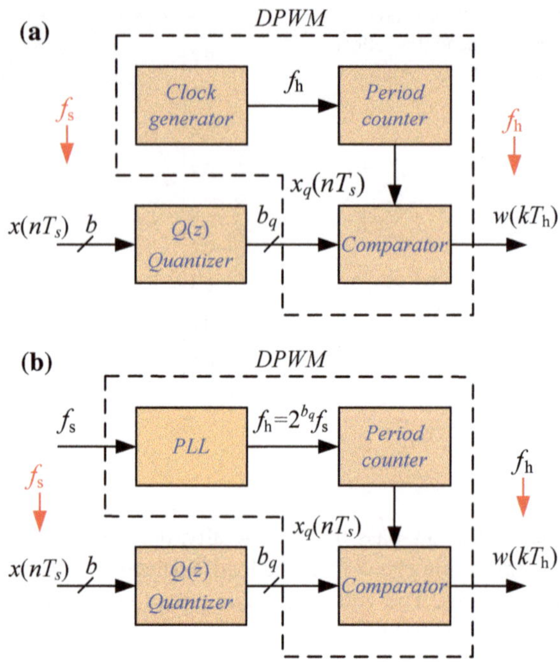

Fig. 6.5 Simplified block diagram of DPWM: **a** asynchronous, **b** synchronous

The output time pulse $w(kT_h)$ is generated by a digital comparator connected to a period counter according to input digital signal $x(nT_s)$. If the digital input signal has a bigger value than the current value in the period counter then the output signal $w(kT_h)$ is high, otherwise it is low. The period counter clock frequency can be expressed as

$$f_h = f_c N_h \quad , \tag{6.1}$$

where: N_h—number of period counter states, or for number of states of period counter that are a power of two

$$f_h = f_c 2^b \quad , \tag{6.2}$$

where: b—number of bits.

For the case in consideration, for typical audio sampling ratio $f_s = 44.1\,\text{kHz}$ and resolution $b = 16\,\text{bit}$, the value of the period counter clock frequency is $f_h \approx 2.89\,\text{GHz}$ and time resolution $T_h = 1/f_h \approx 350\,\text{ps}$. This is too high even for modern standard integrated circuits. Therefore the digital input signal should be quantized. For a given maximum period the counter clock frequency bit rate can be calculated

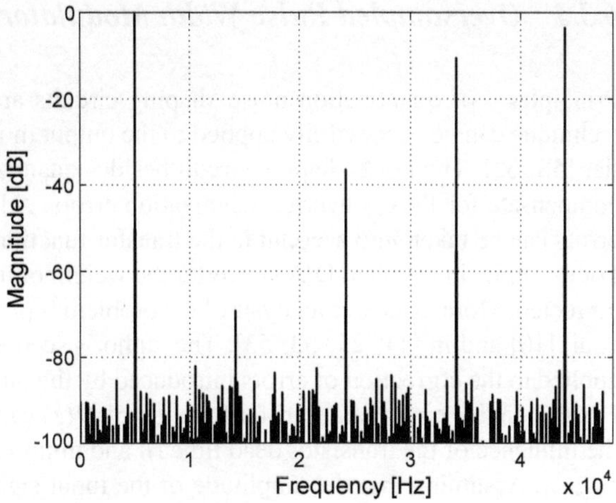

Fig. 6.6 NPWM signal spectrum, for $f = 5\,\text{kHz}$, $f_c = 44100\,\text{Hz}$, $b = 12$ bit

$$b = \text{floor}\left(\frac{log\left(\frac{f_h}{f_c}\right)}{log2}\right) . \tag{6.3}$$

For modern integrated circuits a value of counter clock frequency $f_h = 200\,\text{MHz}$ is usual, with transistor switching frequency $f_c = 44100\,\text{Hz}$, hence the number of bits for the above data is around $b = 12$ bit.

A spectrum segment of digital, naturally sampled (NPWM) simulation for $f_c = f_s = 44100\,\text{Hz}$, and input signal frequency $f = 5\,\text{kHz}$ and digital PWM resolution $b = 12$ bit, is presented in Fig. 6.6. The quantization noise level of signal from Fig. 6.6 is around 67 dB for typical audio band. In the spectrum segment there are: signal, switching frequency and intermodulation components: f, f_c, $f_c \pm 2f$, $f_c \pm 4f$, $f_c \pm 6f$ .... The component $f_c - 6f = 14100\,\text{Hz}$ is in the audio band. This is the main disadvantage of the classical PWM modulation, i.e., the transistor switching frequency f_c must be much higher than the end of the audio band. So oversampled input signal should be used. In that case the transistor switching frequency is increased R times. For increasing input signal sampling rate a signal interpolator should be applied. A typical value of oversampling ratio R for digital class-D audio power amplifier is $R = 8$. The transistor switching frequency is calculated by the equation

$$f_c = Rf_s \quad . \tag{6.4}$$

For typical values of signal sampling rate $f_s = 44100\,\text{Hz}$ and $R = 8$, the transistor switching frequency is equal to $f_c = 352.8\,\text{kHz}$. This value of switching frequency is used by the digital class-D audio power amplifier family PruePath™from Texas Instruments Inc. [72] and it is a compromise value between transistor power losses and quality of output signal.

6.3.1 Oversampled Pulse Width Modulator

Principles of quantization noise shaping circuits are presented in Chap. 2. This technique can be successfully applied to the output modulator of the class-D amplifier [58, 62]. The noise shaping circuit besides quantization noise shaping can also compensate for D/A converter quantization errors. Additionally, known systematic errors can be taken into account in the transfer function of quantization block $Q(z)$. For example, in a typical D/A converter the weight of individual bits can be digitally corrected. More extensive analysis of the problem is presented in the works of Carley et al. [10] and in [11, 25, 33, 53]. The author's second order noise shaping circuit applied to the correction of errors introduced by the pulse amplifier [62] is shown in Fig. 6.7. In this circuit the transfer function of $Q(z)$ can be easily modified to cancel the influence of the transistor dead time t_D and minimum transistors switching time $t_{on(min)}$. Assuming that the amplitude of the input signal $y(k)$ is in the range from -1 to 1 and b-bit of the modulator, the output signal is as follows

$$y_q(k) = \begin{cases} 0 & \text{for } \; |N_n/2y(k)| - N_D < N_{min} \\ N_h/2 & \text{for } \; N_q/2y(k) - Y_D > N_q/2 \\ -N_h/2 & \text{for } \; N_h/2y(k) + N_D < -N_h/2 \\ \text{int}(N_h/2y(k)) - N_D & \text{for } \; N_{min} \leq N_h/2y(k) \leq N_h/2 \\ \text{int}(N_h/2y(k)) + N_D & \text{for } \; -N_{min} \geq N_h/2y(k) \geq -N_h/2 \end{cases} \quad (6.5)$$

where: N_{min}—represents the transistor minimum switch on time $t_{on(min)}$, N_D—represents the transistors dead time t_D.

In the circuit $Q(z)$, b_{in}-bit resolution of the input signal $y(k)$ is reduced to b-bit resolution of the output signal $y_q(k)$. This circuit was successfully introduced to a digital class-D audio power amplifier [58, 62]. Figure 6.8 shows the spectrum of the amplifier output voltage. The diagram shows the falling curve of quantization noise, which is the effect of the noise shaping circuit. In this case SNR reached 80 dB.

In the author's opinion, the presented modulator with noise shaping circuit and with error correction can be employed in the traditional power inverters of other

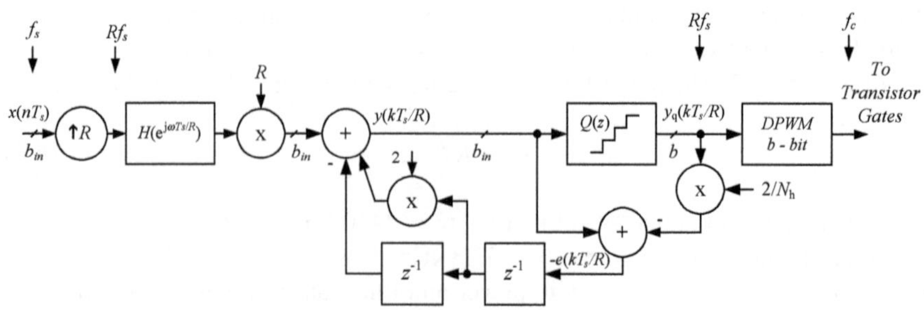

Fig. 6.7 Noise shaping circuit for class-D power amplifier

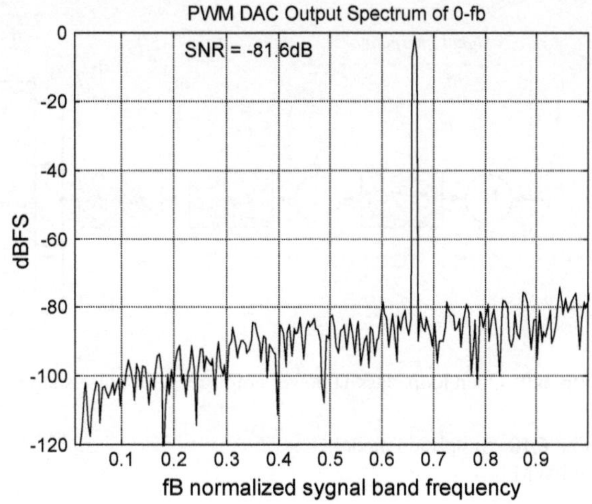

Fig. 6.8 Spectrum of class-D with noise shaping circuit output voltage

power electronics circuits, e.g., APF, uninterruptible power supply (UPS) etc. The additional workload for the processor due to a noise shaping circuit is very small, so it can also be easily implemented in existing control circuits.

6.4 Basic Topologies of Control Circuits for Digital Class-D Power Amplifiers

A complete discussion of all the possible topologies is beyond the scope of this work and the section below considers only the most important topologies.

6.4.1 Open Loop Amplifiers

The open loop (without feedback) digital class-D amplifier is shown in Fig. 6.9. The requirements of an open loop digital class-D amplifier power supply are stricter than a power supply for a classical class B amplifier. Parameters such as power voltage ripple or output impedance become important for this application.

For the simplified case of average output voltage of a class-D amplifier: Assuming a number of simplifications, for a class-D amplifier the average output voltage $U_{L(av)}$ depends linearly on the PWM duty ratio D and the reference supply voltage of amplifier U_{Zref} can be written as

$$U_{L(av)} = DU_{Zref} \quad . \tag{6.6}$$

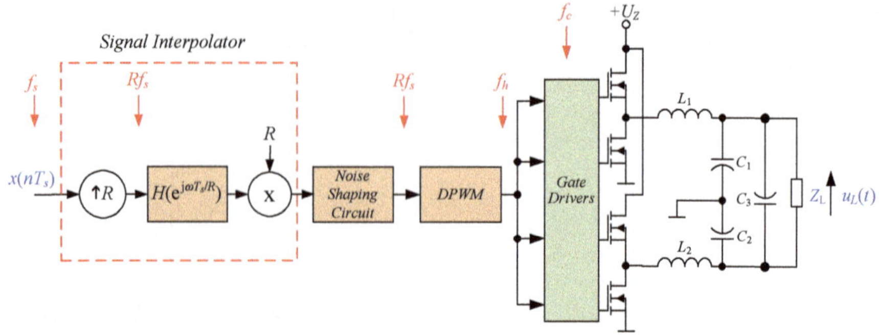

Fig. 6.9 Open loop class-D power amplifier with noise shaping circuit

Fig. 6.10 Simplified model of PWM

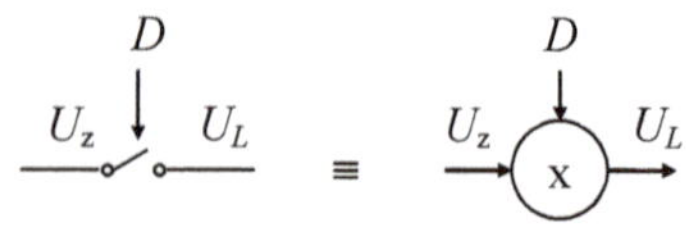

Figure 6.10 shows a simplified model of PWM. It shows that any ripple voltage is transferred by duty ratio D to the output amplifier. Therefore, the open loop amplifier needs to be powered by a high quality regulated voltage source. In the case of the use of a non-stabilized power source, an additional duty ratio correction should be used. To determine the coefficient D a reference supply voltage value U_{Zref} is assumed. However, if the value of the voltage is different from the reference value an error is created. To achieve the same value of output voltage $U_{L(av)}$, a different value of duty ratio should be used

$$U_{L(av)} = D_c U_Z \quad . \tag{6.7}$$

So for the same output voltage $U_{L(av)}$, using Eqs. (6.6 and 6.7), it possible to calculate the duty ratio D_c

$$DU_{Zref} = D_c U_Z \quad , \tag{6.8}$$

$$D_c = D\frac{U_{Zref}}{U_Z} \quad . \tag{6.9}$$

Discussions on power supply parameters for open loop class-D amplifier appear in [8, 9, 37]. The open loop digital class-D power amplifier needs a power supply source with very low impedance for the whole amplifier frequency range, for audio application this is 20 kHz! The output signal THD ratio is dependent on the supply source impedance Z_Z as follows [9]

$$THD = \frac{Z_Z M^2}{4Z_L} \quad , \tag{6.10}$$

where: M—is maximum modulation factor.

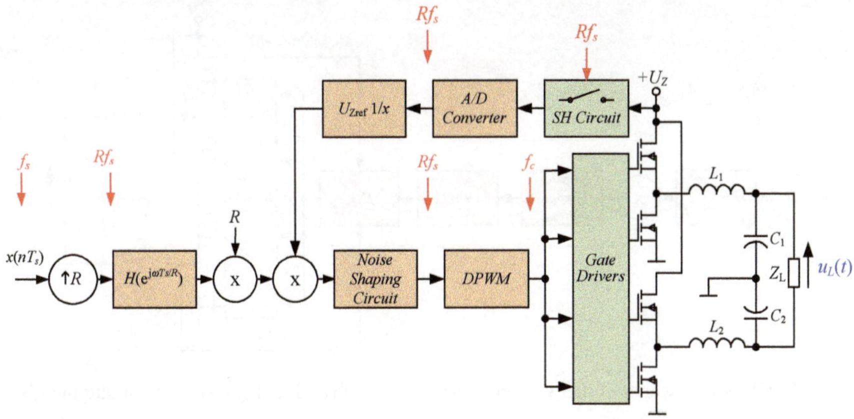

Fig. 6.11 Class-D power amplifier with noise shaping circuit and digital feedback for supply voltage

6.4.2 *Amplifiers with Digital Feedback for Supply Voltage*

Figure 6.11 shows a class-D power with noise shaping circuit and digital feedback for supply voltage correction. In this circuit for measurement of supply voltage a high resolution A/D converter is used. Digital representation of supply voltage is used for calculation of duty ratio according to Eq. (6.9).

6.4.3 *Amplifiers with Analog Feedback for Output Pulses*

Digital correction of output pulses needs a high speed and high quality successive approximation A/D converter with sample-and-hold circuit on the front. Using an analog circuit it is possible to compensate the voltage supply fluctuation and amplitude errors resulting from the non-linear on-state resistance of the MOSFET switches by changing the duty ratio. Figure 6.12 shows this type of circuit. In this circuit analog feedback signals are taken directly from the transistor, so it can also compensate the influence of transistor switching distortions. For this purpose an analogue compensation circuit is proposed by the author [62]. A simplified diagram of such a compensation circuit is shown in Fig. 6.13. The circuit consists of two sets of integrators, comparators and switches. One set an integrates input signal $D_{in}(n)$ and a second one controls power transistors. Input signal is a square wave with duty ratio $D_{in}(n)$ and the output signal is a square wave with duty ratio $D_{out}(n)$. For all components throughout the switching period T_c, it is possible to calculate $D_{out}(n)$ from the equation

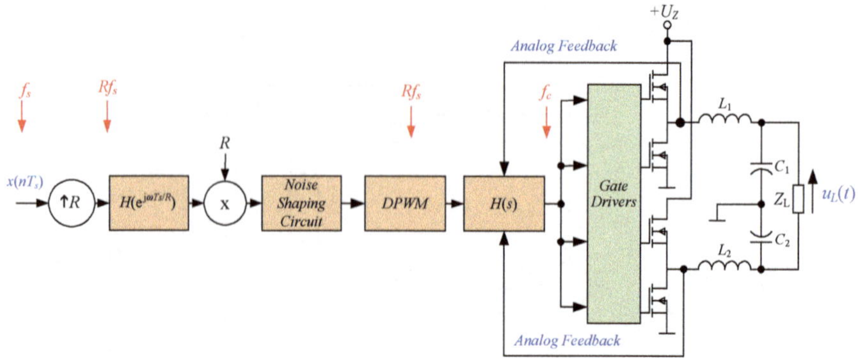

Fig. 6.12 Class-D power amplifier with noise shaping circuit and analog feedback for output pulses

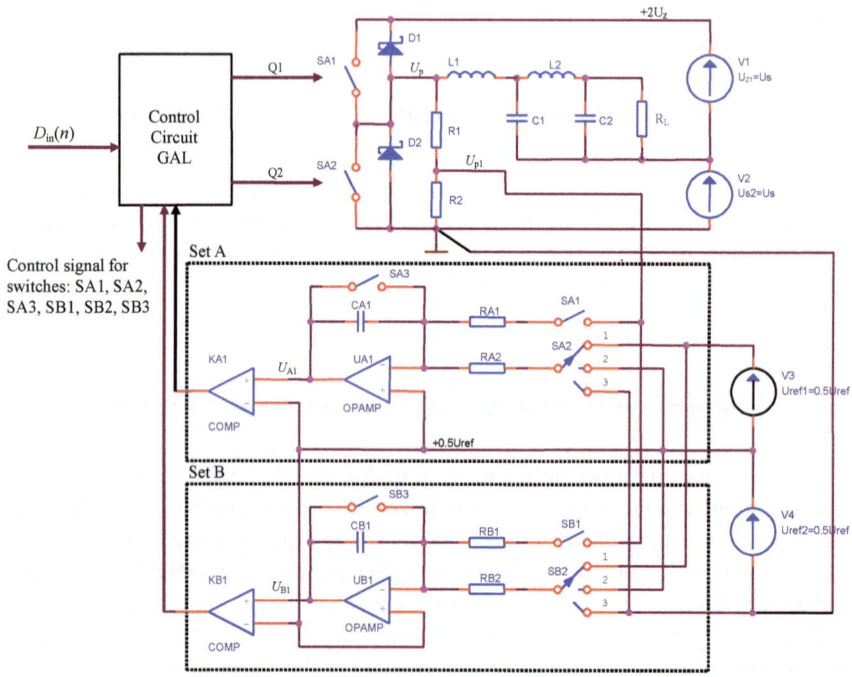

Fig. 6.13 Simplified diagram of analog compensation circuit

$$\begin{aligned} D_{out}(n) = \frac{U_{ref2}}{U_{p1P}(n)} D_{in}(n) - (1 - D_{in}(n)) \frac{U_{ref2}}{U_{p1P}(n)} + \\ + (1 - D_{out}(n-1)) \frac{U_{p1N}(n-1)}{U_{p1P}(n)} \quad , \end{aligned} \tag{6.11}$$

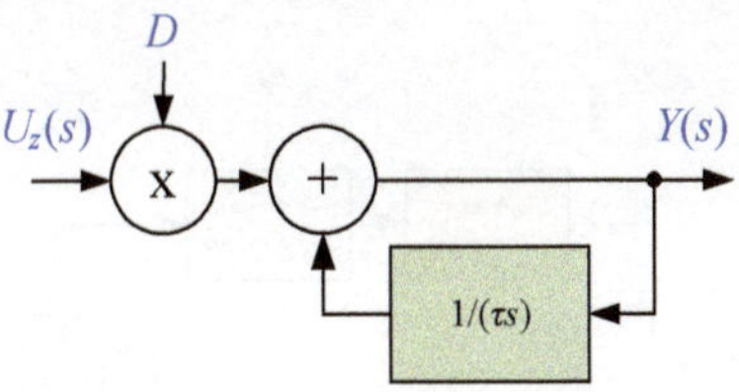

Fig. 6.14 The small signal model of analog compensation circuit

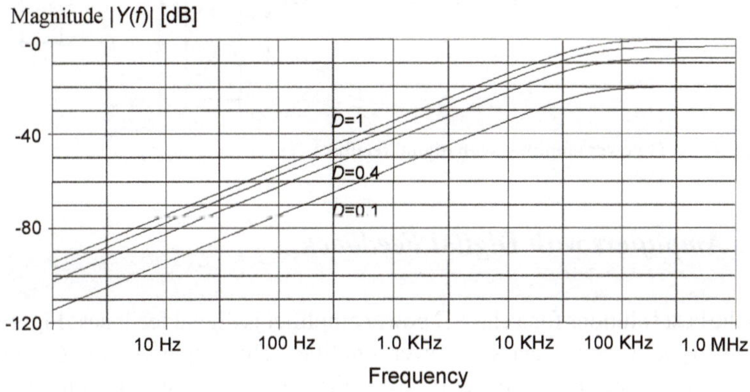

Fig. 6.15 Ripple rejections for different duty ratios *D*

where: U_{ref1}, U_{ref2}—reference voltages, U_{p1P}—output voltage when transistor Q_1 is switched on, U_{p1N}—output voltage when transistor Q_2 is switched on. A simplified small signal model of the compensation circuit and its ripple rejections for different duty ratios $D_{in}(n)$ is shown in Fig. 6.14. Its small signal transfer function is

$$Y(s) = U_z(s)D\left(\frac{\tau s}{1+\tau s}\right), \tag{6.12}$$

where: τ—integrator time constant. Figure 6.15 shows ripple rejections of the compensation circuit for different duty ratios *D*.

Amplifiers with analog feedback for output pulses are employed in some IC commercial solutions. For example, Texas Instruments introduced power IC, TAS5631 300 W stereo PurePath™ digital-input power stage [75]. Analog compensation provides an 80 dB power supply rejection ratio (PSRR), similar to that in an analog class-B power amplifier. Therefore, for such amplifier a simple power supplier can be used.

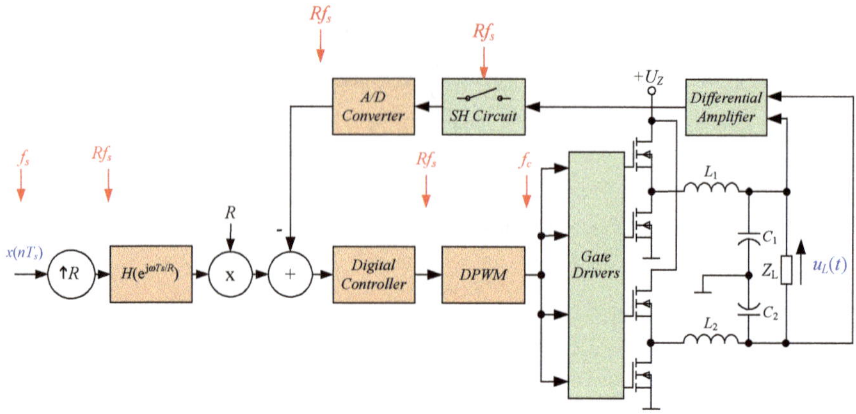

Fig. 6.16 Class-D power amplifier with digital feedback

6.4.4 *Amplifiers with Digital Feedback*

One of the best solutions for a class-D power amplifier is digital feedback. Figure 6.16 shows this type of power class-D power amplifier. The output voltage U_L through the differential amplifier is converted by a high precision fast A/D converter. The quality of the amplifier depends on the A/D converter parameters. The signal from the A/D converter is then used as a feedback signal for the digital controller. These circuits are not frequently described in the literature, however, they are discussed in [16, 40].

Amplifiers with digital feedback have a number of fundamental advantages. For example, in the open loop solution, it is not possible to decrease output impedance. At signal frequency 20 kHz a typical 22 μH output serial inductor has an impedance of 2.8 Ω. It is obvious that different loads have a significant impact on frequency response. The only remedy is to control the output voltage using a correctly designed feedback loop, where even at 20 kHz an output impedance below tens of milliohms can be achieved, similar to that of a class-B analog amplifier. Another advantage of amplifiers with digital feedback is the reduction of distortion products arising from saturation in the output inductors. Of course, in such amplifiers it is possible to compensate the impact of variations in the power supply voltage, errors resulting from the non-linear on-state resistance of the MOSFET switches, current dependent overshoot and current dependent delays of the switching transitions. In these circuits it is possible to compensate most amplifier errors, but design of the digital controller is very difficult. However, the author believes that after eliminating the problems associated with stability, jitter, etc. this solution will be widely used. The high quality audio applications pose the greatest design challenge. In such case a single chip containing the high resolution PWM (with picosecond resolution), 18-bit successive approximation A/D converter, digital PLL (with extremely low jitter), 64-bit DSP and digital audio receiver for input signal should be used.

6.5 Supply Units for Class-D Power Amplifiers

The next solution is to use a power supply with low output impedance and very low ripple amplitude. Such examples are shown in [8, 74]. As has already been discussed, the quality of the amplifier operating in open loop directly depends on the quality of the supply voltage.

The author has developed a high quality power supply unit. It is controlled by a digital controller realized using digital signal processor TMS320F2835 [71, 73]. In the circuit a 12-bit A/D converter built in TMS320F2835 is used. The simplified circuit of the power supply unit is depicted in Fig. 6.17. The A/D converter resolution is insufficient, therefore feedback error is calculated by an additional analog circuit. Hence, the A/D converter converts only feedback error $e_W(t)$ and thus the resolution is sufficient. In the next stage the error signal is processed by a digital PID controller. Signal $u(n)$ controls a high resolution PWM circuit. The PWM output impulses control transistors Q_1 and Q_2 of the DC/DC synchronous buck converter. The converter produces from a 65 V input voltage a lower output regulated voltage, with the voltage range from 0 to 50 V. The output current range is 0–10 A. All digital circuits are realized using digital signal processor TMS320F2835. The feedback error analog signal $e_W(t)$ is calculated thus:

$$e_W(t) = U_{off}\frac{R_3}{R_1}(k_d U(z) - U_{off}) + \frac{R_3}{R_1}(U_{ref} - U_{off}) \quad , \tag{6.13}$$

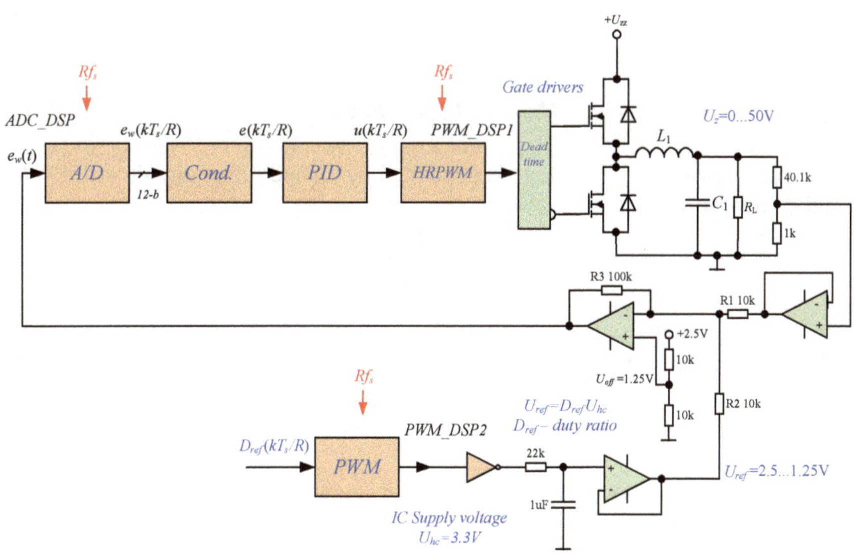

Fig. 6.17 Supply unit for class-D power amplifier

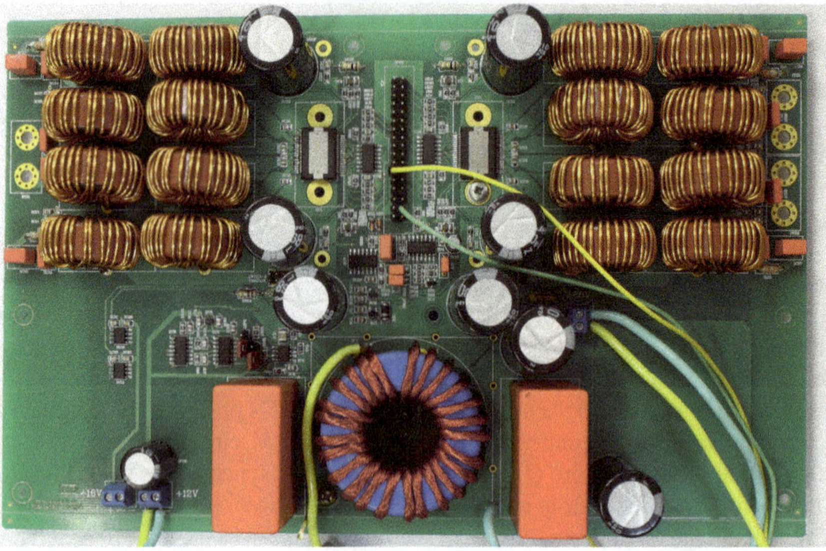

Fig. 6.18 Power stages of a 2 × 100 W digital stereo amplifier with the supply unit

where: τ—integrator time constant and

$$k_d = \frac{R_4}{R_4 + R_5} \quad . \tag{6.14}$$

The controller reference voltage U_{ref} is

$$U_{ref} = 2U_{off} - k_d U_z \quad . \tag{6.15}$$

The supply unit reference voltage is generated by an additional circuit with PWM, and the duty ratio:

$$D_{ref} = \frac{U_{ref}}{U_H C} \quad . \tag{6.16}$$

where: $U_H C$—supply voltage of digital IC 74HC14. The feedback error signal for digital proportional-integral-derivative (PID) controller is:

$$e(t) = -(e_W(t) - U_{off}) = -e_W(t) + U_{off} \quad . \tag{6.17}$$

Figure 6.18 shows author's design of the power stages of a 2 × 100 W TAS5121 [68] digital stereo amplifier with the supply unit. All are controlled by TMS320F2835 DSC.

For laboratory purposes the author has built a hybrid switching mode and linear supply unit for a 2 × 316 W class-D audio power amplifier. The supply unit consists

Fig. 6.19 The 2 × 316 W digital stereo power amplifier with the hybrid supply unit

of a high quality switching mode power supply which transfers power from mains power to DC voltage 60V/20 A with low ripple. Then voltage is controlled by a linear supply unit to an output voltage from 0 to 50 V. Of course this solution is inefficient, but it provides a high quality supply voltage for a class-D audio power amplifier. The amplifier was built using TAS5518-5261K2EVM evaluation module [69]. Figure 6.19 shows the amplifier.

6.6 Click Modulation

Click modulation is a coding technique developed in the 80's by Logan [36] to retrieve information encoded by the zero crossings of certain bipolar signals. Using click modulation it is possible to remove modulation components from the signal band to the high frequency band. Therefore the demodulation process can be easily performed by a low order low-pass LC filter. Click modulation is also called zero position coding (ZePoC). The block diagram of the analog click modulation algorithm is shown in Fig. 6.20. Given a band-limited passband such as signal $f(t)$ with spectral content confined to $(f_L \dots f_H)$, where $0 < f_L < f_H < \infty$, the signal $f(t)$ has a zero

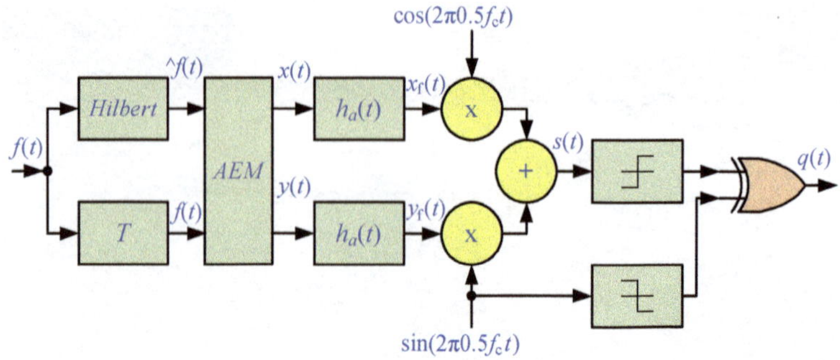

Fig. 6.20 Block diagram of analog click modulation algorithm

value DC component. Input signal is transformed to analytic signal $f_A(t)$ by Hilbert transform: the analytic signal

$$f_A(t) = f(t) + \mathrm{j}\hat{f}(t) \quad , \tag{6.18}$$

where

$$\hat{f}(t) = f(t) * \frac{1}{\pi t} \quad , \tag{6.19}$$

symbol "*" represents convolution in time domain. In the next stage analytic signal is converted through an analytic exponential modulator (AEM):

$$z(t) = \mathrm{e}^{-\mathrm{j}f_A(t)} = \mathrm{e}^{\hat{f}(t) - \mathrm{j}f(t)} \quad , \tag{6.20}$$

where

$$z(t) = x(t) + \mathrm{j}y(t) \quad , \tag{6.21}$$

and

$$x(t) = \mathrm{e}^{\hat{f}(t)} \cos(f(t)) \quad , \quad y(t) = -\mathrm{e}^{\hat{f}(t)} \sin(f(t)) \quad . \tag{6.22}$$

The signal $z(t)$ is also analytic. In the following stage it is filtered by the low-pass filter $h_a(t)$. Discussion of the filter parameter is available in [44, 67, 81]. The real value of signal $s(t)$ defined by

$$\begin{aligned} s(t) &= \mathrm{Re}\{z(t)\mathrm{e}^{-\mathrm{j}2\pi f_c t}\} = \\ &= x(t)\cos(2\pi 0.5 f_c t) + y(t)\sin(2\pi 0.5 f_c t) \quad . \end{aligned} \tag{6.23}$$

Finally, the binary signal with separated baseband $q(t)$ is prepared from $s(t)$ by

$$q(t) = -\frac{\pi}{2}\{\mathrm{sgn}(s(t))\} \cdot \{\mathrm{sgn}(\sin(2\pi 0.5 f_c t))\} \quad . \tag{6.24}$$

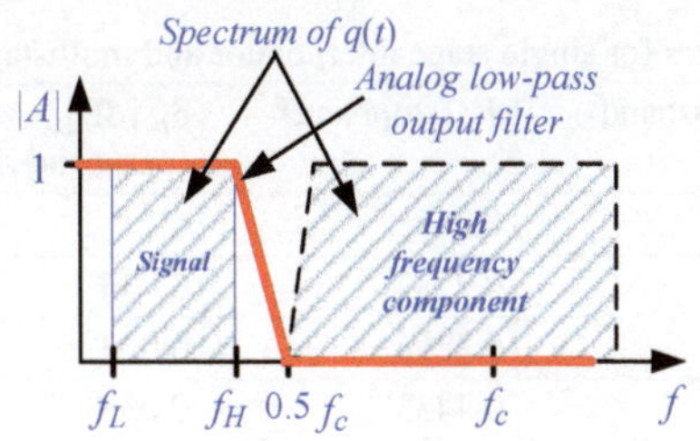

Fig. 6.21 Spectrum of click modulator output signal $q(t)$ and analog output low-pass LC filter frequency response

The spectrum of click modulator output signal $q(t)$ is shown in Fig. 6.21. The spectrum consists of two bands: signal band and high frequency modulation component band. The high frequency band is suppressed by output LC filter.

6.7 Interpolators for High Quality Audio Signals

Because of the high dynamic range of audio signal (120 dB) designing an interpolator for audio applications is a big challenge. This section includes implementations of a single stage and a multistage interpolator for high quality audio signals.

As an illustrative example a cascaded interpolator for a class-D power audio amplifier is used. Parameters chosen by the author for this interpolator are: passband ripple $\delta_p < 0.1$ dB, oversampling ratio $R = 8$, passband from 4 to 20000 Hz, signal-to-noise and distortion ratio $SINAD < 90$ dB.

First a single stage interpolator based on IIR and FIR filters is presented. Secondly, is discussed a multistage interpolator based on bireciprocal modified lattice wave digital filters. Followed one based on two-path (polyphase) digital filters. The interpolators are implemented in a floating point digital signal processor SHARC. The results of these implementations are presented and compared.

The parameters for a single stage interpolator and a multistage interpolator are shown in Table 6.2. In the multistage interpolator it is possible to reduce requirements for stages 2 and 3 by means of the suppression introduced in the stopband by an output analog low-pass filter.

6.7.1 Single Stage Interpolators

The author has designed and implemented a single-stage interpolator with the parameters shown in Table 6.2 in a digital signal processor SHARC. The following types of interpolators have been analyzed:

Table 6.2 Design parameters for single stage interpolator and multistage interpolator

Stage	F_p (passband)	F_z (stopband)	δ_p [dB] (passband riple)	δ_z [dB] (stopband riple)
Single stage	0.0567	0.0683	0.1	−90
1	0.2267	0.2732	0.033	−90
2	0.1134	0.3866	0.033	−90
3	0.0567	0.4433	0.033	−90

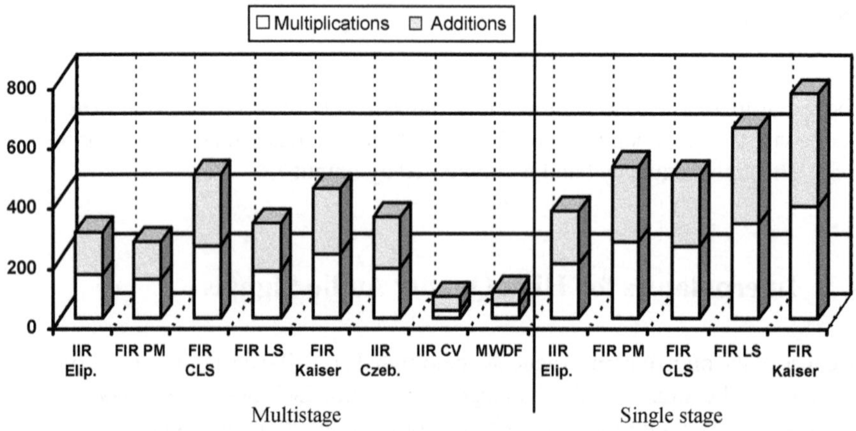

Fig. 6.22 Quantity of arithmetical operations for interpolation of one sample (for R = 8), results of implementations of single stage and multistage versions of the interpolators realized on SHARC DSP

- an interpolator with an elliptic filter IIR (Elip),
- interpolators with polyphase FIR filters: Parks-McClellan (PM), Kaiser window (Kaiser), least squares (LS), and constrained least squares (CLS).

Figure 6.22 shows the quantity of arithmetical operations necessary for one sample interpolation (where $R = 8$). Interpolators with FIR filters have a polyphase structure with periodically time-varying coefficients (considered in Chap. 3). From the filters analyzed the IIR filters required the least number of arithmetical operations, however the polyphase FIR filters showed a similar efficiency.

6.7.2 *Multistage Interpolators*

The multistage interpolator with parameters in Table 6.2 was designed and realized using a digital signal processor SHARC. The following types of interpolators have been analyzed:

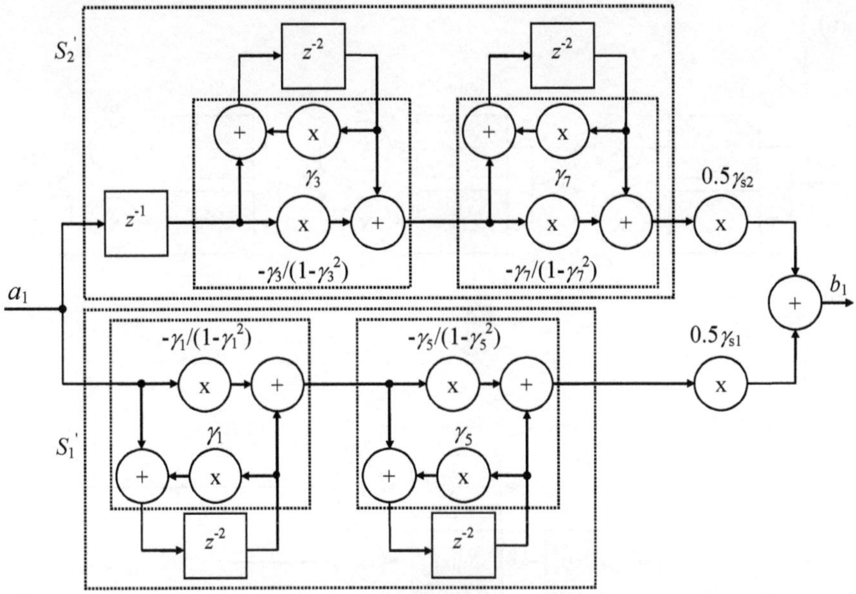

Fig. 6.23 Block diagram of interpolator used 9-order bireciprocal lattice modified wave digital filter

- interpolators with FIR filters and the FIR Parks-McClellan filter (PM FA) employing a stopband characteristic of an analog low-pass filter (Fig. 3.42),
- interpolators with classical IIR filters, elliptic (IIR Elip.) and Czebyshev (IIR Czeb.)
- interpolators with bireciprocal lattice modified wave digital filters (MWDF),
- interpolators with two-path (polyphase) filters (IIR CV).

Bireciprocal lattice modified wave digital filters were also used in the interpolator design [12, 22, 23, 26, 61]. A single stage interpolator for $R = 2$ used 9-order bireciprocal lattice modified wave digital filter is shown in Fig. 6.23. A block diagram of the interpolator for $R = 8$ is depicted in Fig. 6.24b. Modified wave digital filters are very efficient for implementation with modern floating point signal processors, especially for applications where a wide dynamic range of the signal is important. The author applied bireciprocal lattice wave digital elliptic filters for this realization. Filter coefficients are designed with author's program prepared in the the MATLAB® environment, based on the methods presented in Chap. 3. The interpolator was realized with SHARC DSP using modified wave digital filters. The structure of the interpolator is depicted in Fig. 6.24a. The resulting value of the coefficient γ_{sw1} is given by the following equation

$$\gamma_{sw1} = \gamma_{s12}\gamma_{s22}\gamma_{s32} \quad , \tag{6.25}$$

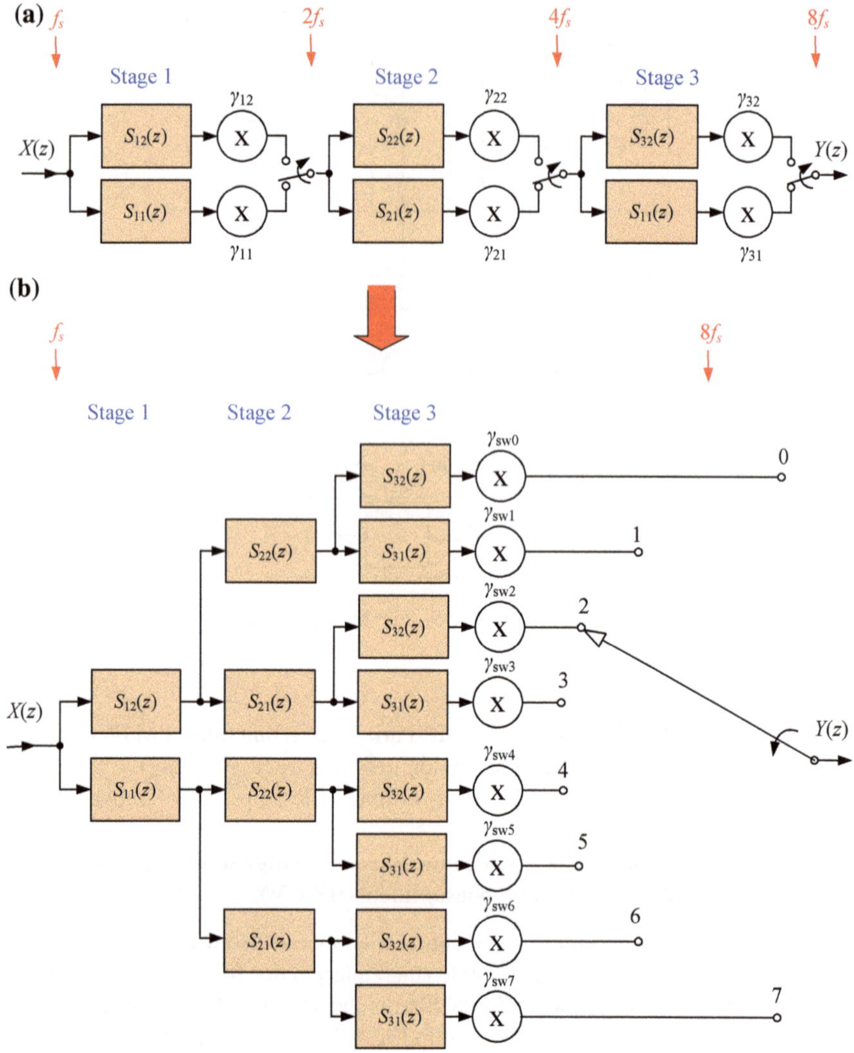

Fig. 6.24 Cascaded version of the interpolator for $R = 8$ (**a**), version of the interpolator with a single switch and the resulting multipliers (**b**)

where: γ_{s12}, γ_{s22}, γ_{s32} are the resultant coefficients of upper branches for stages 1, 2, 3, respectively. The author modified the structure of this interpolator (Fig. 6.24a) to the equivalent circuit depicted in Fig. 6.24b. This solution is very useful for realization by modern digital signal processor with parallel instruction set because this structure allows for the implementation of parallel computing. Frequency response of the cascaded interpolator realized with SHARC DSP for $R = 8$ is shown in Fig. 6.25. The interpolator achieves the signal-to-noise and distortion ratio *SINAD* near to

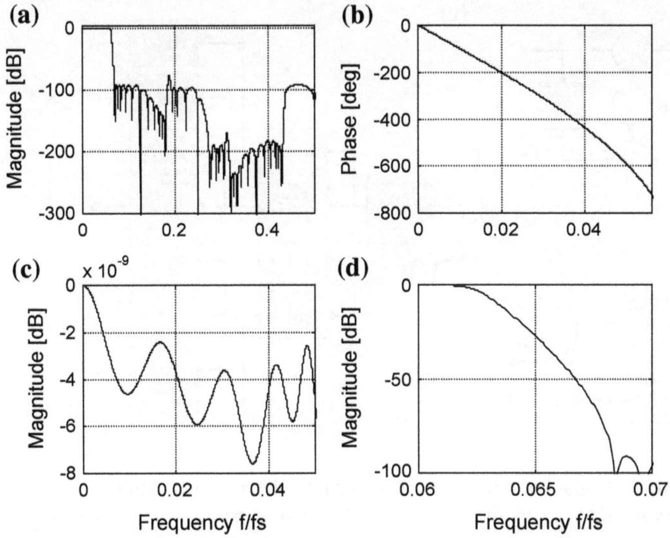

Fig. 6.25 Frequency response of cascaded interpolator realized by SHARC DSP for $R = 8$: **a**, **c**, **d** magnitude response, **b** phase response

-90 dB and the passband ripple $\delta_p \approx 8 \times 10^{-9}\, dB$. To compute the response for one input sample it needs 50 multiplications and 42 additions.

Among the polyphase IIR filters, special attention was paid to a two-path (polyphase) filter designed according to methods introduced by Venezuela and Constantindes [80]. This filter consists of two branches with allpass filters (Fig. 6.26a). Block diagrams of these filters are shown in Fig. 6.26b, c. The two-path filters have very high performance and they are easily implemented and computationally efficient. The quantity N of allpass filter stages depends on: the stopband ripple δ_z and the relative frequency of transition bandwidth ΔF and is given by [31]

$$N = \frac{\delta_z}{72\Delta F + 10} \,. \tag{6.26}$$

Among the analyzed filters (Fig. 6.22) the multistage interpolator based on a polyphase two-path filter required the smallest number of arithmetical operations for implementation with a SHARC digital signal processor. For applications, for which a linear phase response is important, a multistage interpolator with a Parks-McClellan FIR filter requires the smallest number of arithmetical operations. Using the symmetry of FIR filter coefficients, it is possible to decrease the number of arithmetic operations. Good results are also obtained for multistage interpolators based on modified wave digital filters.

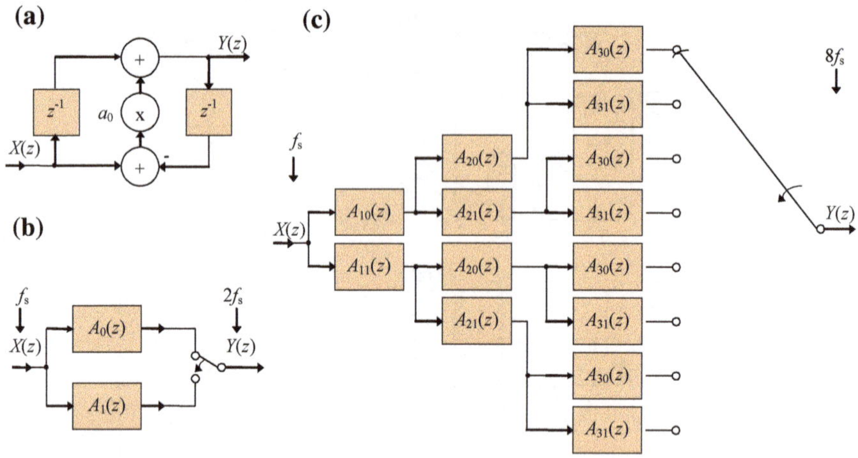

Fig. 6.26 Block diagram of interpolator realized by polyphase two-path filters: **a** allpass section, **b** interpolator for $R = 2$, **c** multistage version of the interpolator for $R = 8$ with a single switch

6.8 Class-D Audio Power Amplifiers

Devices to play music are one of the most common devices used by people. The main part of these devices is the power amplifier. Problems of analog power audio amplifiers are widely described in the literature, among the many publications the following can be cited [5, 15, 21, 41, 46, 47, 49–51]. Digital signal processing algorithms for audio applications are described, inter alia, by Zolcer et al. [84, 85], Ledger and Tomarakos [34], Orfanidis [42, 43], Bateman and Paterson-Stephens [6]. Another very important part of devices to play music are the electroacoustic transducers, usually the loudspeakers. Among the many publications on the loudspeakers the following can be recommended [3, 13, 14, 21]. The main problems of finding loudspeaker parameters and designing loudspeaker system boxes were successfully solved by Thiele and Small [52–57, 76–78].

The frequency characteristics of the impedance magnitude of a typical loudspeaker is not constant, as shown in Fig. 6.27. This shows that the loudspeaker is not an easy type of load for an audio power amplifier and for a crossover network. The loudspeakers have a local maximum of impedance at the resonant frequency. For higher frequencies there is an increase of impedance due to the impact of coil inductance. Therefore it is difficult to design a good passive crossover for the flat frequency response of the whole speaker system, independently of the speaker impedances. To solve this problem, individual impedance compensation networks are necessary for particular speakers. Typically, the impedance of the loudspeaker is defined for a frequency of 1 kHz, though this is not sufficient to properly design a passive crossover network. A simplified diagram of a power amplifier output circuit is depicted in Fig. 6.28, Z_S is the output impedance and Z_L load impedance. The power amplifier

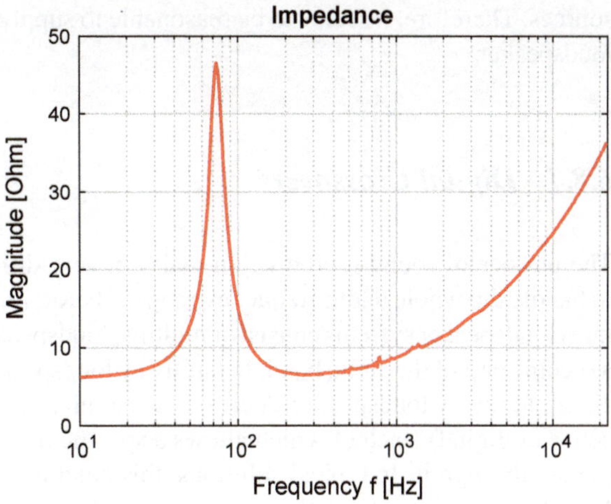

Fig. 6.27 Impedance of 8 Ω midrange-woofer loudspeaker

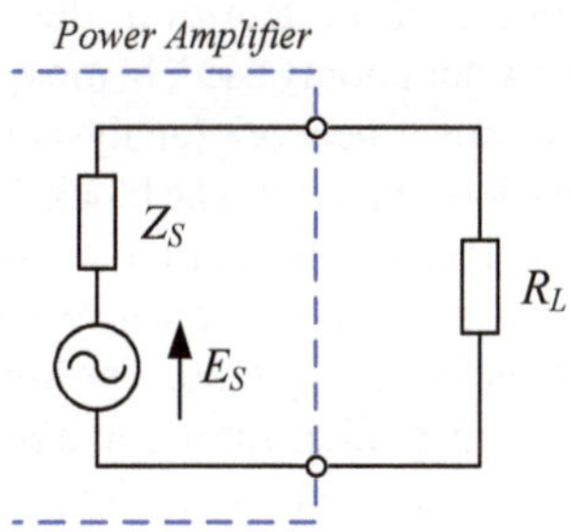

Fig. 6.28 Simplified diagram of power amplifier output circuit

output impedance is usually less than 0.1 Ω. For a power audio amplifier a damping factor is defined thus

$$D_F = \frac{Z_L + Z_S}{Z_S} \quad . \tag{6.27}$$

The damping factor is not often specified in the amplifier. A high value of damping factor is better for controlling a complex load such as a loudspeaker. A typical value of damping factor range for a classical analog power audio amplifier with direct coupled output stage and negative feedback is from 50 to 2000. For an amplifier without feedback the range is 0.1–10. It should be noted, between the speaker and the amplifier are connected the speaker cable and crossover. They add their impedance to the output impedance of the amplifier. Therefore impedance of crossover and speaker cable should be minimized and this significantly increases the cost.

Today the majority of audio sources are digital. Using CD and DVD players, DAT, MP3 players, digital audio processors, digital TV, digital broadcasting systems, Internet audio receivers and so on, there is direct access to the digital signal sources. Therefore, it seems to be reasonable to supply a digital signal directly to the loudspeaker.

6.8.1 *Digital Crossovers*

The physics of sound reproduction makes it very difficult for a single loudspeaker to handle the whole audio frequency range. Therefore, for high fidelity applications, most loudspeaker systems consist of multiple loudspeakers, each of them reproducing a specific part of the audio band. In multiway loudspeaker systems there is a crossover circuit for every loudspeaker. A crossover circuit is a set of electrical filters (passive, active or digital), each of which allows a specific portion of the frequency spectrum to pass through it. In typical solutions, this band is divided into two or three parts. Most loudspeakers can work satisfactorily well in the frequency range of about 10:1 only. Thus, the whole audio band (20 Hz–20 kHz) should be covered by at last three loudspeakers. However, due to the lower price, two-way loudspeaker systems are very commonly used. In most current loudspeaker systems, a passive (typically RLC) crossover network (analysis filter bank) is placed between the power amplifier and the loudspeakers. The block diagram of such a classical system is shown in Fig. 6.29a. Another function of the crossover is to equalize different sensitivities of particular loudspeakers. A typical two-way speaker system with 2-order passive crossover is depicted in Fig. 6.30. The simplest crossover network consists of a low-pass and a high-pass filter for use in a two-way loudspeaker system.

In the author's opinion, a much better solution is to use the active loudspeaker system, in which the speaker is connected directly to the amplifier. Such a system is shown in Fig. 6.29b. It possesses many advantages over the passive realization. The active system is more accurate and its design is simpler because there is no loudspeaker impedance influence on crossover parameters. For example, the low output impedance of a power amplifier suppresses the speaker resonance phenomena. Moreover, the size of active or digital filters is smaller and the cost is less than that for their passive equivalents.

Using digital circuits it is easy to introduce additional time delays into individual signal paths in order to correct the delay differences. This allows the speakers to have their acoustic centres aligned even though they are mounted at a particular distance to each other. However, in the author's opinion the classic sets consisting of a good class-AB analog amplifier and a well-designed loudspeaker system sounds excellent.

A digital version of the active crossover is proposed and discussed in this section. It possesses many advantages over the analog realization. It should be stressed, however, that currently the overall system can be more expensive for the active realization because of the cost of separate power amplifiers, which are required in this case in each band, because the filters should be separated from individual loudspeakers by the power amplifiers. This will, however, change in the near future with lower costs of electronic components. A block diagram of a two-way digital active loudspeaker system with digital class-D power amplifier is shown in Fig. 6.31.

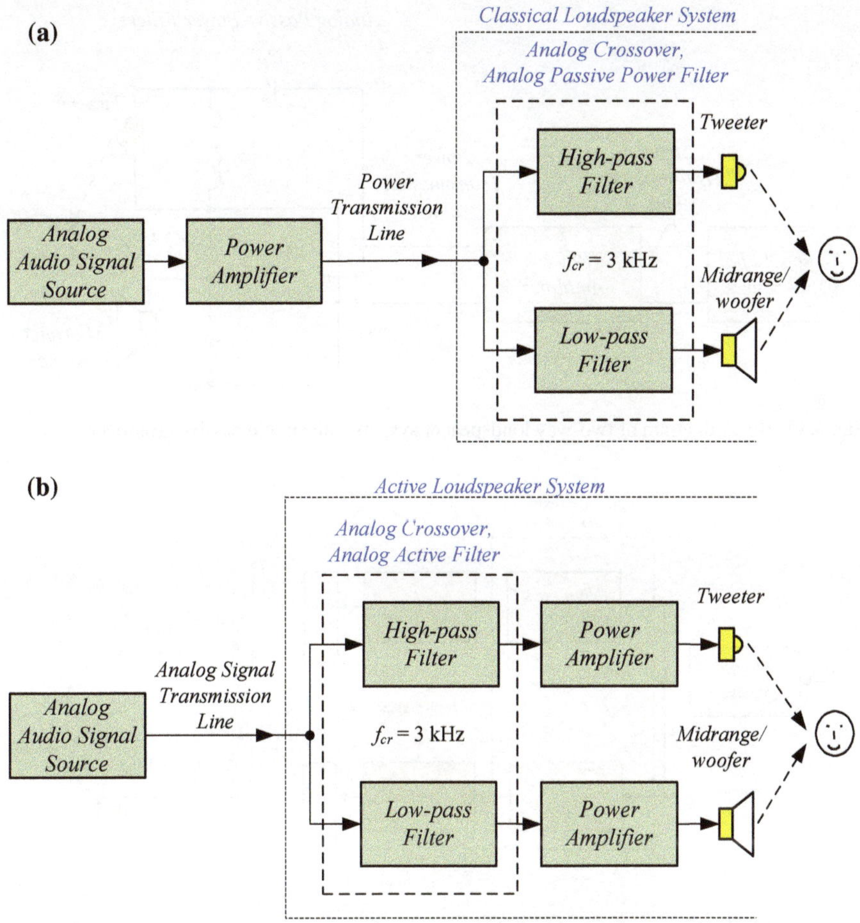

Fig. 6.29 Block diagram of analog two-way loudspeaker systems: **a** passive, **b** active

The digital audio input signal S/PDIF or AES/EBU (in the CD player standard, i.e., $b = 16$bit with sampling rate $f_s = 44.1$ kHz) is divided into two channels, left and right, by a digital audio interface receiver (DAI). A typical audio band has a range of 20 Hz–20 kHz. The signal of the channel is divided into two subbands, high-pass for tweeter and low-pass for midrange/woofer. For a two-way loudspeaker system a typical value of crossing frequency f_{cr} is in the range from 2 to 3 kHz. In the next stages digital pulse width modulators (DPWM) produce pluses controlling pulse power amplifier transistors. The transistor switching frequency f_c can be different for both power amplifiers, it can be lower for midrange/woofer. The loudspeakers are connected to a pulse amplifier through an LC low-pass filter used for suppressing modulation components.

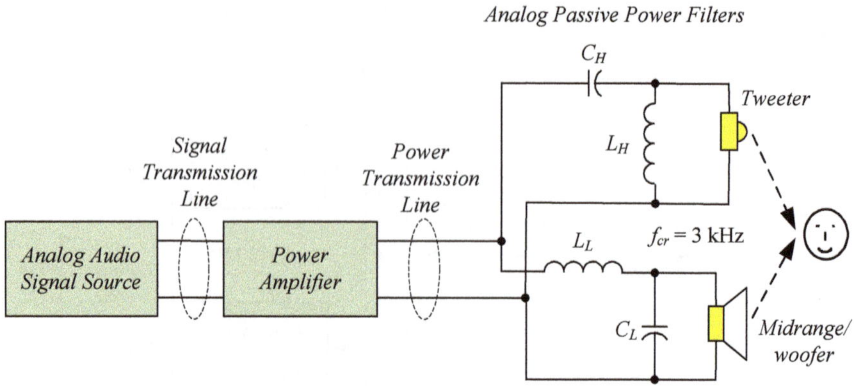

Fig. 6.30 Block diagram of two-way loudspeaker system with simple passive crossover

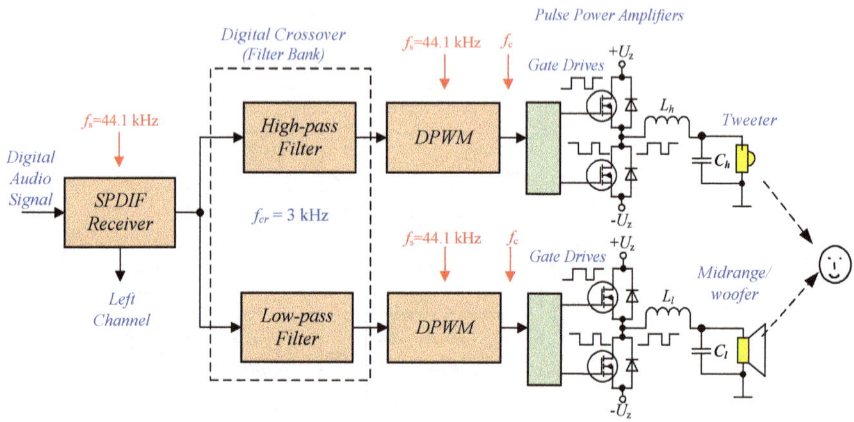

Fig. 6.31 Block diagram of two-way digital active loudspeaker system with digital class-D power amplifiers

6.9 Loudspeaker Measurements

The process of converting an electrical signal into an acoustic wave is very complex and hence very difficult to describe mathematically or simulate. However, loudspeakers are characterized by great variability and their operation depends on the size and shape of the enclosure and the crossover parameters. Therefore, during the design process reliance exclusively on calculations and simulations is not possible, and verification of the results with measurements is needed. This necessitates an acoustic chamber and suitable measuring equipment. When it comes to the acoustic chamber an anechoic chamber would be best, but this is big and expensive (especially at low frequencies). So often the use of acoustic chambers only provides noise isolation from the environment. Similarly, good measuring equipment is expensive, with

Fig. 6.32 The acoustic chamber

the most comfortable measuring equipment to use coming from companies such as Audio Precision, Bruel and Kjaer etc. However, you can use cheaper solutions, such as the Clio system from the company of Audiomatica.

The author uses an acoustic chamber with an acoustic insulation and Clio measurement system [4]. The walls and ceiling of the chamber are covered with a 11 cm layer of mineral wool. Figure 6.32 shows the chamber during a measurement of a loudspeaker box. The chamber is located in the Institute of Electrical Engineering at University of Zielona Gora. The author was also involved in the design and construction of the chamber and now it is under his supervision.

The measurement circuit is shown in Fig. 6.33a. Thanks to its advanced signal processing methods, using Clio it is possible to measure frequency characteristics of acoustic systems without an anechoic chamber. The measurements are taken using LogChirp or MLS (**M**aximum **L**ength **S**equence) signal and time gating for canceling reflected acoustic waves, only direct response from the loudspeaker is measured. Using the time gating technique it is possible to cancel the influence of the reflected energy. The block diagram of the measurement algorithm is depicted in Fig. 6.33b. The loudspeaker impulse response can be divided into three regions: delay region, meter-on region and reflections region. In the analysis only the meter-on region is used and remaining regions are filled by zeros. The floor reflection geometry is shown in Fig. 6.34. The time gating can be calculated by the formula

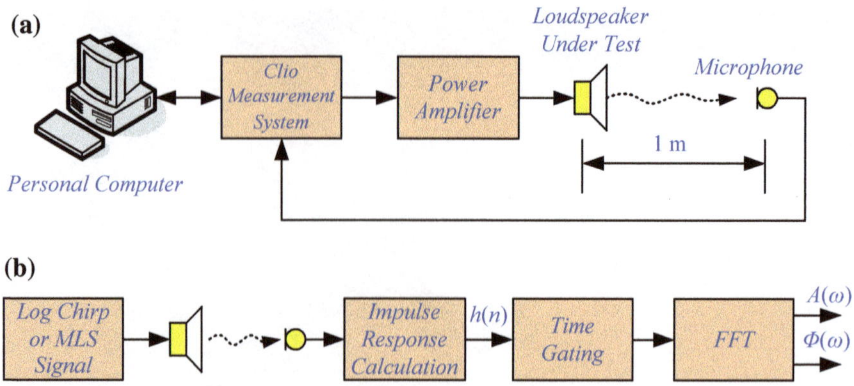

Fig. 6.33 Loudspeaker measurement: **a** circuit, **b** algorithm

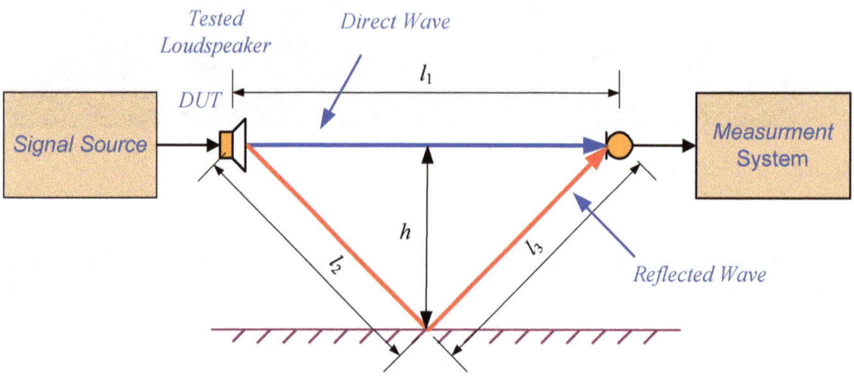

Fig. 6.34 The floor reflection geometry

$$\Delta t = \frac{2\sqrt{h^2 - \frac{l_1^2}{4}} - l_1}{v}, \tag{6.28}$$

where: v—speed of sound, l_1—distance between loudspeaker and microphone, h—distance between loudspeaker and nearest reflecting surface (floor). The lowest measured frequency is equal to

$$f_{min} = \frac{1}{\Delta t}. \tag{6.29}$$

Reflected energy from nearby walls, floor and ceiling arrives at the test microphone later than direct waves, as shown in Fig. 6.35. For this particular case, the measurement is valid for the samples between the time points $t_{min} = 2.5 \times 10^{-3}$ s and $t_{max} = 5.8 \times 10^{-3}$ s, which gives the value of $\Delta t = 3.3 \times 10^{-3}$ s Thus, the value of the lowest measured is $f_{min} = 303.03$ Hz. The impulse response (Fig. 6.35) is

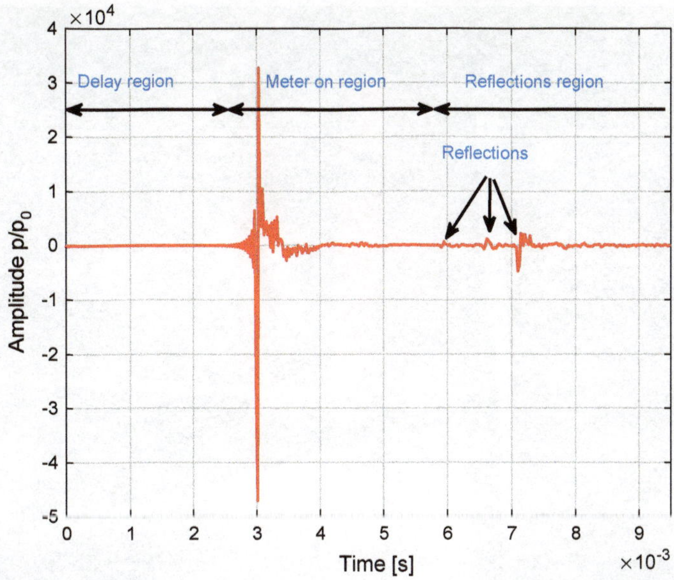

Fig. 6.35 Tweeter impulse response calculated from LogChrip signal response

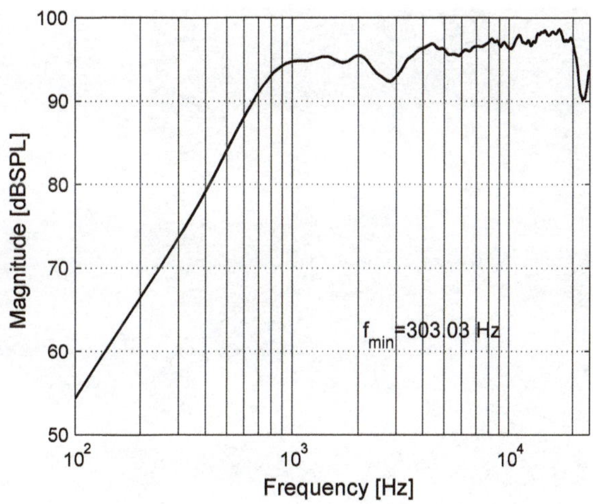

Fig. 6.36 Tweeter frequency response calculated from impulse response

used for calculating the frequency response of the loudspeaker. It has been shown in Fig. 6.36.

To measure lower frequencies a large anechoic acoustic chamber or outdoor measurements outside in the open air should be used. The open-air loudspeaker measurement system made by the author at the Institute of Electrical Engineering backyard is shown in Fig. 6.37.

Fig. 6.37 The open-air the loudspeaker system measurement

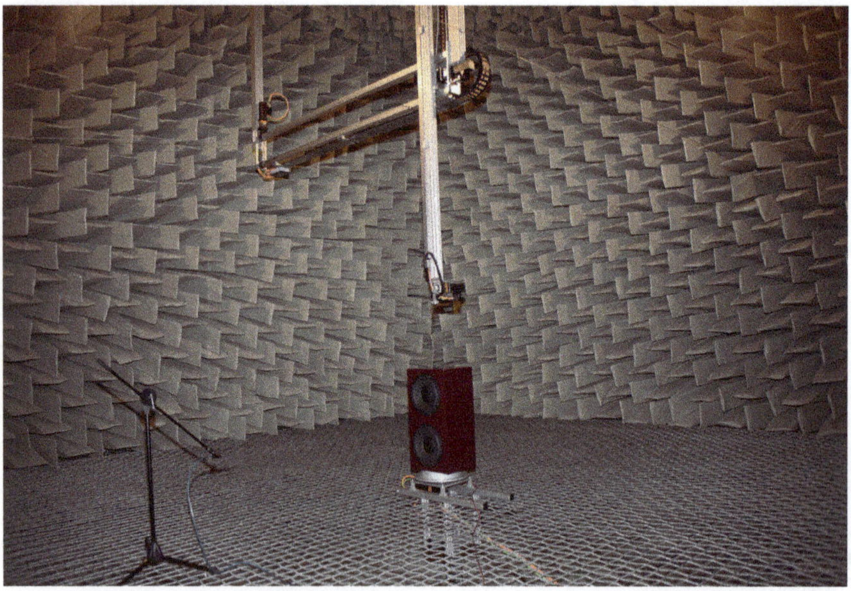

Fig. 6.38 The anechoic acoustic chamber

The near-field measurement method for low frequencies may also be used and then measurements for near-field and far-field can be combined to give the entire frequency characteristics. This evaluation method requires great care in order to obtain reliable results. Loudspeaker measurement problems are fairly well described by D'Appolito [3].

Currently, the author can use a much better acoustic chamber. It is an anechoic acoustic chamber (dimensions: 10 m × 10 m × 10 m) in the Electroacoustics Laboratory in the Science and Technology Park of the University of Zielona Gora in Nowy Kisielin. Figure 6.38 shows a low-frequency loudspeaker system with passive radiator under the tests in the chamber. The author was also involved in the design and construction of the chamber, and he is currently the head of Electroacoustics Laboratory. Thanks to the chamber and the measurement instruments installed in it, it is possible to perform very precise acoustic measurements, especially in the low frequency range.

6.10 Class-D Power Amplifier with Digital Click Modulator

The author has designed and built digital two-way loudspeaker system with digital click modulator [60]. The enclosure was built by friendly company Audiotholn from Zielona Gora, it is very stiff and very carefully done. The enclosure is a 8.5-l vented box, with the ceramic cone tweeter 26HD1/A8 [19] and HexaCone kevlar midrange/woofer 5-880/25 Hex [20], both from Eton. The HexaCone cone is a honeycomb Nomex structure that makes them both extremely light and very stiff. The digital active loudspeaker system was tested in an acoustic chamber. The designed loudspeaker system under the tests is shown in Fig. 6.39.

The author proposes the use of the digital click modulator. Problems with implementation of the digital CM are represented among others by [30, 32, 44, 48, 60, 66, 67, 81]. Application of the CM allows the reduction of the switching frequency of transistors in comparison to the classical PWM. However, implementation of digital CM encounters difficulties in realization for low-frequency signal. For instance, the Hilbert transform requires for a low signal frequency, such as 20 Hz, a very high order FIR filter—from several hundreds to thousands. Such high order FIR filters result in a large computational work load and introduce a very long delay in the signal. Therefore, the author decided to use two different modulation techniques for the lower band and upper band of the signal. Thus, for the lower band, classical PWM, and for upper band, click signal modulation are applied respectively. The input signal is divided by crossover into two bands (−60 dB): $f_{low} = 20 - 6000$ Hz, $f_{high} = 500 - 20000$ Hz. A block diagram of the proposed solution is presented in Fig. 6.40.

By using two types of modulators in the system, it is possible to use a very low transistor switching frequency $f_c = 44100$ Hz. The system also employs a delay block for correction of different signal delays in low and high channels. This difference is a result of the application of different modulation algorithms, and various positions of speakers in the box. The pulse power amplifier used was the integrated

Fig. 6.39 The digital active loudspeaker system under the tests

circuit TAS5121 from Texas Instruments [68]. The amplifier for each band achieves an output power 100 W with a supply voltage equal to 30 V and a load impedance equal to 4 Ω.

6.10.1 Digital Crossovers

In the digital crossover system that is taken into consideration there is only an analysis filter bank. The synthesis filter bank results from acoustic wave addition of particular loudspeaker signals. A simplified block diagram of such circuit is shown in Fig. 6.40. In the considered digital active crossovers two solutions are discussed: strictly complementary (SC) finite impulse response (FIR) filter bank [24, 58, 60, 79, 82, 83] and Linkwitz-Riley (LR) infinite impulse response (IIR) filter bank [35].

As an example, a strictly complementary analysis filter bank based on the Kaiser FIR filter was designed. The design parameters of the filter are described in Table 6.3. The MATLAB program for calculating response of the (SC) digital crossover is shown in Listing 6.1. The function **[b,a] = stricomp(b,a)** for coefficients calculating of strictly complementary FIR filter is shown in Listing 3.5.

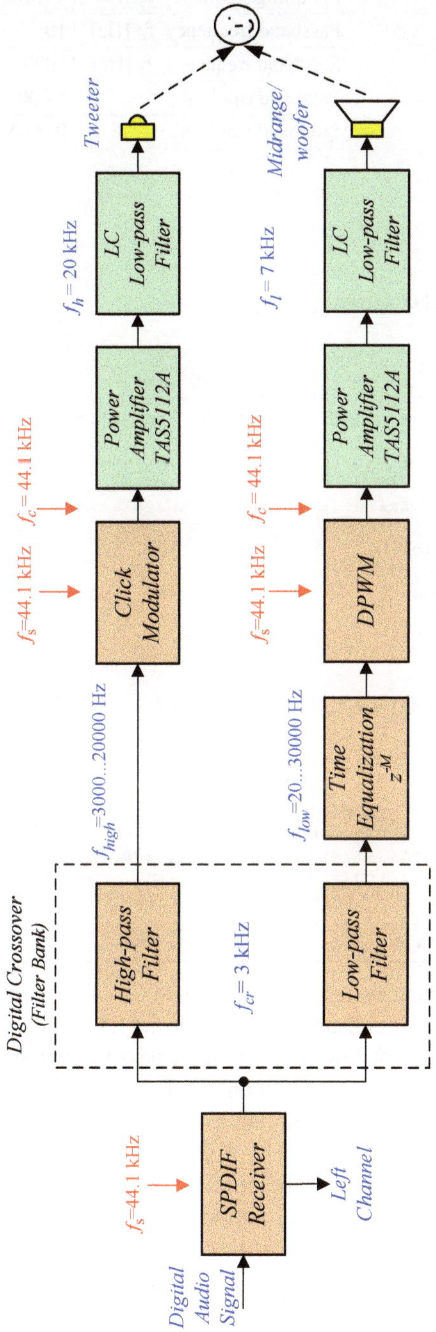

Fig. 6.40 Block diagram of proposed two-way digital active loudspeaker system with digital class-D power amplifiers

Table 6.3 Design parameters of the Kaiser FIR filter

Parameter	Value
Sampling frequency f_s [Hz]	44100
Passband frequency f_p [Hz]	10
Stopband frequency f_z [Hz]	6000
Passband ripple R_p	0.0005
Stopband ripple R_z	0.0005

Listing 6.1 The SC digital crossover

```
clear all; close all;
grub_lin=2; roz_fon=10;
N=2^11; fs=44100;
fg=10; fz=6000; Rp=0.0005; Rz=0.0005;
[n,Wn,beta,typ] = kaiserord([fg fz],[1 0],[Rp Rz],fs)
if rem(n,2)~=0
   n=n+1;   %Filter order should be even
end
bd = fir1(n, Wn, typ, kaiser(n+1,beta), 'noscale');
ad=1; ag=1;
[bg,ag]=stricomp(bd,ad);
input=zeros(1,N); input(1,1)=1;
out_L=filter(bd,ad,input); out_H=filter(bg,ag,input);
out_sum=out_L+out_H;
tt=(0:N-1)/fs; ff=(0:N-1)*fs/N;
figure('Name','Responses of SC crossover','NumberTitle','off');
subplot(211),
plot(tt,out_L,'+',tt,out_H,'*',tt,out_sum,'o','linewidth',grub_lin);
grid on; set(gca,'FontSize',[roz_fon],'FontWeight','n'),xlabel('Time [s]'),
ylabel('Amplitude'), title('(a)'); legend('Low','High','Sum');
set(gca,'ylim',[-0.2 1.0]);set(gca,'xlim',[0 n/fs]),
subplot(212),
plot(ff,20*log10(abs(fft(out_L))+eps),ff,20*log10(abs(fft(out_H))+eps), ...
    ff,20*log10(abs(fft(out_sum))+eps),'linewidth',grub_lin);
set(gca,'xlim',[0 0.5*fs]);set(gca,'ylim',[-100 10]);
set(gca,'FontSize',[roz_fon],'FontWeight','n'), xlabel('Frequency [Hz]');
ylabel('Magnitude [dB]');title('(b)'); grid on;
legend('Low','High','Sum');
print('Freq_SC_crossover.pdf','-dpdf');
```

The frequency characteristics of the designed 38-order 2-channel strictly complementary analysis filter bank are shown in Fig. 6.41. Crossing point coordinates of magnitude characteristics of low-pass and high-pass filters are −6 dB and 3000 Hz respectively. In Fig. 6.41a impulse responses of the filter bank are depicted; especially interesting is the sum of the responses of the two filters. The sum is the same as the input impulse, but is only delayed by $N/2$ samples. The frequency characteristic of the output sum is flat. Another type of filter bank well suited for audio crossover applications is a filter bank based on Linkwitz-Riley filters [35] (also called squared Butterworth). The Linkwitz-Riley crossover achieves:

- sum of the outputs with flat frequency response,
- absolutely flat amplitude response throughout the passband with a steep 24 dB/octave roll off rate after the crossover point,

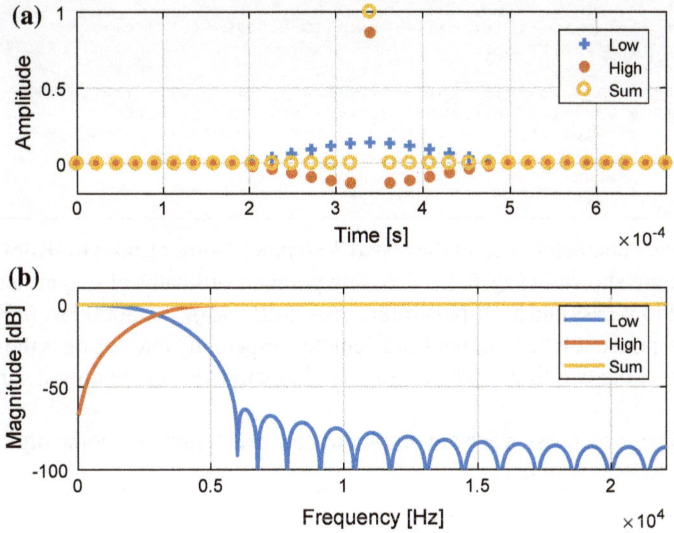

Fig. 6.41 Responses of the designed 2-channel strictly complementary analysis filter bank: **a** impulse responses: low-pass (+), high-pass (*) and sum of responses (o), **b** frequency responses

- outputs in-phase at the crossover frequency,
- phase relationship of the outputs allowing time correction for drivers that are not in the same acoustic plane,
- zero phase difference between drivers at crossover frequency,
- all drivers always wired the same (in phase).

The usual implementation of fourth order Linkwitz-Riley crossovers is a simple series connection of two second-order Butterworth filters (two for the high-pass channel, and two more for the low-pass channel). In this particular case a 4-order Linkwitz-Riley analysis filter bank with crossing frequency f_{cr}=3000 Hz was designed. The MATLAB program for calculating response of the (LR) digital crossover is shown in Listing 6.2.

Listing 6.2 The Linkwitz-Riley digital crossover

```
% Linkwitz-Riley digital crrosover
clear all; close all; grub_lin=2; roz_fon=10;
fs=44100; N=2^12;
fcr=3000; % crrosing frequency
[bd ad]=butter(2,2*fcr/fs);[bg ag]=butter(2,2*fcr/fs,'high');
input=zeros(1,N); input(1,1)=1; % impulse
out_L=filter(bd,ad,input); out_L=filter(bd,ad,out_L);
out_H=filter(bg,ag,input); out_H=filter(bg,ag,out_H);
out_sum=out_L+out_H;
t=(0:N-1)/fs;
ff=(0:N-1)*fs/N;
figure('Name','Responses of LR crossover','NumberTitle','off');
subplot(211),
plot(t,out_L,'+',t,out_H,'*',t,out_sum,'o','linewidth',grub_lin);
grid on; set(gca,'FontSize',[roz_fon],'FontWeight','n'),xlabel('Time [s]'),
```

```
ylabel('Amplitude'), title('(a)'); axis([0 0.0005 -0.3 1.0]);
legend('Low','High','Sum');
subplot(212),
plot(ff,20*log10(abs(fft(out_L))+eps),ff,20*log10(abs(fft(out_H))+eps), ...
    ff,20*log10(abs(fft(out_sum))+eps),'linewidth',grub_lin);
axis([0 0.5*fs -100 10]); set(gca,'FontSize',[roz_fon],'FontWeight','n'),
xlabel('Frequency [Hz]'); ylabel('Magnitude [dB]');title('(b)'); grid on;
legend('Low','High','Sum');
print('Freq_LR_crossover.pdf','-dpdf');
```

Frequency characteristics of the tested 2-channel 4-order Linkwitz-Riley analysis filter bank are shown in Fig. 6.42. Crossing point co-ordinates of magnitude characteristics of low-pass and high-pass filters are −6 dB and $f_{cr} = 3000$ Hz. In Fig. 6.42a impulse responses of the filter bank are depicted; especially interesting is the response of the sum of the two filters. The frequency characteristics of the output sum are flat too.

Both designed crossovers are implemented using floating point digital signal processor ADSP-21364.

6.10.2 *Realization of Digital Click Modulator*

The click modulator is a big challenge for the processor, for example, [65, 67] used three processors with computational power of 233 MMACs and two FPGAs. Therefore, the author decided to use a powerful floating point digital signal processor

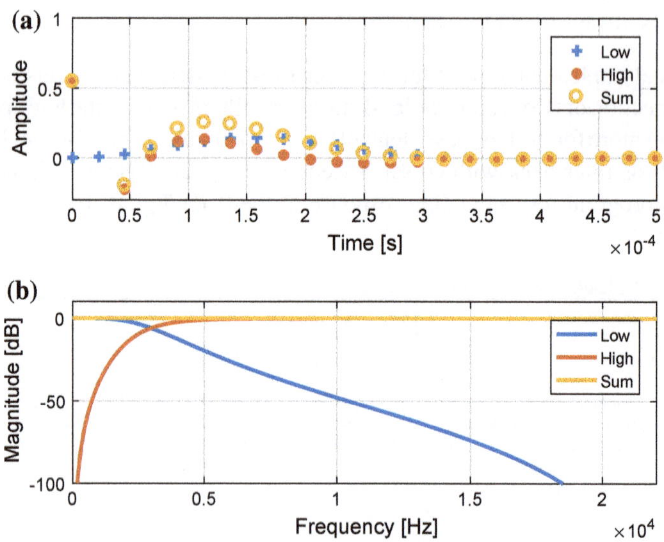

Fig. 6.42 Responses of the designed 2-channel LR analysis filter bank: **a** impulse responses: low-pass (+), high-pass (*) and sum of responses (o), **b** frequency responses

(DSP) ADSP-21364 from Analog Devices [1, 2]. The efficiency of the DSP processor used for the realization of the modulator is sufficient to support the entire algorithm at full speed (44.1 kHz).

The block diagram of the realization of the laboratory experimental circuit is shown in Fig. 6.43 [60]. In this circuit, for simplicity, a digital audio input is used. The signal is in Sony/Philips Digital Interconnect Format (S/PDIF) standard and it uses 75-ohm coaxial RCA connector. The 16-bit stereo digital audio input signal is received by an ADSP-21364 digital audio receiver called also a digital audio interface (DAI). The digital signal has a sampling rate $f_s = 44.1$ kHz. The main part of the modulator is realized using floating point digital signal processor ADSP-21364 with $f_{clk} = 300$ MHz clock frequency, delivering 300 million floating point instructions per second. It is possible to calculate the quantity of available processor operations per input sample $L_{DSP} = \text{floor}(f_{clk}/f_s) = 6802$. The digital PWM is realized with ADSP-21364 counters and it has a 12-bit resolution. The counters work with a frequency of $f_M = 300$ MHz. The switching frequency of the pulse amplifier transistors is $f_c = 44.1$ kHz. The data to PWM is fed with frequency $f_c = 44.1$ kHz.

The block diagram for the digital realization of the click modulator algorithm is presented in Fig. 6.43. Based on the linear phase response of the whole algorithm, finite impulse response filters (FIR) have to be used. For Hilbert transform of the input signal a FIR filter is applied. A practical FIR implementation of the Hilbert transform will exhibit bandpass characteristics. The bottleneck of this algorithm is the low-frequency performance. The Hilbert FIR filter was designed using the MATLAB Signal Processing Toolbox, as shown in Listing 6.3.

Listing 6.3 The Hilbert FIR filter

```
N_H=100; % Order
F=[0.02 0.98]; % Frequency Vector
A=[1 1]; % Amplitude Vector
W=1; % Weight Vector
b=remez(N_H,F,A,W,'hilbert');
```

The frequency response of the Hilbert FIR filter (red line) and delay line (blue line) is depicted in Fig. 6.44. Realization of ADSP-21364 code for such FIR filter is described in Listing 6.4, with every filter tap executed in a single processor machine cycle 3.333 ns (for 300 MHz clock).

Listing 6.4 The ADSP-21364 code for FIR filter

```
/* load sample from circular buffer
in data memory and coefficient from
 circular buffer in program memory */
f2=dm(i0,m0), f4=pm(i8,m8);
/* loop initialization */
lcntr=TAPS-1, do (pc,1) until lce;
/* calculate filter tap */
f8=f2*f4,f12=f8+f12,f2=dm(i0,m0),
f4=pm(i8,m8);
/* calculate last tap */
f8=f2*f4 , f12=f8+f12;
/* last accumulation */
f12 = f8+f12;
```

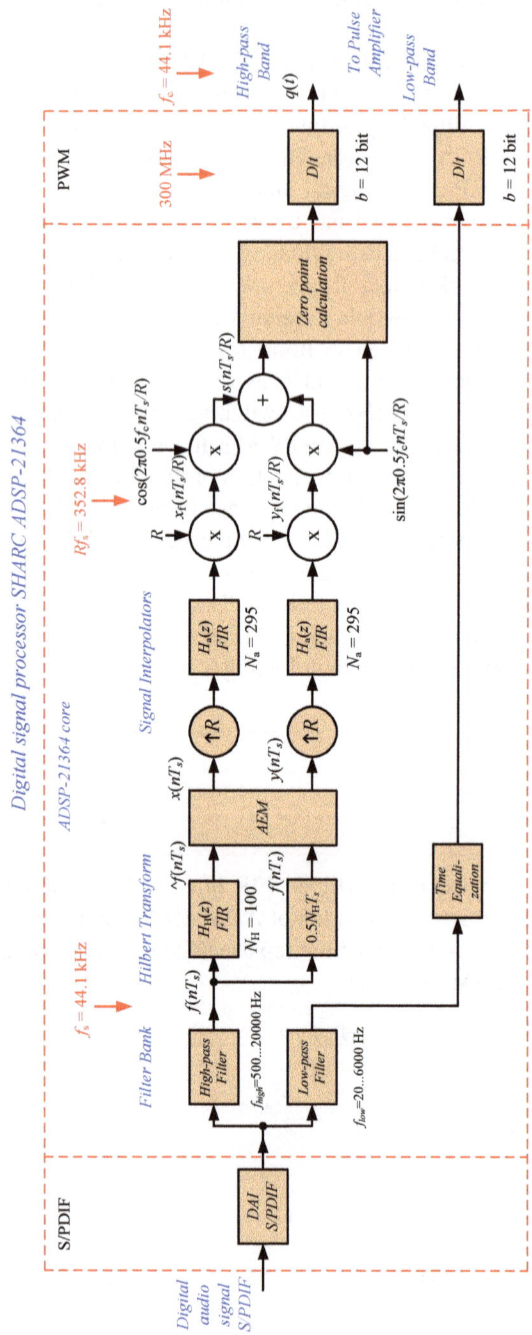

Fig. 6.43 Realization block diagram of the ALS

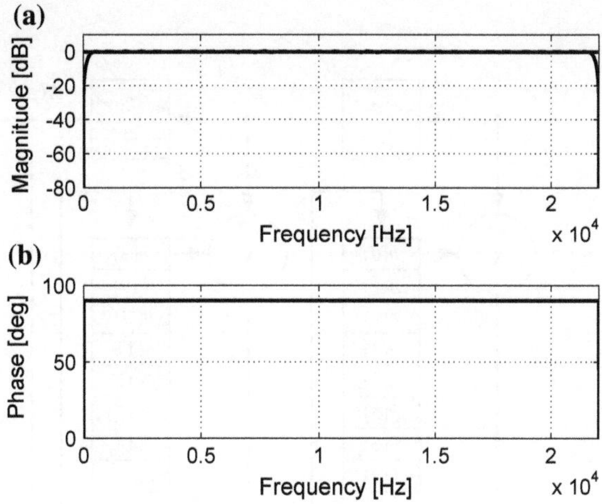

Fig. 6.44 Frequency responses of the Hilbert FIR filter and delay line: **a** amplitude responses, **b** difference of phase responses

The digital algorithm determined moments of the zero crossing of signal $s(t)$; to increase the accuracy of the digital process, signal sampling rate should be increased R times. The chosen interpolator ratio is $R = 8$, which is a compromise between modulation accuracy and computational complexity. The sampling rate is increased before the filter $H_a(z)$, so that the filter also fulfills the role of the interpolation filter. Similar to the Hilbert filter is the design of the $H_a(z)$ FIR filter. In practice, the low-pass filter should sufficiently attenuate the stop-band to suppress the unwanted images of the baseband. Finally the signal amplitude is increased R times to compensate amplitude losses. In the designed modulator the finite impulse response filter FIR has to be used according to its linear phase response. The FIR filter was designed using the MATLAB Signal Processing Toolbox as shown in Listing 6.5.

Listing 6.5 The MATLAB FIR filter design

```
fs=44100*8; % sampling frequency
Na=295; % filter order
Fpass=20000; % passband Frequency
Fstop=27000; % stopband Frequency
Wpass=1; % passband Weight
Wstop=1; % stopband Weight
b=firls(N,[0 Fpass Fstop fs/2]/(fs/2),..
[1 1 0 0],[Wpass Wstop]);
```

The filter order is $N_{int} = 295$. The interpolator requires $(N_a + 1)R$ multiplications and additions per one input sample.

It is possible to decrease the quantity of arithmetic operations by elimination of multiplication and addition for zero value samples. The block diagram of such solution is shown in Fig. 6.45. This is a FIR based signal interpolator for $R = 8$ with periodically switched coefficients and filter order $N_a = 295$. In this case the interpolator requires $N_a + 1$ multiplications and additions per one input signal sample. This kind of filter structure is easy and efficiently realized by the DSP.

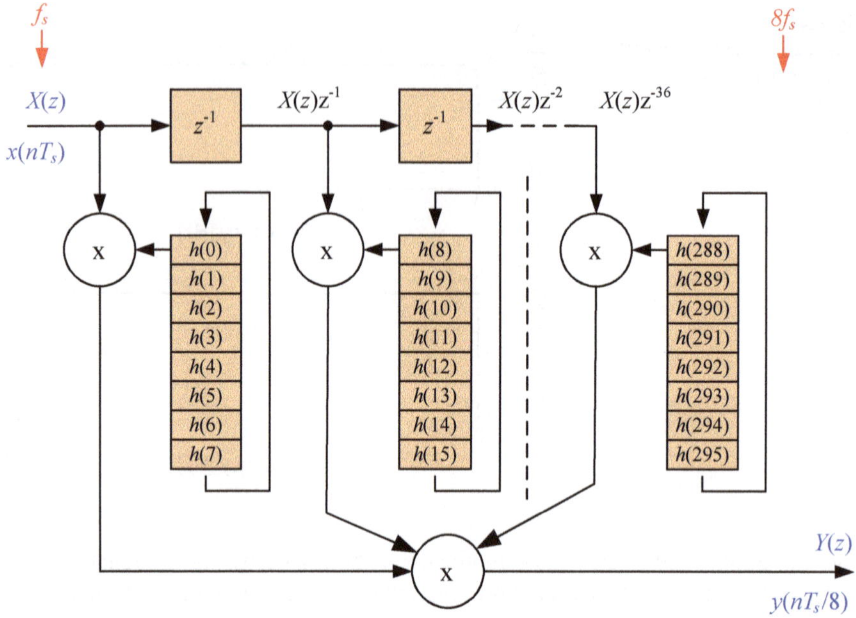

Fig. 6.45 Block diagram of FIR based signal interpolator for $R = 8$ with periodically switched coefficients and filter order $N_{int} = 295$

The signal $s(kT_s/R)$ has a sampling rate equal to 352.8 kHz. This time resolution is still too low to perform high quality audio signal. Therefore zero crossing point has to be calculated with higher accuracy. The time counters work with a clock frequency equal to 300 MHz. The zero crossing point is calculated using linear interpolation. This process is shown in Fig. 6.46.

The weak point of the algorithm is the need to use FIR filters with very high orders, which causes high DSP workload and signal latency. Instead of FIR filters, the author successfully applied a linear-phase IIR filter, which is described in Chap. 3. This results in a significant reduction of DSP workload for the same results.

6.10.3 Experimental Results

Experimental results for the sinusoidal input signal for the realized click modulator are presented in Fig. 6.47. Presented is the spectrum of the output signal for an input signal of 5 kHz. Modulation components are moved from a signal band of 500 Hz–20 kHz, some harmonics in the signal band on −80 dB level are connected with limited time resolution.

Acoustic measurements were made using the computer controlled system Clio from Audiomatica [4]. Thanks to its advanced signal processing methods, using the

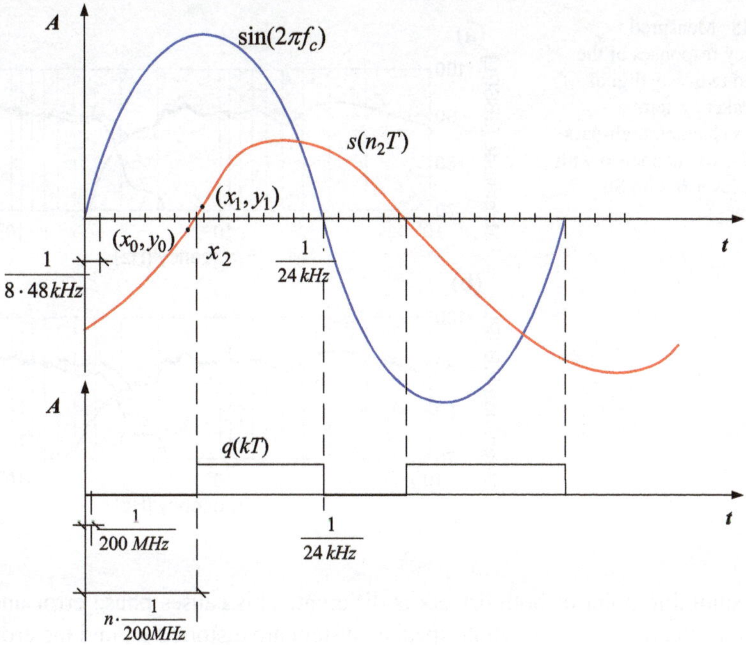

Fig. 6.46 Zero crossing calculation

Fig. 6.47 Experimental results of digital audio amplifier output signal with click modulator for $f_c = 44.1$ kHz and sinusoidal input signal $f = 5$ kHz

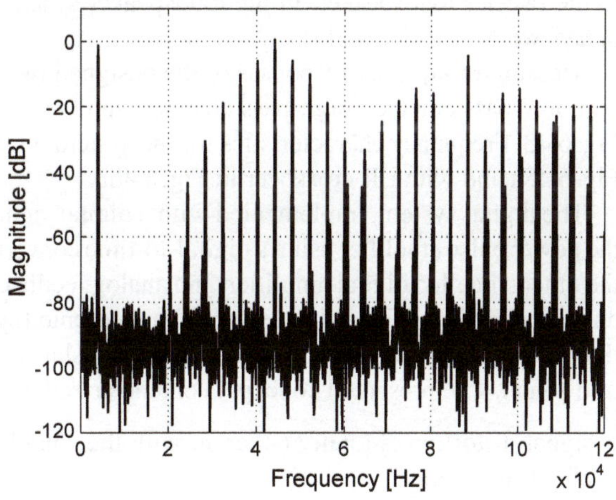

Clio system, it is possible to measure frequency characteristics of acoustic systems without an anechoic chamber.

Acoustic waves generated by two drivers must be coincident. This means the drivers have to radiate from exactly the same point in space and time. According to different sizes of tweeter and woofer, in a typical loudspeaker system, the positions of driver acoustic centres are not located on the same plane. Therefore, the distance

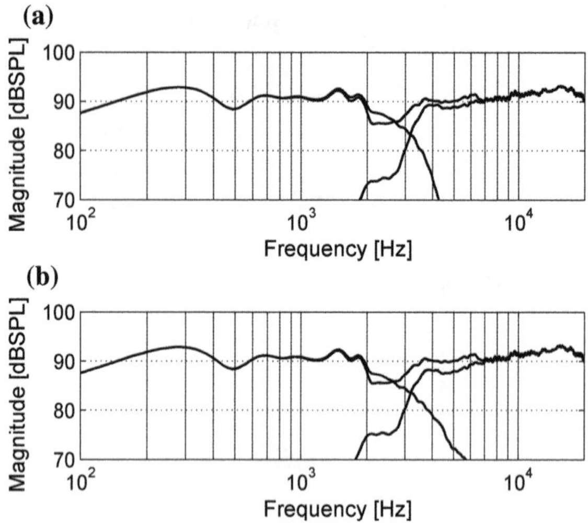

Fig. 6.48 Measured frequency responses of the designed two-way digital loudspeaker system: low-pass channel, high-pass channel, two channels **a** with LR crossover **b** with SC crossover

to the summing point of both drivers is different. This causes phase error and the amplitude characteristics of whole speaker system are distorted around the crossing frequency. A simple digital delay circuit can be used to equalize time-aligns, thus harmonizing the phase of both drivers, and reducing lobing error by adding delay to the tweeter loudspeaker. In the loudspeaker systems under consideration digital delays were used (Fig. 6.43).

Measured frequency responses of the designed two-way digital loudspeaker system, low-pass channel, high-pass channel, and two channels together are shown in Fig. 6.48. Frequency characteristics for the system with SC crossover are depicted in Fig. 6.48a and with LR crossover in Fig. 6.48b.

The digital system, implemented with a digital signal processor, directly controls the power pulse amplifier using a digital-to-time converter with noise shaping. Unlike the other so-called digital amplifiers, no analog feedback or analog signal processing amplification is involved at any stage of the presented system. The resulting system is thus a high power D/A converter device that translates the digital information directly into sound. The presented concept is characterized by numerous advantages:

- signal distortion is totally coherent with the sound (music), no ringing or decay effects can appear,
- transient intermodulation distortion cannot occur,
- distortion is the same under steady state and dynamic conditions.

The proposed digital crossovers based on a strictly complementary filter bank and Linkwitz-Riley filter bank are well suited for a digital active loudspeaker box. The results of both filter banks are similar. The main advantage of click modulation is low switching frequency close to the upper signal band limit and high efficiency of energy conversion. The main disadvantage of click modulation is complication of the control algorithm. It is a big challenge even for the fastest digital signal processors. Another difficult problem is output pulse time resolution. Fortunately, the speed of modern digital signal processors and microcontrollers is continuously growing. Using two types of modulation allowed the use of very low switching transistor frequency while maintaining good performance of the system. Using low switching frequency reduces power losses and reduces EMC interference. The designed active loudspeaker system covers the whole audio band, theoretically from 20 Hz to 20 kHz, though practically, low frequency is higher according to woofer/midrange loudspeaker and box parameters.

6.11 Digital Audio Class-D Power Amplifier with TAS5508 DSP

The author has designed and built a high quality stereo three-way digital loudspeaker system. The system consists of a digital crossover and a digital class-D power audio amplifier. It is based on TAS5508-5121K8EVM class-D digital power amplifier evaluation module. Figure 6.49 shows the loudspeaker system in an acoustic chamber. The TAS5508-5121K8EVM class-D digital power amplifier evaluation module is shown in Fig. 6.50.

While, the block diagram of one channel of the loudspeaker system is depicted in Fig. 6.51. The system consists of a digital audio interface (DAI) on the front. In DAI

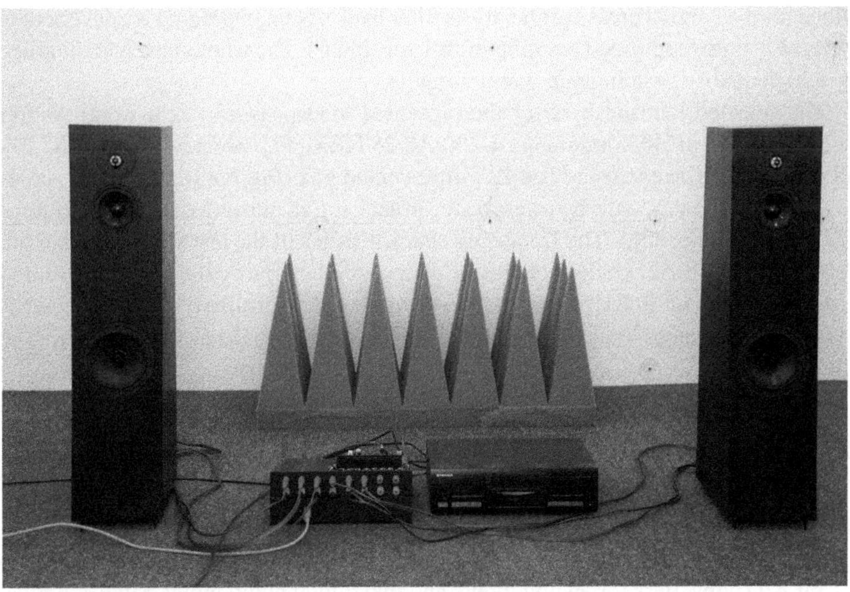

Fig. 6.49 The three-way loudspeaker system

Fig. 6.50 The TAS5508-5121K8EVM module

digital audio signal is divided into two channels. Then signal is divided into three subbands by a digital crossover (analysis filter bank). Next, the signal is converted by DPWM to pulse controlled amplifier transistor signals. The whole system is supplied by a high quality switch mode power supplier.

The selected loudspeakers for the presented loudspeaker system were: woofer 7-200/A8/32 [18] HEX, midrange 4-200/A8/25 HEX [17], tweeter 26HD1/A8 [20], all from Eton. The enclosure is a 12.5-litre vented box (Fig. 6.49).

The loudspeakers in the box were tested in acoustic chamber (located in Institute of Electrical Engineering). The frequency characteristics of the tested loudspeaker are depicted in Fig. 6.52. Under conditions in acoustic laboratory the lowest measured frequency is about 268 Hz. The graph shows that the sensitivity of the tweeter is higher than the other loudspeakers, but it can be easily equalized.

6.11.1 TAS5508-5121K8EVM

The proposed system consists of the TAS5508-5121K8EVM evaluation module board (EVM) and Input PC Board [70]. The Input PC Board has three stereo audio 24-bit A/D converters for analog inputs and two digital audio inputs SPDIF: optical Toslink and Coaxial. The system is controlled by a personal computer using a USB interface.

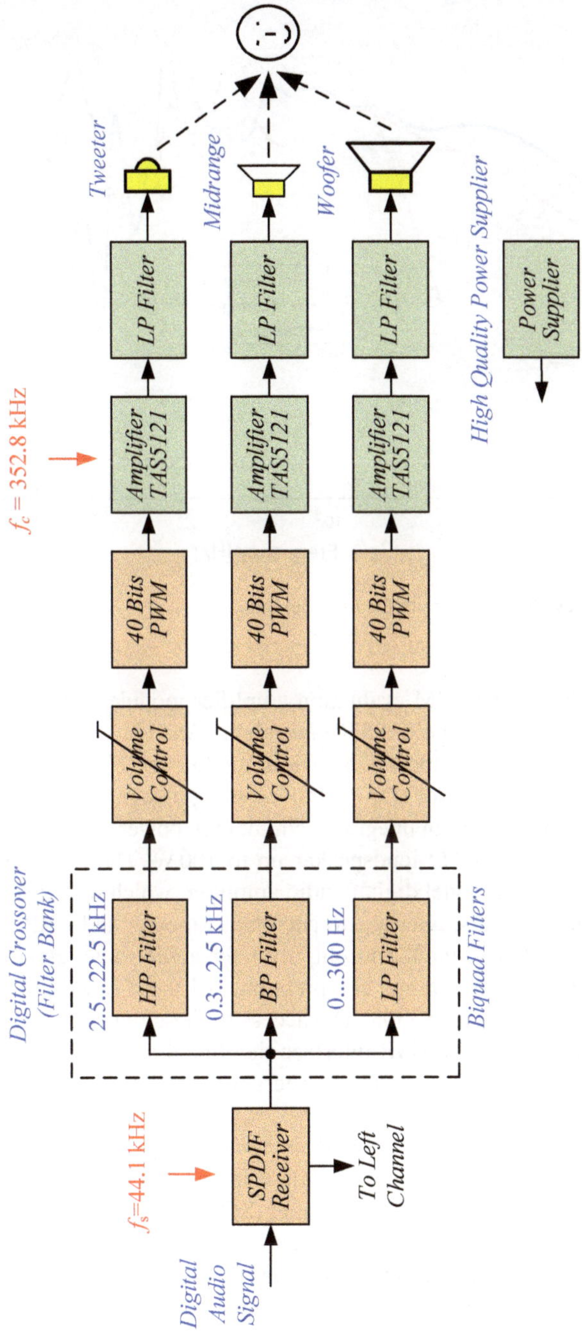

Fig. 6.51 The block diagram of the three-way loudspeaker system

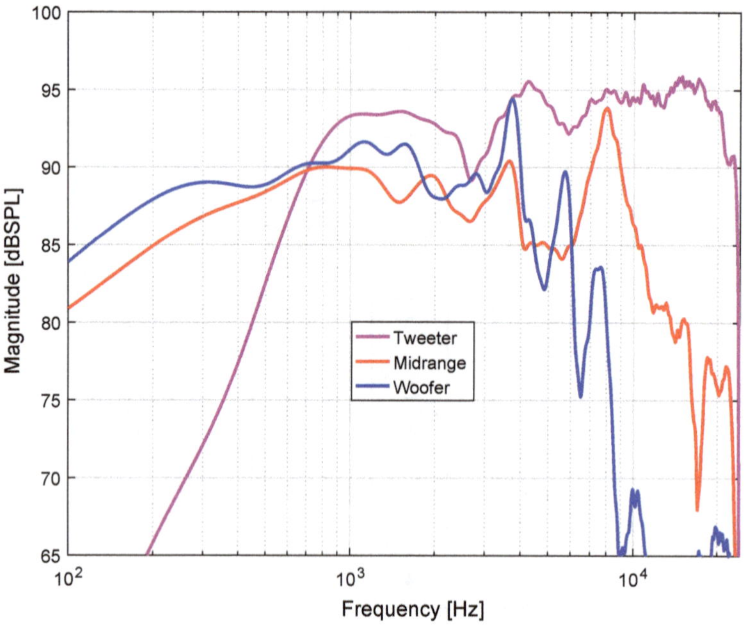

Fig. 6.52 Frequency characteristics of the loudspeakers

The TAS5508-5121K8EVM evaluation amplifier module consists of two main types of integrated circuits: one TAS5508 and eight TAS5121. The TAS5508 is a high performance 32-bit (24-bit input) pulse width modulator (PWM) and 48-bit multi channel digital audio processor (DAP). It accepts input signal sample rate from 32 to 192 kHz. The TAS5121 is an integrated circuit high power digital amplifier power stage designed to drive a 4 Ω loudspeaker up to 100 W. The EVM and Input PC Board is a complete 8-channel digital audio amplifier, which includes digital audio inputs, analog audio inputs, interface to personal computer and DAP features such as digital volume control, input and output mixer audio mute, equalization, tone controls, loudness, dynamic range compression. All these function are controlled by special registers in the TAS5508. The access to these registers is possible using the USB interface. Using the USB interface the digital amplifier is connected to a personal computer. The content of the registers is controlled by TAS5508 graphical interface software.

Analog or digital audio signal SPDIF is converted by an Input PC Board and then is transmitted to the evaluation module using an I2S interface. Then it passes through an input mixer to selected biquad filter (SOS) groups. The biquad filter group consists of seven filters. The filter coefficients b_0, b_1, b_2, $-a_1$, $-a_2$ (Fig. 6.53) are 28-bit, using a 5.23 number format. This means that there are 5 bits to the left of the decimal point and 23 bits to the right of the decimal point. From the SOS filter group the signal is sent to the output mixer. In the next stage digital signals are amplified

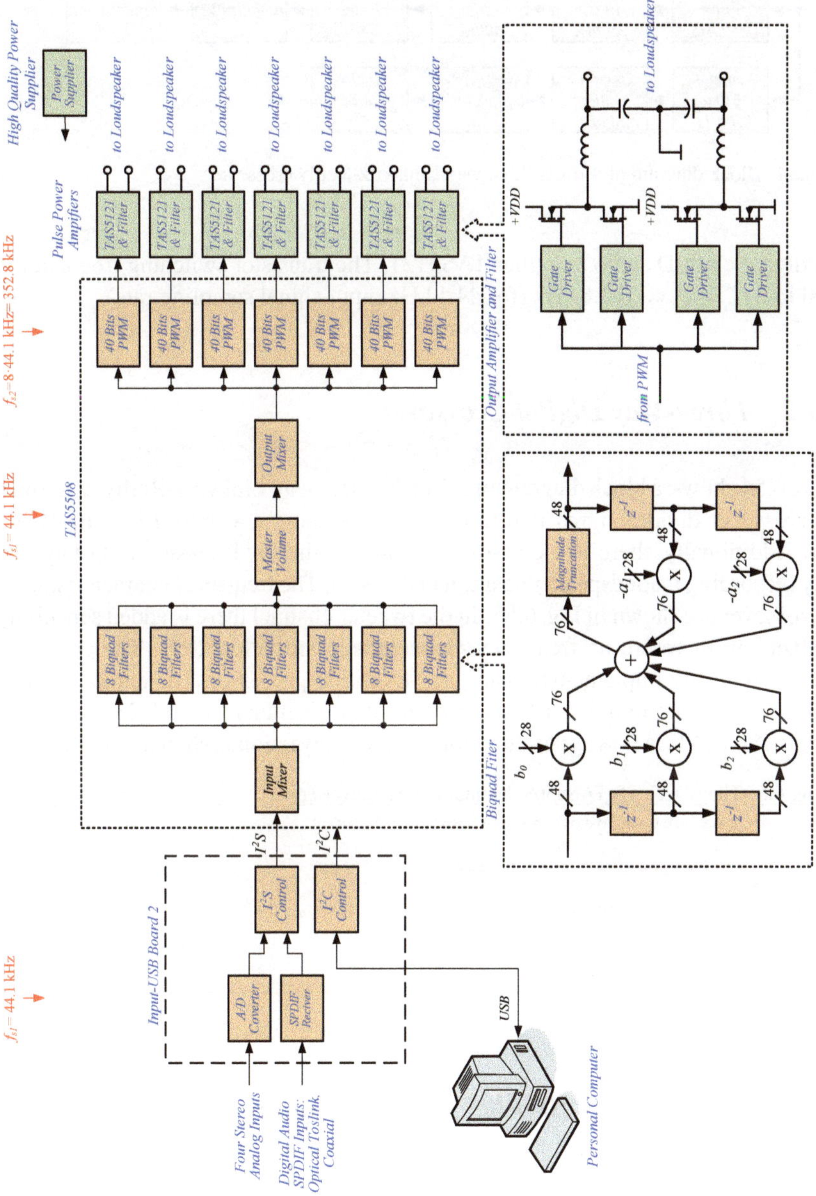

Fig. 6.53 Block diagram of TAS5508 evaluation module

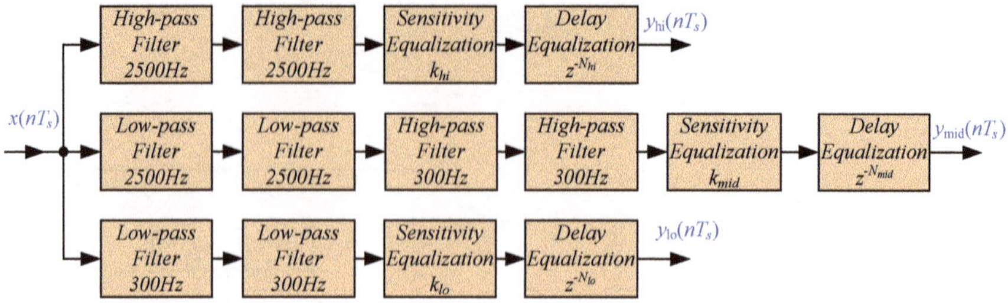

Fig. 6.54 Block diagram of 4-order three way Linkwitz-Reilly crossover

by a digital class-D audio amplifier TAS5121. The transistor switching frequency is equal to $8f_s = f_c = 352.8\,\text{kHz}$ (for 44.1 kHz input signal sampling rate).

6.11.2 *Three-Way Digital Crossover*

Figure 6.54 shows a block diagram of a 4-order three way Linkwitz-Reilly crossover. The crossover divides signal into three subbands using a 4-order Linkwitz-Reilly filter. Additionally, there is a crossover equalizing the loudspeaker sensitivity and delay (according to loudspeaker position in the box). The frequency characteristics of the crossover are shown in Fig. 6.55. In the tweeter channel there is added sensitivity equalization, therefore its frequency characteristic are below 0 dB. The crossover filters were easily implemented using the TAS5508 processor. The author wrote a MATLAB program to transfer floating-point filter coefficients into 5.23 fixed point format. The MATLAB program function for this conversion is shown in Listing 6.6.

Listing 6.6 The MATLAB TAS5508 decimal to hex converter

```
function hex_out=f_dec2hex_TAS5508(in_dec)
% Quantize to 23 bits and round to the nearest integer
quant = abs(round((2^23)*in_dec));
if in_dec <  % i.e. negative
   quant = bitcmp(quant,28)+1; %quant = quant+2^27
end
% Convert the decimal number to hex
hex_out=dec2hex(quant,8);
```

6.11.3 *Experimental Results*

All measurements were made in an acoustics chamber using the Clio system. The simplified block diagram of the measurement system is depicted in Fig. 6.56. Figure 6.57 shows frequency characteristics of the loudspeaker system channels loudspeaker and crossover tweeter, midrange, woofer and the whole system. Similar to the case of measuring the loudspeaker lower frequency the frequency is limited by the conditions

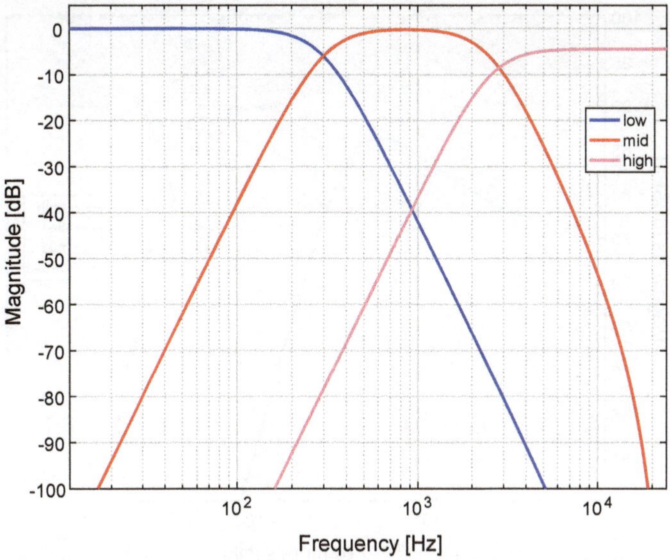

Fig. 6.55 Frequency characteristics of 4-order three way Linkwitz-Reilly crossover

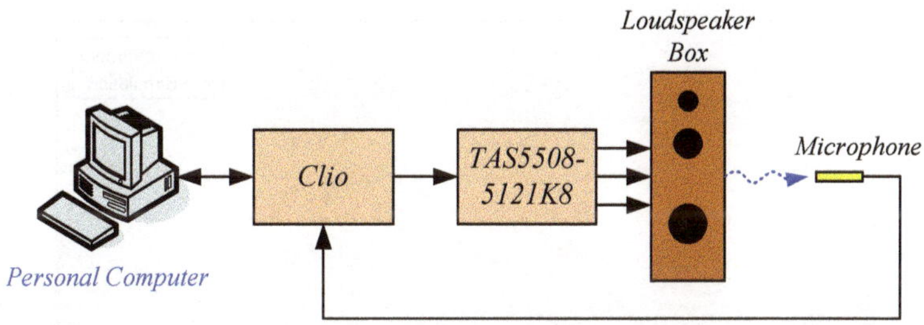

Fig. 6.56 Measurement circuit for three way loudspeaker system

of the measurement and is 268 Hz. A comparison of loudspeaker system simulations with the experimental results obtained from the measurement of the entire system was also made. In simulation tests, the measured impulse responses of the loudspeakers were used. Figure 6.58 shows such frequency characteristics. The characteristics curves show consistency of simulation and experimental results. The presented system is also very useful for an electroacoustic laboratory, especially for experiments with crossovers for speaker boxes. Using laptop computer and the MATLAB program for designing the filter it is very easy to change filter parameters.

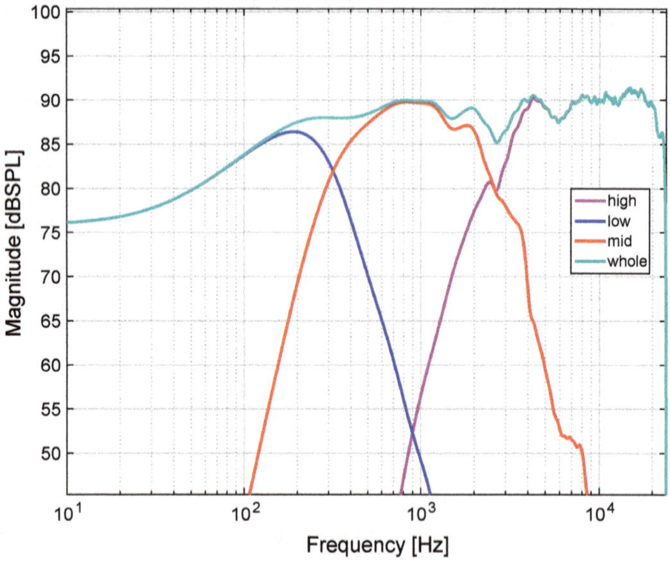

Fig. 6.57 Frequency characteristics of the measured loudspeaker system, channels: tweeter, midrange, woofer, whole system

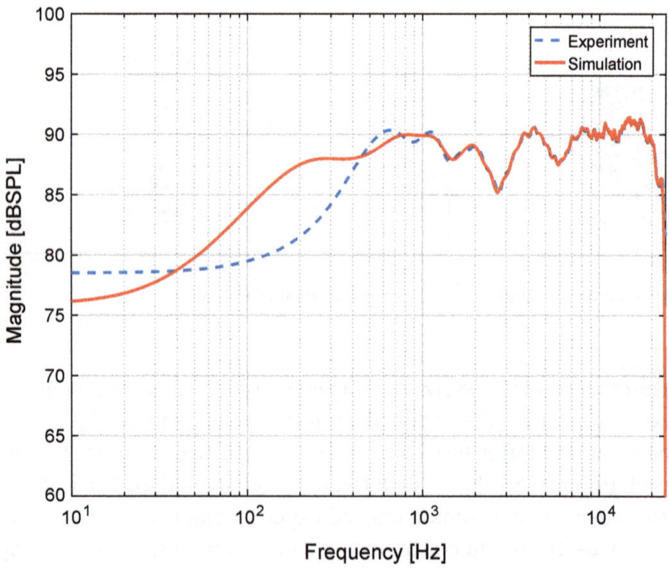

Fig. 6.58 Comparison of simulation and experimental results of the loudspeaker system

6.12 Conclusions

The proposed noise shaping circuit for the digital class-D amplifier PWM makes it possible to increase the quality of the D/A conversion. In future research more efficient noise shaping circuits should be investigated. Special attention also should be paid to class-D amplifiers with full digital feedback. But this task is very difficult in audio applications with respect to the creation of audible transient suboscillation. The results of such circuits should be carefully verified by subjective tests.

Open loop digital class-D amplifiers are free from such problems but they require very high quality power supply voltage. For an audio application the power supply source impedance should be low, up to 20 kHz! So the author has presented an analog circuit for supply voltage fluctuation and transistor amplitude errors resulting from non-linear on-state resistance compensation.

In the digital class-D amplifier signal oversampling is required, therefore, there have also been considered signal interpolators. Such interpolators allow for increasing sampling frequency whilst maintaining substantial separation of signal from noise.

Finally, two-way and three-way loudspeaker systems, designed by the author, have been presented, where a signal from input to output is digitally processed. The proposed two-way loudspeaker system with employed click modulator for the higher band and ordinary PWM for the lower band makes it possible to keep transistor switching frequency equal to 44.1 kHz. Using the low switching frequency reduces transistor power losses and reduces EMC interference. The designed active loudspeaker system covers the whole audio band, theoretically from 20 Hz to 20 kHz.

In the author's opinion, in the near future, digital active loudspeaker systems with digital input will become more and more popular, especially for home cinema systems. Another advantage of such systems is the possibility to control individual loudspeaker characteristics and give overload protection.

The algorithms and applications presented by the author in this Chapter are a little bit off topic for traditional power electronics applications. However, the scope of applications of power electronics is now quickly expanding to more and more areas. The problems presented in this Chapter can be successfully applied in typical power electronics circuits. In particular, a noise shaping circuit for compensating systematic errors of inverter output stages can be used. This type of circuit can decrease the influence of dead time and minimum switch-on time. In the author's opinion this compensation should be very successful for multilevel inverters too. The additional workload on the processor for a noise shaping circuit is very small, so it can also be easily implemented in existing control circuits.

References

1. Analog Devices (2005) ADSP-2136x SHARC processor hardware reference. Rev 1.0. Analog Devices Inc.
2. Analog Devices (2007) ADSP-21364 Processor EZ-KIT lite evaluation system manual. Rev 3.2, Analog Devices Inc.
3. D'Appolito J (1998) Testing loudspeaker. Audio Amateur Press, Peterborough
4. Audiomatica (2005) Clio electrical and acoustical test. User's Manual, Audiomatica
5. Barbour E (1998) The cool sound of tubes. IEEE Spectr 35(8):24–35
6. Bateman A, Paterson-Stephens I (2002) The DSP handbook: algorithms, applications and design techniques. Prentice Hall, New York
7. Bresch E, Padgett WT (1999) TMS320C67-based design of a digital audio power amplifier introducing novel feedback strategy. In: Texas Instruments DSPS Fest 99
8. Bruunshuus T (2004) Implementation of power supply volume control. Application report, SLEA038, Texas Instruments Inc.
9. Bruunshuus T (2004) Power supply considerations for AV receivers. Application report, SLEA028, Texas Instruments Inc.
10. Carley RL, Schreier R, Temes GC (1997) Delta-sigma ADCs with multibit internal conveters. In: Norsworthy SR, Schreier R, Temes GC (eds) Delta-sigma data converters, theory, design and simulation. IEEE Press
11. Cataltepe T, Kramer AR, Larson LE, Temes GC, Walden RH (1992) Digitaly corrected multi-bit $\Sigma\Delta$ data converters. In: Candy JC, Temes GC (eds.) Oversampling delta-sigma data converters theory, design, and simulation, IEEE proceedings ISCAS'89, May 1989. IEEE Press
12. Dabrowski A, Sozanski K (1998) Implementation of multirate modified wave digital filters using digital signal processors. In: XXI Krajowa Konferencja Teoria Obwodów i Układy Elektroniczne, KKTUIE98, Poznan
13. Dickason V (2000) The loudspeaker design cookbook. Audio Amateur Press, Peterborough
14. Dobrucki A (2007) Electroacoustic transducers. WNT, Warszawa (in Polish)
15. Duncan B (1996) High performance audio power amplifier for music performance and reproduction, Newnes
16. Esslinger R, Gruhler G, Stewart RW (2004) Feedback strategies in digitally controlled class-D amplifiers. In: Conference proceedings, AES 114th convention, Amsterdam, The Netherlands, 22–25 March 2003. Audio Engineering Society
17. Eton (2012) Midrange loudspeaker 4–200/A8/25 HEX. Data sheet, Eton GmbH
18. Eton (2012) Midrange loudspeaker 7–200/A8/32 HEX. Data sheet, Eton GmbH
19. Eton (2012) Midrange loudspeaker 5–880/25 Hex. Data sheet, Eton GmbH
20. Eton (2012) Tweeter loudspeaker 26HD1/A8. Data sheet, Eton Gmbh
21. Everest F (2000) Master handbook of acoustics. McGraw-Hill, New York
22. Fettweis A (1982) Transmultiplexers with either analog conversion circuits, wave digital filters, or SC filters—a review. IEEE Trans Commun 30(7):1575–1586
23. Fettweis A (1989) Modified wave digital filters for improved implementation by commercial digital signal processors. Sig Process 16(3):193–207
24. Flige N (1994) Multirate digital signal processing. Wiley, New York
25. Galton I (1997) Spectral shaping of circuit errors in digital-to-analog converters. IEEE Trans Circ Syst II Analog Digital Sig Proc 44(10):789–797
26. Gazsi L (1985) Explicit formulas for lattice wave digital filters. IEEE Trans Circ Syst 32(1):68–88
27. Goldberg JM, Sandler MB (1994) New high accuracy pulse width modulation based digital-to-analogue convertor/power amplifier. IEE Proc Circ Devices Syst 141(4):315–324
28. Gwee BH, Chang JS, Adrian V (2007) A micropower low-distortion digital class-D amplifier based on an algorithmic pulsewidth modulator. IEEE Trans Circ Syst I Regul Pap 52(10):2007–2022
29. Holmes DG, Lipo TA (2003) Pulse width modulation for power converters: principles and practice. Institute of Electrical and Electronics Engineers, Inc.

30. Kostrzewa M, Kulka Z (2005) Time-domain performance investigations of the click modulation-based PWM for digital class-D audio power amplifiers. In: Signal processing 2005, IEEE conference proceedings, pp 121–126
31. Krukowski A, Kale I, Morling R, Hejn K (1994) A Design technique for polyphase decimators with binary constrained coefficients for high resolution A/D converters. In: IEEE international symposium on circuits and systems (ISCAS'94), pp 533–536
32. Kuncewicz L (2009) Design and realization of PWM with click modulation algorithm. Master's thesis, University of Zielona Gora, Poland (in Polish)
33. Larson LE, Cataltepe T, Temes G (1992) Multibit oversampled - A/D converter with digital error correction. In: Candy JC, Temes GC (eds) Oversampling delta-sigma data converters, theory, design and simulation. IEEE electronics letters, 24, August 1988. IEEE Press
34. Ledger D, Tomarakos J (1998) Using the low cost, high performance ADSP-21065L digital signal processor for digital audio applications. Revision 1.0, Analog Devices, Norwood, USA
35. Linkwitz SH (1976) Active crossover networks for non-coincident drivers. J Audio Eng Soc 24(1):2–8
36. Logan BF (1984) Click modulation. AT&T Bell Lab Tech J 63(3):401–423
37. Madsen K, Soerensen T (2005) PSRR for PurePath digital™ audio amplifiers. Application report, SLEA049, Texas Instruments Inc.
38. Midya P, Roeckner B (2010) Large-signal design and performance of a digital PWM amplifier. J Audio Eng Soc 58(9):739–752
39. Mosely ID, Mellor PH, Bingham CM (1999) Effect of dead time on harmonic distortion in Class-D audio power amplifiers. IEEE Electron Lett 35(12):950–952
40. Mouton T, Putzeys B (2009) Digital control of a PWM switching amplifier with global feedback. In: Conference proceeding, AES 37th international conference, Hillerod, Denmark, 28–30 August 2009. Audio Engineering Society
41. Nielsen K (1998) Audio power amplifier techniques with energy efficient power conversion. PhD thesis, Departament of Applied Electronics, Technical University of Denmark
42. Orfanidis SJ (1996) ADSP-2181 experiments. http://www.ece.rutgers.edu/orfanidi/ezkitl/man.pdf. Accessed Dec 2012
43. Orfanidis SJ (2010) Introduction to signal processing. Prentice Hall Inc., New Jersey
44. Oliva A, Paolini E, Ang SS (2005) A new audio file format for low-cost, high-fidelity, portable digital audio amplifiers, Texas Instruments
45. Pascual C, Song Z, Krein PT, Sarwate DV, Midya P, Roeckner WJ (2003) High-fidelity PWM inverter for digital audio amplification: spectral analysis, real-time DSP Implementation and results. IEEE Trans Power Electron 18(1):473–485
46. Putzeys B (2008) A universal grammar of class D amplification, Tutorial, 124th AES convention. http://www.aes.org. Accessed 26 June 2012
47. Putzeys B, Veltman A, Hulst P, Groenenberg R (2006) All amplifiers are analogue, but some amplifiers are more analogue than others. Convention paper 353, 120th convention 2006 May. France, Audio Engineering Society, Paris, pp 20–23
48. Santi S, Ballardini M, Setti Rovatti RG, (2005) The effects of digital implementation on ZePoC codec. ECCTD III:173–176. IEEE
49. Self D (2002) Audio power amplifier design handbook. Newnes
50. Self D (2008) Linear audio power amplification. In: Tutorial, 124th AES convention. http://www.aes.org. Accessed 26 June 2012
51. Slone GR (1999) High-power audio amplifier construction manual. McGraw-Hill, New York
52. Small R (1973) Vented-box loudspeaker systems. J Audio Eng Soc Part I 21:363–372
53. Small R (1973) Closed-box loudspeaker systems. J Audio Eng Soc Part II 21:11–18
54. Small R (1973) Vented-box loudspeaker systems. J Audio Eng Soc Part III 21:635–639
55. Small R (1973) Vented-box loudspeaker systems. J Audio Eng Soc Part II 21:549–554
56. Small R (1972) Direct-radiator loudspeaker system analysis. J Audio Eng Soc 20:383–395
57. Small R (1972) Closed-box loudspeaker systems. J Audio Eng Soc Part I 20:798–808
58. Sozanski K (1999) Design and research of digital filters banks using digital signal processors. PhD thesis, Technical University of Poznan (in Polish)

59. Sozanski K (2007) Subwoofer loudspeaker system with acoustic dipole. Elektronika : Konstrukcje, Technologie, Zastosowania 4:21–26
60. Sozanski K (2010) Digital realization of a click modulator for an audio power amplifier. Przeglad Elektrotechniczny (Electric Rev) 2010(2):353–357
61. Sozanski K (2002) Implementation of modified wave digital filters using digital signal processors. In: Conference proceedings, 9th international conference on electronics, circuits and systems, ICECS 2002, pp 1015–1018
62. Sozanski K, Strzelecki R, Fedyczak Z (2001) Digital control circuit for class-D audio power amplifier. In: Conference proceedings, 2001 IEEE 32nd annual power electronics specialists conference, PESC 2001, pp 1245–1250
63. Sozanski K (2015) Selected problems of digital signal processing in power electronic circuits. In: Conference proceedings SENE 2015. Lodz Poland
64. Sozanski K (2016) Signal-to-noise ratio in power electronic digital control circuits. In: Conference proceedings: Signal processing, algorithms, architectures, arrangements and applications, SPA 2016. Poznan University of Technology, pp 162–171
65. Stefanazzi L, Chierchie F (2014) Low distortion switching amplifier with discrete-time click modulation. IEEE Trans Ind Electron 61(7):3511–3518
66. Streitenberger M, Bresch H, Mathis W (2000) Theory and implementation of a new type of digital power amplifiers for audio applications. In: ICAS 2000. IEEE, vol I, pp 511–514
67. Streitenberger M, Felgenhauer F, Bresch H, Mathis W (2002) Class-D audio amplifiers with separated baseband for low-power mobile applications. In: Conference proceedings, ICCSC'02.IEEE, pp 186–189
68. Texas Instruments (2004) TAS5121 Digital amplifier power stage. Texas Instruments Inc.
69. Texas Instruments (2007) TAS5518-5261K2EVM. User's guide, SLAA332A, Texas Instuments Inc.
70. Texas Instruments (2007) TAS5508-5121K8EVM evaluation module for the TAS5508 8-channel digital audio PWM processor and the TAS5121 digital amplifier power output stage. User's guide, SLEU054b.pdf, Texas Instrumentss Inc.
71. Instruments Texas (2008) TMS320F28335/28334/28332, TMS320F28235/28234/28232 digital signal controllers (DSCs). Texas Instruments Inc., Data manual
72. Texas Instruments (2010) TAS5508C 8-channel digital audio PWM processor. Data manual, SLES257, Texas Instruments Inc.
73. Texas Instruments (2010) C2000 Teaching materials, tutorials and applications. SSQC019, Texas Instruments Inc.
74. Texas Instruments (2010) A 600W, universal input, isolated PFC power supply for AVR amplifiers based on the TAS5630/5631. Reference Design, SLOU293 Texas Instruments Inc.
75. Texas Instruments (2012) TAS5631B 300 W stereo/ 600 W mono PurePath™ HD digital-input power stage. Data sheet, SLES263C, Texas Instruments Inc.
76. Thiele N (1971) Loudspeakers in vented boxes. J Audio Eng Soc Part I 19:382–392
77. Thiele N (1971) Loudspeakers in vented boxes. J Audio Eng Soc Part II 19:471–483
78. Thile N, Small R (2008) Loudspeaker parameters, Tutorial. In: AES 124th convention. http://www.aes.org
79. Vaidyanathan PP (1992) Multirate systems and filter banks. Prentice Hall Inc., New Jersey
80. Venezuela RA, Constantindes AG (1982) Digital signal processing schemes for efficient interpolation and decimation. IEE Proc Part G(6):225–235
81. Verona J (2001) Power digital-to-analog conversion using sigma-delta and pulse width modulations. In: ECE1371 Analog Electronics II, ECE University of Toronto 2001(II), pp 1–14
82. Zielinski TP (2005) Digital signal processing: from theory to application. Wydawnictwo Komunikacji i Lacznosci, Warsaw (in Polish)
83. Zielinski TP, Korohoda P, Rumian R (eds) (2014) Digital signal processing in telecommunication, basics, multimedia transmission. Wydawnictwo Naukowe PWN, Warsaw (in Polish)
84. Zolzer U (2008) Digital audio signal processing. Wiley, New York
85. Zolzer U (ed) (2002) DAFX—digital audio effects. Wiley, New York

Chapter 7
Conclusion

7.1 Summary of Results

In this book the author has presented his research, analysis and completed projects in the field of signal processing. It began by discussing analog signal acquisition through conversion to digital form, then examined the methods of its filtration and separation, and finally focused on pulse control of output inverters. To start with an analysis was made of the most common sources of errors during conversion of analog signal into its digital form. This process is unquestionably very important for the quality of the entire digital control system. The presented discussion has given a deeper understanding of the selection of control system parameters.

The author has focused on two applications for the considered methods of digital signal processing: an active power filter and a digital class-D power amplifier. In this book the author's original solutions for both applications have been analyzed and implemented. Both applications require precise digital control circuits with a very high dynamic range of control signals. Hence, these applications have provided very good illustrations for the considered methods.

The scope of the monograph has included the following selected digital signal processing methods:

- Selected digital signal processing algorithms useful for a power electronics control circuits: special attention has been paid to implementation aspects using digital signal processors. The author has presented an overview of the characteristics of microprocessors useful for implementing digital signal processing.
- Wave digital filters: the properties of such filters give excellent results in implementation. Thus, this book has included descriptions of wave digital filters and a wave digital filter with the author's modifications. The monograph has analyzed the critical path of such filters, and selected a circuit with a shorter critical path. There has also been shown an effective application of wave digital filters in multirate circuits. In spite of the advantageous features of the type of filters described in Chap. 3, they are not commonly used. The presented methods and circuit have been used in a selected application.

K. Sozański, *Digital Signal Processing in Power Electronics Control Circuits*,
Power Systems, DOI 10.1007/978-1-4471-7332-8_7

- Linear-phase IIR filters based on non-causal IIR filter: these filters have been shown to be a good alternative to the high-order FIR filter, as they require less arithmetic operations. In this book the design of such filters has been presented. Additionally these filters have been used to build signal interpolators.

The most important issues concerning active power filter control circuits which have been considered in the monograph are:

- A review and analysis of selected algorithms based on DFT transform useful for the implementation of control systems APF: sliding DFT, sliding Goertzel and moving DFT algorithms have been considered. In 2003, the author was one of the first to introduce the sliding DFT algorithm to APF control systems. The usefulness of these algorithms has been confirmed through simulation and experimental studies.
- A review and analysis of selected filter banks for signal separation useful in applications for power electronics: The filter banks have been applied to a selective harmonics compensation algorithm. Instantaneous power theory and moving DFT have been considered. These algorithms have been confirmed through simulation studies.
- Dynamic distortion in APF: its presence makes it impossible to fully eliminate line harmonics. In some cases, the line current THD ratio for systems with APF compensation can reach a value of a dozen or so percent. Hence, the problems of active power filter dynamics have been investigated. Power loads can be divided into two main categories: predictable loads and noise-like loads. Most loads belong to the first category. For this reason it is possible to predict current values in subsequent periods, after a few periods of observation. The author has proposed simplified APF models suitable for analysis and simulation of this phenomena. The author has found a solution to these problems. For predictable line current changes, the author has designed a modification using a predictive circuit to reduce dynamic compensation errors. The author's experimental results have confirmed the usefulness of the compensation method with predictive circuit. This modification for APF control algorithms is very simple and the additional computational workload is very small. Therefore it is shown to be very easy to implement it in an existing APF digital control circuit based on a digital signal processor, microcontroller or programmable digital circuit (FPGA, CPLD, etc.), thereby improving the quality of harmonic compensation. In addition, the research has shown that current prediction circuits may be applied to other power electronics devices, such as serial APF, power conditioners, high quality AC sources, UPS etc.
- Unpredictable line current changes: the author has developed a multirate APF. The presented multirate APF has shown a fast response to sudden changes in the load current. Therefore using multirate APF it is possible to decrease THD ratio of line current even for unpredictable loads.

For the control circuit of digital class-D power amplifiers the following topics have been included:

- Signal interpolators for high quality audio signal with high signal-to-noise ratio.

- A second order noise shaping circuit for compensating quantization noise and systematic errors of inverter output stages: this can increase the output voltage signal-to-noise ratio. This type of circuit is shown to be able to decrease the influence of dead time and minimum switch-on time. The additional workload on the processor for a noise shaping circuit is shown to be very small, making it easy to implement in existing control circuits. The circuit can be also applied in other power electronics devices. In the author's opinion this compensation should be very fruitful for multilevel inverters too.
- Problems of supply voltage fluctuation influence on a class-D amplifier.
- Analysis and design of an analog circuit for supply voltage fluctuation and transistor amplitude errors resulting from non-linear on-state resistance compensation.
- The design of a two-way loudspeaker system with digital click modulator: when such a system is used for the higher band and ordinary PWM for the lower band it has been shown to make it possible to keep transistor switching frequency equal to 44.1 kHz. Using the low switching frequency reduces transistor power losses and reduces EMC interference. The designed active loudspeaker system covers the whole audio band.

The greater part of the presented methods and circuits in this book are the original work of the author. The results of simulation and experimental studies have been achieved by the original work of the author. For some algorithms listings from MATLAB® or in C language have been presented. In the author's opinion the presented methods and circuits can be successfully applied to the whole range of power electronics circuits.

7.2 Future Work

The scope of future research planned for selected signal processing algorithms related to specific applications:

- The design of an efficient linear-phase IIR filter with modified lattice wave digital filters and two-path filters: in the next phase signal interpolators with such filters will be researched.
- The analysis and design of a digital crossover using filter banks with loudspeaker characteristics equalization: in this crossover the use of wave digital filters will be studied.
- Research of methods for power quality analysis and fault detection [2].

The scope of future research on active power filter control circuits:

- Continued investigation of APF dynamics: the work on APF control circuits based on iterative learning control algorithms, repetitive control algorithm [3] and wavelets will be continued.
- The investigation of closed loop APF control circuits with predictive current circuits.

- Improvement of control circuit properties: to this end the application of a digital signal processing circuit fully synchronized with power line frequency will be studied. For this purpose, a digital phase locked loop circuit will be developed.

The scope of future research on digital class-D power amplifiers:

- High quality audio signal digital circuits: such circuits need to be well synchronized with a low level of jitter [1]. Research and construction of a fully synchronized digital circuit with low-noise digital phase locked loop circuit will be undertaken.
- Signal-to-noise ratio: to increase this factor research of more efficient high-order noise shaping circuits will be investigated.
- A class-D amplifier with full digital feedback [4–6]: special attention will be paid to such an amplifier, which should have very low output impedance. However, this task is very difficult in audio applications with respect to the creation of audible transient suboscillation. Therefore, such a circuit will be investigated and designed. The results of such circuits will be carefully verified by tests.
- An analysis of woofer functioning with the class-D digital amplifier: the purpose of this analysis is to determine the position of the speaker cone with respect to the electromotive force generated in the coil. This will lead to better control of the speaker in the low frequency range.

References

1. Azeredo-Leme C (2011) Clock jitter effects on sampling: a tutorial. IEEE Circ Syst Mag 3:26–37
2. Bollen MHJ, Gu IYH, Santoso S, McGranaghan MF, Crossley PA, Ribeiro MV, Ribeiro PF (2009) Bridging the gap between signal and power. IEEE Sig Process Mag 26(4):12
3. Buso S, Mattavelli P (2015) Digital control in power electronics, 2nd edn. Morgan & Claypool, San Rafael
4. Mouton T, Putzeys B (2009) Digital control of a PWM switching amplifier with global feedback. In: AES 37th international conference, Hillerod, Denmark, 28–30 August 2009. Audio Engineering Society
5. Midya P, Roeckner B (2010) Large-signal design and performance of a digital PWM amplifier. J Audio Eng Soc 58(9):739–752
6. Putzeys B (2008) A universal grammar of class D amplification, Tutorial. 124th AES convention. www.aes.org. Accessed Dec 2012

Index

A
AC analysis, 169
Acoustic chamber, 304
Active crossover, 302
Active loudspeaker system, 300
Active power filter (APF), 5, 6, 14, 199
AD7608, 72
A/D conversion, 57
A/D converter, 50, 68, 70
ADS1278, 73
ADS8364, 71
ADS8568, 74
ADSP-21364, 310
ADSP21367, 159
Advance time T_A, 245
AES/EBU, 17, 300
Aliasing, 11, 47
Allpass section, 94
Amplifier with analog feedback, 287
Amplifier with digital feedback, 290
Analog input, 23
Anechoic acoustic chamber, 304, 309
Anti-causal circuit, 5, 107
APF control circuit, 243
Application specific integrated circuits (ASIC), 146

B
Barrel shifter, 150
Bilinear transform, 90, 182
Butterworth filter, 208, 310

C
Cascaded signal decimator, 119
Cascaded signal interpolator, 115, 295
Causal circuit, 5, 107
Circular addressing, 149
Circular buffer, 149
Clark transformation, 218, 225
Class-D power amplifier, 5, 16, 277
Click modulation (CM), 293, 314
Click modulator (CM), 309
Closed loop Hall effect sensor, 36
Code Composer Studio, 192
Coherent sampling, 47
Common mode capacitance, 26
Common mode rejection (CMR), 26
Common mode transient rejection (CMTR), 26
Common mode voltage, 25
Complex number in MATLAB, 169
Control circuit, 200
Critical path, 86, 89, 97, 100
Current measurement, 32
Current sensor with Hall sensor, 36
Current transducer with air coil, 38
Current transformer, 33
Current transformer with magnetic modulation, 38

D
Damping factor, 300
DC analysis, 169
DC/DC converter, 183
Dead time, 281, 285
Decimation filter, 119
Delayed branch, 152
Delta sigma modulator (DSM), 70, 73
Design of digital IIR filters, 90
DFT algorithm, 129
DFT filter bank, 129

K. Sozański, *Digital Signal Processing in Power Electronics Control Circuits*, Power Systems, DOI 10.1007/978-1-4471-7332-8

Digital audio interface (DAI), 17, 303, 314, 321
Digital class-D power amplifier, 277, 278, 300, 321
Digital crossover, 302, 310
Digital filter, 84
Digital filter bank, 126, 254, 255, 302, 310, 326
Digital hysteresis current controller, 180
Digital pulse width modulator (DPWM), 16, 279, 281, 284, 309, 321
Digital signal controller (DSC), 75, 155
Digital signal processor (DSP), 146
Diode reverse recovery charge Q_{rr}, 278
Dirac delta, 6
Discrete-time circuit, 6, 84
Dither, 63
Dynamic range of signal, 59
Dynamics of shunt APF, 233

E
Effective number of bits (ENOB), 66
Electromagnetic interference (EMI), 278

F
Fast Fourier transform (FFT), 130
Field-programmable gate array (FPGA), 146, 154
Filter specification, 84
Finite impulse response (FIR) filter, 85
First harmonic detector, 206

G
Galvanic isolation, 23
Goertzel algorithm, 131

H
Hall sensor, 35
Hardware-controlled loop, 150, 152
Hilbert FIR filter, 314
Hilbert transform, 294

I
Infinite impulse response (IIR) filter, 87, 90, 182
Instantaneous reactive power theory (IPT), 201, 222
Intermodulation component, 281
Interpolation filter, 115, 295
Interrupt, 152
Isolation amplifier, 27, 32

J
Jitter, 54

K
Kirchhoff's current law (KCL), 174
Kirchhoff's voltage law (KVL), 175, 177
Kronecker delta, 6

L
Laplace transform, 174, 177
Lattice bireciprocal wave digital filter (LB-WDF), 121, 139
Lattice wave digital filter (LWDF), 83, 94, 96, 139, 209
Linear time-invariant circuit (system) (LTI), 6, 84, 87
Linear-phase IIR filters, 107
Linkwitz-Riley filter, 310
LogChrip, 304
Loop analysis, 174
Low-pass filter(LPF), 201

M
MAC multiply-accumulate operation, 8, 86, 90, 111, 114, 115, 123, 124, 126, 148, 153, 154, 160
MATLAB, 43, 49, 55, 83, 90, 96, 111, 117, 119, 143, 169, 171, 173, 176, 178, 180, 182, 204
Matrix in MATLAB, 169
Maximum length sequence (MLS), 304
Metal-oxide-semiconductor field-effect transistor (MOSFET), 278
Midrange, 309
Million floating-point instruction per second (MFLOPS), 157
Million instruction per second (MIPS), 157
Modified lattice wave digital filter (ML-WDF), 83, 100
Modified wave digital filter (MWDF), 297
Moving average, 8, 85, 227
Moving DFT algorithm (MDFT), 136, 215
Multichannel system, 50
Multiplication and accumulation, 148
Multirate circuit, 3, 13, 83, 115
Multirate control circuit, 13, 201
Multirate shunt APF, 258

N
Nodal analysis, 174
Noise shaping technique, 60, 284
Non-causal circuit, 5, 107
Nonlinear circuit, 43
Number of bits, 5, 12

O
Open loop amplifier, 285
Open loop Hall effect current sensor, 35
Oversampled pulse width modulator, 284
Oversampling, 13
Oversampling ratio, 115

P
Parallel architecture, 152
Park transformation, 218, 225
Passive crossover, 302
Passive loudspeaker system, 300
Phase locked loop (PLL), 47, 201
Pipeline, 152
Power electronics system, 1
Power supply rejection ratio (PSRR), 289
P-q algorithm, 218, 222
Predictive circuit, 244, 251
Printed circuit board (PCB), 278
Proportional-integral-derivative (PID) controller, 292
Pseudorandom signal, 63
PSIM, 184, 185, 192, 206
PWM, 281

Q
Quadrature mirror filter (QMF), 139
Quantization error, 57
Quantization step, 58

R
RC circuit, 170
Real-time control system, 10
Resistive shunt, 32
Resolution, 58
Rogowski coil, 38, 40

S
Sample–and–hold (SH) circuit, 45, 71, 74, 75, 77
Sampling rate, 5, 11, 45
Saturation arithmetic, 150
Second-order section (SOS), 89, 96, 160
Sense resistor, 32
Sequential sampling, 50
SHARC, 159
Shunt active power filter, 184, 200
Signal acquisition, 49
Signal acquisition time, 49
Signal conditioning, 23
Signal decimator, 115, 119
Signal headroom, 59
Signal interpolator, 115, 295
Signal quantization, 57
Signal signal downsampler, 119
Signal signal upsampler, 115, 295
Signal to noise ratio (SNR), 12, 42, 46, 55, 58, 59, 61, 63, 65, 66
Simulink, 183
Simultaneous sampling, 12, 50, 70–72
Simultaneous sampling A/D converter, 70
Slew rate, 25
Sliding DFT algorithm (SDFT), 131, 184, 210
Sliding Goertzel algorithm (SGDFT), 136, 215
S/PDIF, 300
Square root, 213
Strictly complementary (SC) filter bank, 128, 310
Successive approximation (SA), 69, 71, 72, 74, 77
Synchronization, 47, 201

T
TAS5508, 321
Three-way digital crossover, 326
TMS320C6xxx family, 157
TMS320F28335, 75, 155, 192
TMS320F28377D, 231
TMS320F2837xD, 77, 157
TMS320F28xx family, 155
Total harmonic distortion (THD) ratio, 15, 43, 233
Track–and–hold (TH) circuit, 45, 71, 72
Transformer, 33
Transient analysis, 177
Tweeter, 309
Two-path (polyphase) filter, 299
Two-port adaptor, 94

U
Uninterruptible power supply (UPS), 285

V

Very long instructions word (VLIW), 157

W

Wave digital filter (WDF), 9, 94, 139, 209
Weighted harmonic distortion (WTHD) ratio, 45
Woofer, 309

Lightning Source UK Ltd.
Milton Keynes UK
UKHW02n1302221018
330975UK00003B/42/P